Springer
Tokyo
Berlin
Heidelberg
New York
Barcelona
Hong Kong
London
Milan
Paris
Singapore

H. Osada (Ed.)

Bioprobes

Biochemical Tools for Investigating Cell Function

With 112 Figures

 Springer

Hiroyuki Osada
Professor
Antibiotics Laboratory, RIKEN
Hirosawa 2-1, Wako, Saitama 351-0198, Japan

QP
579
.9
.M64
B55
2000

ISBN 4-431-70247-4 Springer-Verlag Tokyo Berlin Heidelberg New York

Library of Congress Cataloging-in-Publication Data

Bioprobes : biochemical tools for investigating cell function / H. Osada(ed.).
    p. cm.
  Includes bibliographical references and index.
  ISBN 4431702474 (hard cover : acid-free paper)
  1. Molecular probes. 2. Cytology—Methodology. I. Osada, H.(Hiroyuki),1954-

QP519.9.M64 B55 1999
571.6'028—dc21

99-049857

Printed on acid-free paper

Printing and binding: Hirakawa-kogyo, Japan
SPIN: 10676798   5 4 3 2 1

Preface

In 1929, Alexander Fleming serendipitously observed an interesting phenomenon: a fungal contaminant from the air had grown on an agar plate and inhibited the growth of *Staphylococcus aureus*. In the Second World War, Howard W. Florey and his colleagues developed penicillin as a chemotherapeutic medicine based on Fleming's observation. This success story ignited the exploitation of magic bullets from microorganisms. Selman A. Waksman defined antibiotics as being produced by microorganisms and inhibiting the growth of microorganisms. At the beginning of antibiotic research, most antibiotics were included within the limits of his definition. However, the definition became wider and wider to include various biological and pharmacological substances such as immunosuppressants and, nowadays, antitumor agents. Moreover, it is well known that penicillin and streptomycin played important roles not only in the medical field but also in the biochemical field, leading to the elucidation of the biochemical mechanisms of bacterial cell wall and ribosome, respectively. Cycloheximide could not be used for chemotherapy because of its toxicity, but it is a very useful reagent to investigate protein synthesis in eukaryotic cells. Such compounds showing selective toxicity are useful as biochemical tools.

In this book, we use the term "bioprobe" for agents that are useful to reveal molecular mechanisms of eukaryotic cells. With respect to selective toxicity and screening technology, bioprobe research is similar to antibiotic research; however, there is more emphasis on the function of bioprobes as a biochemical tool compared with research in antibiotics. Not only organic compounds isolated from natural sources but also synthetic compounds could be termed bioprobes. In this book, however, the authors discuss bioprobes that for the most part are isolated from microorganisms.

The book has six chapters. After a brief introduction of the development of bioprobes, the biological fields in which bioprobes affect the molecular targets are surveyed. These reviews cover the broadly advancing areas of the cell cycle, differentiation, apoptosis, and immunological responses. "Bioprobes at a Glance" contains information on the most important bioprobes that are useful in biological studies. The book thus presents the application of bioprobes isolated from microbial metabolites which assisted the investigation of the complex biochemical processes of the cell cycle, differentiation, apoptosis, and immunological responses. Recently, scientists from two disciplines previously independent, namely, organic chemists interested in the isolation of inhibitors and molecular biologists interested in the biological function of enzymes, have been meeting in the same stream of this bioprobe research.

We have attempted to provide substantial references to show who originally isolated each compound and determined its chemical structure and who revealed the molecular target. We hope that this book will be valuable to both natural product chemists and cell biologists, eventually resulting in progress in bioprobe research.

HIROYUKI OSADA, Ph.D.
July 3, 1999

Contents

1
Trends in Bioprobe Research

Hiroyuki Osada

1 Introduction

The term "Bioprobe" in this book is reserved for small molecular weight compounds mainly isolated from microbial secondary metabolites which are useful not only for biochemical research but also as a source of successful drugs with diverse activities. In general, bioprobes were discovered as inhibitors of specific functions of mammalian cells and afterwards their molecular targets in eukaryotic cells were determined (Fig. 1).

Some bioprobes such as tunicamycin [1] and cyclosporin A [2] were discovered by cell-based screening. Afterwards, the target molecules that are inhibited by the compounds were clarified. FK506 [3] and ML-236B [4] later developed as Tacrolimus and pravastatin (mevalotin), respectively, were discovered by more target-oriented screening systems. From the beginning of screening, the target enzyme is limited in these sophisticated screenings. These discoveries accelerated bioassay guided screening in pharmaceutical companies because these compounds are important not only as bioprobes but also as human therapeutic agents. However, many pharmaceutical companies recently stopped natural products screening and chose combinatorial chemistry for high throughput screening. Natural products screening is thought to be a slow and costly strategy to discover lead candidates and combinatorial chemistry is thought to be a more rapid and attractive approach. Major problems in the field of natural products screening are the steps for compound purification and identification, which are time-consuming. However, natural products are like a treasure box containing interesting lead compounds with novel structures and useful biological activities. Chemical biology and chemical genetics are fascinating new fields which rely on the new bioprobes. Efforts to obtain new bioprobes are described in the following sections.

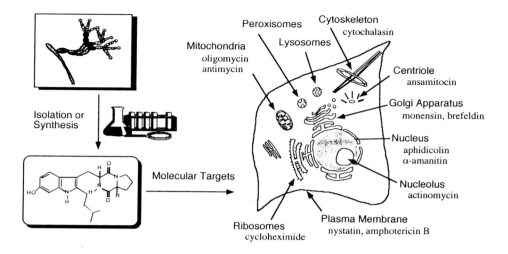

Fig. 1. Bioprobes are small molecules that are useful in investigating biological functions

2 Sources of Bioprobes

2.1 Microorganisms

Microorganisms are most important as a reservoir of chemical diversity. Although nobody can tell why microorganisms produce such a wide variety of compounds, microorganisms may produce chemical signals for microbial interactions or defense substances to fight other organisms. Traditional antibiotics are antimicrobial compounds, which were isolated from the fermentation broth of soil microorganisms. Bioprobes acting on mammalian cells are also isolated from soil microorganisms, and eukaryotic microorganisms especially are a major source of bioprobes. Radicicol [5], cyclosporin A [2], cytochalasin [6] and so on are fungal metabolites. These compounds might have important roles in fungal physiology or be used as defensive weapons for survival.

In this book, antibacterial or antifungal compounds are not referred to. However, bioprobes that suppress viral growth are known. For example, tunicamycin was originally discovered as an antiviral antibiotic produced by *Streptomyces lysosuperificus* [7], and then used as an inhibitor of *N*-glycosylation in mammalian cells [8].

Fungi produce many terpenoid compounds, but the actinomycetes seldom produce them. On the contrary, polyketide compounds are common metabolites in actinomycetes compared to fungi. Combinatorial biosynthesis might be a new strategy to produce chemical diversity [9].

Microtubules, consisting of tubulin α, β heterodimers with microtubule associated proteins (MAPs and tau), are known to be the main component of spindles in the mitotic apparatus of eukaryotic cells, and are also involved in many other basic and essential cell functions. Pironetin, produced by *Streptomyces* sp. binds to tubulin and inhibits microtubule polymerization [10].

2.2 Marine Organisms

Marine organisms, e.g., sponges and bryozoa, are important sources of bioprobes, some of which are highly toxic. Bioprobes that act on protein phosphorylation are often isolated from marine organisms [11]. Okadaic acid is a protein phosphatase 2B specific inhibitor, which was isolated from *Halichondoria okadai*. Calyculin A is a protein phosphatase 1 inhibitor, which was isolated as a potent antitumor metabolite from the marine sponge *Discodermia calyx*. Bryostatin 1, a metabolite of the colonial bryozoan, *Bugula neritina,* is a protein kinase C (PKC) activator which exhibits potent and selective activity against leukemia cell lines and excellent in vivo antitumor activity [12, 13]. Many metabolites of marine organisms have been developed as bioprobes, however, in general it was difficult to supply enough amount of materials for in vivo evaluation.

Dolastatin 10 from *Dolabella auricularia* and halichondrin B from sponges are inhibitors of glutamate- and microtubule-associated protein-dependent polymerization of tubulin [14]. Tryprostatin A is also an inhibitor of microtubule-associated protein-dependent polymerization of tubulin [15]. The producer fungus strain taxonomically belongs to *Aspergillus fumigatus* which is easily isolated from soil. However, this producer strain was isolated from sea bottom sediment. Members of another series of marine products, the spongistatins from a *Spongia* sp., also inhibit tubulin polymerization by binding to the vinca alkaloid domain of tubulin and have exceptional cytotoxic properties [16].

As described above, most bioactive marine products are extremely low in natural abundance; however, the structural complexity and the high potency of the marine products make them attractive as bioprobes.

2.3 Plants

There are a number of plant metabolites that interfere with microtubule function to cause mitotic arrest of eukaryotic cells [17–19]. These antimitotic agents are used for medicinal purposes as well as for biochemical tools for investigating the regulatory mechanism of microtubule networks. Most such antimitotic agents, with a few exceptions, bind to β-tubulin, among them, colchicine [20], vinblastine [21] and taxol [22] have been of major importance in biochemical studies of microtubules and in studies of their intracellular functions. The former two both inhibit microtubule assembly but their binding sites on β-tubulin are different; the colchicine-site and vinblastine-site, and many microtubule inhibitors bind to either site. Taxol binds to tubulin at a site other than the colchicine- and vinblastine-site, and promotes microtubule assembly.

DNA topoisomerase I (topo I) has been identified as a principal target of a plant alkaloid, camptothecin, and its derivative (CPT-11) [23]. Camptothecin derivatives irinotecan, topotecan and 9-aminocamptothecin were developed after identification of the mechanism of action of camptothecin, an alkaloid from the Chinese tree *Camptotheca acuminata*, as topo I inhibition. Irinotecan and topotecan are water soluble camptothecin derivatives. Topotecan is an active principal directly acting on the target molecules in cells, but irinotecan is a pro-drug that undergoes de-esterification in vivo to yield a more potent metabolite [24, 25].

The michellamines A-C, novel naphthyl isoquinoline alkaloids, were isolated as anti-HIV agents from the tropical plant, *Ancistrocladus korupensis* [26]. Michellamine B inhibited the enzymatic activities of reverse transcriptases from HIV and also syncytium formation [27].

3   Development of Bioprobes

Development of bioprobes starts from screening, which involves testing crude extracts of microbial fermentation broth based on bioassays. Where activity of sufficient interest is detected, the active components are isolated by a process of bioassay-guided purification. The physicochemical properties of the active components should be determined at the earliest possible stage and compared with those of known compounds to eliminate duplicate isolation. For novel compounds, sufficient pure material for structure elucidation and evaluation shall be obtained by large-scale fermentation of microorganisms.

3.1 Cell-Based Screening

A bioassay to detect desired compounds is the most important process to develop bioprobes. Many bioassays using intact cells or genetically engineered cell lines to express specific proteins or receptors have been reported [28–31]. Although many useful bioprobes were isolated by cell-based screening, it is a potential disadvantage of the screening that it is difficult to distinguish specific inhibitors from undesired toxic compounds.

The bleb forming assay using K-562, a human chronic myeloid leukemia cell line, is a convenient detection method for inhibitors of serine/threonine kinases and phosphatases and suitable for the 96-well screening procedure [29]. When K-562 cells were treated with phorbol 12,13-dibutyrate (PDBu) or teleocidin that are activators of PKC, many blebs appeared on the cell surface of K-562 within 10 minutes. This appearance of blebs is inhibited by staurosporine and H7 that are known to be protein kinase C (PKC) inhibitors. The bleb forming assay satisfied the criteria (simplicity and specificity) required for preliminary screening of activators or inhibitors of PKC.

A completely different approach is provided by chemical screening for novel metabolites. This involves searching for compounds with particular chemical characteristics using procedures such as thin layer chromatography combined with specific colorimetric methods. Staurosporine (*see* p.50) was discovered by chemical screening to detect alkaloids and later it is revealed to be a potent PKC inhibitor [32]. A HPLC equipped with a photodiode array system is a powerful apparatus to detect UV-visible compounds. Staurosporine derivatives, RK-286C [33], RK-286D [34], RK-1409 [35] and RK-1409B [36] were detected by a photodiode array system. UV-profiles of these compounds were similar to that of staurosporine but the retention time of these compounds in HPLC is different from that of staurosporine. Compounds are purified on a multi milligram scale to evaluate biological properties and to determine chemical structure.

3.2 Combinatorial Biosynthesis

As the biosynthetic pathways of bioactive natural products such as the polyketides have been recently revealed, it is possible to produce a series of natural chemical diversity using protein-engineering strategies. Polyketides can be designed rationally by recombinant assembly of enzyme subunits to produce unnatural bioprobes [9]. An unusual polyketide, EM18 was prepared using *Streptomyces* strains engineered to express combinations of appropriate enzymatic subunits from naturally occurring polyketide synthases [37]. A similar approach is being studied more widely for compounds other than polyketides [38]. DNA technology is also being investigated for the production of interesting bioprobes from organisms that are difficult to cultivate, such as marine organisms. Moreover, new technologies enable us to isolate the aimed genes from the environmental community of microorganisms. With polymerase chain reaction (PCR) primers derived from conserved regions of known ketosynthase or acyl carrier protein genes specifying the formation of polyketides, it is possible to clone genes even from uncultured microorganisms [39]. Since substantially less than 1% of the microorganisms present in nature are thought to be cultivatable by standard methods, this technology is a potential way to gain access to a more extensive range of microbial molecular diversity and to biosynthetic pathways whose products can be tested for biological applications.

4 Targets for Bioprobes

4.1 Cell Proliferation and Tumors

Microbial metabolites contain a wide variety of compounds that regulate cell proliferation and tumor growth by regulating the specific proteins required for cell proliferation, as reviewed in Chapter 2. Bioprobes for the investigation of cell proliferation were mainly screened as cell cycle inhibitors [30, 40]. Protein kinase inhibitors, staurosporine and related compounds have been characterized as cell cycle inhibitors [41–43]. Many kinds of bioprobes for cell cycle inhibition have been developed. Cyclin-dependent kinases (Cdk), which are activated by the binding with cyclin and simultaneously by its own phosphorylation/dephosphorylation, play important roles as engines in the cell cycle. Tubulins are considered to be one of the most important proteins of the cell division machinery. Therefore, cdk inhibitors [44, 45] and tubulin binders [15, 46] are useful bioprobes for investigating cell cycle mechanisms and also as possible anticancer drugs. Since the function of proteins controlling the cell cycle is also regulated by phosphorylation and dephosphorylation, inhibitors of protein kinases and phosphatases are considered not only as biochemical tools but also as possible antitumor agents [47–49].

4.2 Differentiation

During embryonic development, precursor cells differentiate to functional matured cells under the control of cytokines and hormones. As tumor cells including leukemia and neuroblastoma are thought to be disrupted by differentiation programs, it is worth while exploiting differentiation inducers of cells. In Chapter 3, a screening strategy to discover novel differentiation inducers from microbial metabolites and their biological activities are described.

Blood cells, neuron, bone, muscle cells, liver cells, and epithelial cells are differentiated from precursor cells. The hemopoietic stem cells are differentiated to functional blood cells by a differentiation determinant. Interleukines are important for lymphocytes to maintain cell growth and differentiation. Neurotrophins, including nerve growth factor (NGF), brain derived neurotrophic factor (BDNF) and so on are major cytokines maintaining both cell survival and differentiation. Many low-molecular-weight organic compounds that induce the differentiation of the eukaryotic cells are mainly isolated from microorganisms.

With the identification of the role of the trk family in the high affinity neurotrophin receptors, the dissection of the subsequent signal transduction pathways becomes possible. Rat pheochromocytoma, PC12 cells are well known to be differentiated to neuronal cells specifically by the addition of exogenous NGF and bFGF [50]. Tyrosine phosphorylation of trk leads to modification of a cascade of protein serine/threonine kinases, known as mitogen activated kinase (MAP) or external signal response kinase (ERK) in PC12 cells. Experiments with neutralizing antibodies, expression of temperature sensitive or dominant negative proteins suggest that the ras-MAP cascade is necessary for differentiation of PC12 cells [51, 52].

One noteworthy example in this area is development of many neuritogenic compounds. Lactacystin (see p.50) was isolated as a neuritogenic compound and now recognized as a proteasome inhibitor [53, 54]. K252a blocks NGF action through the inhibition of tyrosine kinase activity of trk [55, 56]. A fungal metabolite, epolactaene and its derivatives induce neurite outgrowth in human neuroblastoma cells, SH-SY5Y through a ras-MAP independent pathway [57, 58].

4.3 Apoptosis

Apoptosis or programmed cell death is a fundamental process in the development of multicellular organisms. A family of cysteine proteases, termed caspases, which are homologous to the *Caenorhabditis elegans* death gene ced-3, are a common and critical component of the cell death pathway. Overexpression of these proteases leads to apoptosis of various cell types. Although some of these proteases seem to constitute a multiple protease cascade, the identified targets for proteolytic cleavage by caspases are few, and the role of individual targets in mediating particular apoptotic events is not entirely understood [59].

Recent studies have indicated that reactive oxygen species (ROS) including singlet oxygen, superoxide, hydrogen peroxide, hydroxyl radical and nitric oxide, are involved in the signaling pathway for apoptosis [60]. Overexpression of Bcl-2, an inhibitory protein of apoptosis, decreased ROS generation and increased resistance to apoptotic killing by hydrogen peroxide (H_2O_2), menadione and depletion of glutathione [61]. However, whether ROS are essential for apoptosis induction remains to be investigated [62].

In Chapter 4, the signal transduction of apoptosis and the involvement of ROS are discussed based on data using tyrosine kinase inhibitor and so on. The cellular response to diverse external stimuli is controlled by complex phosphorylation cascades. The extracellular signal-regulated protein kinase (ERK) cascade is a prominent component of the mitogen-activated protein (MAP) kinase family that in particular plays an integral role in both growth factor and stress signaling. In addition to classical MAP kinases, novel members of the MAP kinase superfamily including JNK (c-Jun N-terminal kinase)/SAPK (stress-activated protein kinase) and p38 MAPK were reported to be activated by various environmental stresses such as UV irradiation or exposure to agents [63–66]. JNK/SAPK, p38 MAPK and ERK are characterized by having the Thr-Pro-Tyr (TPY), the Thr-Gly-Tyr (TGY) and the Thr-Glu-Tyr (TEY) sequences, respectively, as the dual phosphorylation motif. Each MAP kinase group has a distinct substrate specificity and is regulated by a separate signal transduction pathway. Specific bioprobes (inhibitors or activators) affecting the specific cascade are now under development.

4.4 Immunological Response

A number of microbial products have been introduced or are being evaluated as immunosuppressants (see p.104), as reviewed in Chapter 5. Cyclosporin A is used as the main immunosuppressant drug in this area and together with FK506 from *Streptomyces tsukubaensis* and rapamycin from *Streptomyces hygroscopicus* forms the immunophilin class of immunosuppressants, which are linked by similar but nonidentical mechanisms of action [2, 67]. FK506 (Tacrolimus) [3, 68] has been introduced for the treatment of transplant rejection but, although more potent in vitro, appears to have a lower therapeutic window than cyclosporin A. Rapamycin [69, 70] is currently in development for use in combination with cyclosporin A. 15-Deoxyspergualin (see p.105) is a synthetic analogue of the polyamine fermentation product spergualin, originally isolated from *Bacillus laterosporus* [71, 72]. It has strong immunosuppressive properties and has been introduced for use in the treatment of kidney transplant rejection and is being evaluated for use in combination with other immunosuppressive drugs. The mechanism of action of 15-deoxyspergualin is thought to involve binding to a member of the Hsp70 family of heat shock proteins, leading to blocking of nuclear translocation of some key proteins, including the transcription factor NF-κB [73].

4.5 Cholesterol and Lipid Biosynthesis

Due to page limitations, this book cannot cover all the important compounds that are used in medical and agricultural fields. Several bioprobes acting on cholesterol- and lipid-biosynthesis were isolated from microbial metabolites and some of them are useful for clinical use.

ML-236B (compactin) which was isolated from the soil microorganism, *Penicillium citrinum*, is an inhibitor of cholesterol synthesis [4, 74]. Although this compound itself showed hypocholesterolemic activity in several animal species, microbial hydroxylation of sodium ML-236B (compactin) carboxylate enhanced the in vivo activity. The soluble cytochrome P450 from *Streptomyces carbophilus* is used in this process [75]. The obtained pravastatin (mevalotin) is a potent inhibitor of 3-hydroxy-3 methylglutaryl coenzyme A (HMG-CoA) reductase that is a key enzyme for cholesterol synthesis. It has been reported that pravastatin (mevalotin) decreased the level of high density lipoprotein (HDL) in the serum.

A number of inhibitors of squalene synthase have been reported; the squalestatins isolated from a *Phoma* sp. and the zaragozic acids from various fungi including *Sporormiella intermedia* [76, 77]. These compounds are potent inhibitors of squalene synthase, and exhibit potent activity in in vivo models of cholesterol synthesis. Acyl-CoA:cholesterol acyltransferase (ACAT) is another important target which is inhibited by microbial screening. The pyripyropenes were isolated from *Aspergillus fumigatus* as ACAT inhibitors [78].

5 Trends and Prospects

Development of new bioprobes that inhibit specific biological functions illuminates the new frontier for biologists interested in the mechanistic aspects and implications of molecular interactions. Two examples have generated much productive recent research; discovery of the molecular target of lactacystin and leptomycin. Lactacystin (*see* p.50) was isolated from a *Streptomyces* sp. as a differentiation inducer of neuroblastoma cells [53]. Later, the 20S proteasome was identified as a specific cellular target for lactacystin [54], and now, lactacystin is used in a wide range of studies to reveal the biological meaning of proteasome function in mammalian cells [79]. The other example is identification of the molecular target of leptomycin B. Leptomycin B (*see* p.26) was originally isolated as an antifungal antibiotic exhibiting strong inhibitory activity against *Schizosaccharomyces* and *Mucor* [80]. It is now revealed that leptomycin B binds the CRM1 protein which is responsible for nuclear transport in mammalian cells [81]. Leptomycin B is now considered to be the most important biochemical tool for the investigation of nuclear export signal in mammalian cells [82, 83].

Bioprobes also have an influence on the pharmaceutical industry as the golden egg of medicine. The wide variety of chemotypes from microbial metabolites is potentially useful as lead compounds, which are modified by combinatorial chemistry.

References

1. Takatsuki A, Arima K, Tamura G (1971) Tunicamycin, a new antibiotic. I. Isolation and characterization of tunicamycin A. J Antibiot 24:215–223
2. Shevach EM (1985) The effects of cyclosporine A on the immune system. Ann Rev Immunol 3:397–423
3. Kino T, Hatanaka H, Hashimoto M, Nishiyama M, Goto T, Okuhara M, Kohsaka M, Aoki H, Imanaka H (1987) FK-506, a novel immunosuppressant isolated from a streptomyces. I. fermentation, isolation, and physico-chemical and biological characteristics. J Antibiot 40:1249–1255
4. Endo A, Kuroda M, Tsujita Y (1976) ML-236A, ML-236B, and ML-236C, new inhibitors of cholesterogenesis produced by Penicillium citrinium. J Antibiot 29:1346–1348
5. Kwon HJ, Yoshida M, Fukui Y, Horinouchi S, Beppu T (1992) Potent and specific inhibition of p60^{v-src} protein kinase both in vivo and in vitro by radicicol. Cancer Res 52:6926–6930
6. Cribbs DH, Glenney-Jr JR, Kaulfus P, Weber K, Lin S (1982) Interaction of cytochalasin B with actin filaments nucleated or fragmented by villin. J Biol Chem 257:395–399
7. Takatsuki A, Tamura G (1971) Tunicamycin, a new antibiotic. III Reversal of the antiviral activity of tunicamycin by aminosugars and their derivatives. J Antibiot 24:232–238
8. Gahmberg CG, Jokinen M, Karhi KK, Andersson LC (1980) Effect of tunicamycin on the biosynthesis of the major human red cell sialoglycoprotein, glycophorin A, in the leukemia cell line K562. J Biol Chem 255:2169–2175
9. Hutchinson CR (1998) Combinatorial biosynthesis for new drug discovery. Curr Opin Microbiol 1:319–329
10. Kondoh M, Usui T, Nishikiori T, Mayumi T, Osada H (1999) Apoptosis induction via microtubule disassembly by an antitumour compound, pironetin. Biochem J 340:411–416
11. Ishihara H, Martin BL, Brautigan DL, Karaki H, Ozaki H, Kato Y, Fusetani N, Watabe S, Hasimoto K, Uemura D, Hartshorne DJ (1989) Calyculin A and okadaic acid: inhibitors of protein phosphatase activity. Biochem Biophys Res Commun 159:871–877
12. Sako T, Yuspa SH, Harald CL, Pettit R, Blumberg PM (1987) Partial parallelism and partial blockade by bryostatin 1 of effects of phorbol ester tumor promoters on primary mouse epidermal cells. Cancer Res 47:5445–5450
13. Wender PA, Cribbs CM, Koehler KF, Sharkey NA, Harald CL, Kamano Y, Pettit GR, Blumberg PM (1988) Modeling of the bryostatins to the phorbol ester pharmacophore on protein kinase C. Proc Natl Acad Sci USA 85:7197–7201
14. Pettit GR, Kamano Y, Fujii Y, Herald CL, Inoue M, Brown P, Gust D, Kitahara K, Schmidt JM, Doubek DL, Michel C (1981) Marine animal biosynthetic constituents for cancer chemotherapy. J Nat Prod 44:482–485
15. Usui T, Kondoh M, Cui C-B, Mayumi T, Osada H (1998) Tryprostatin A, a specific and novel inhibitor of microtubule assembly. Biochem J 333:543–548
16. Bai R, Taylor GF, Cichacz ZA, Herald CL, Kepler JA, Pettit GR, Hamel E (1995) The spongistatins, potently cytotoxic inhibitors of tubulin polymerization, bind in a distinct region of the vinca domain. Biochemistry 34:9714–9721

17. Owellen RJ, Hartke CA, Dickerson RM, Hains FO (1976) Inhibition of tubulin-microtubule polymerization by drugs of the Vinca alkaloid class. Cancer Res 36:1499–1502

18. Himes RH, Kersey RN, Heller-Bettinger I, Samson FE (1976) Action of the vinca alkaloids vincristine, vinblastine, and desacetyl vinblastine amide on microtubules in vitro. Cancer Res 36:3798–3802

19. Schiff PB, Fant J, Horwitz SB (1979) Promotion of microtubule assembly in vitro by taxol. Nature 227:665–667

20. Sherline P, Leung JT, Kipnis DM (1975) Binding of colchicine to purified microtubule protein. J Biol Chem 250:5481–5486

21. Owellen RJ, Jr AHO, Donigian DW (1972) The binding of vincristine, vinblastine and colchicine to tubulin. Biochem Biophys Res Commun 47:685–691

22. Kumar N (1981) Taxol-induced polymerization of purified tubulin. Mechanism of action. J Biol Chem 256:10435–10441

23. Tsuruo T, Matsuzaki T, M. M, Saito H, Yokokura T (1988) Antitumor effect of CPT-11, a new derivative of camptothecin, against pleiotropic drug-resistant tumors in vitro and in vivo. Cancer Chemother Pharmacol 21:71–74

24. Slichenmyer WJ, Rowinsky EK, Donehower RC, Kaufmann SH (1993) The current status of camptothecin analogues as antitumor agents. J Natl Cancer Inst 85:271–291

25. Rothenberg ML (1997) Topoisomerase I inhibitors: review and update. Ann Oncol 8:837–855

26. Manfredi KP, Blunt JW, Cardellina-II JH, McMahon JB, Pannell LL, Cragg GM, Boyd MR (1991) Novel alkaloids from the tropical plant Ancistrocladus abbreviatus inhibit cell killing by HIV-1 and HIV-2. J Med Chem 34:3402–3405

27. Boyd M, Hallock Y, Cardellina-II J, Manfredi K, Blunt J, McMahon J, Buckheit R, Jr., Bringmann G, Schaffer M, Cragg G (1994) Anti-HIV michellamines from Ancistrocladus korupensis. J Med Chem 37:1740–1745

28. Nakamura A, Nagai K, Suzuki S, Ando K, Tamura G (1986) A novel method of screening for immunomodulating substances, establishment of an assay system and its application to culture broths of microorganisms. J Antibiot 39:1148–1154

29. Osada H, Magae J, Watanabe C, Isono K (1988) Rapid screening method for inhibitors of protein kinase C. J Antibiot 41:925–931

30. Osada H, Cui C-B, Onose R, Hanaoka F (1997) Screening of cell cycle inhibitors from microbial metabolites by a bioassay using a mouse cdc2 mutant cell line, tsFT210. Bioorg Med Chem 5:193–203

31. Tomoda H, Omura S (1990) New strategy for discovery of enzyme inhibitors: screening with intact mammalian cells or intact microorganisms having special functions. J Antibiot 43:1207–1222

32. Omura S, Sasaki Y, Iwai Y, Takeshima H (1995) Staurosporine, a potentially important gift from a microorganism. J Antibiot 48:535–548

33. Osada H, Takahashi H, Tsunoda K, Kusakabe H, Isono K (1990) A new inhibitor of protein kinase C, RK-286C (4'-demethylamino-4'-hydroxystaurosporine). I. Screening, taxonomy, fermentation and biological activity. J Antibiot 43:163–167

34. Osada H, Satake M, Koshino H, Onose R, Isono K (1992) A new indolocarbazole antibiotic, RK-286D. J Antibiot 45:278–279

35. Osada H, Koshino H, Kudo T, Onose R, Isono K (1992) A new inhibitor of protein kinase C, RK-1409 (7-oxostaurosporine). I. Taxonomy and biological activity. J Antibiot 45:189–194

36. Koshino H, Osada H, Amano S, Onose R, Isono K (1992) A new inhibitor of protein kinase C, RK-1409B (4'-demethylamino-4'-hydroxy-3'-epistaurosporine). J Antibiot 45:1428–1432

37. Alvarez MA, Fu H, Khosla C, Hopwood DA, Bailey JE (1996) Engineered biosynthesis of novel polyketides: Properties of the whiE aromatase/cyclase. Nature Biotech 14:335–338

38. Martin JF (1998) New aspects of genes and enzymes for beta-lactam antibiotic biosynthesis. Appl Microbiol Biotech 50:1–15

39. Seow KT, Meure G, Gerlitz M, Wendt-Pienkowski E, Hutchinson CR, Davies J (1997) A study of iterative type II polyketide synthases, using bacterial genes cloned from soil DNA: A means to access and use genes from uncultured microorganisms. J Bacteriol 179:7360–7368

40. Meijer L (1996) Chemical inhibitors of cyclin-dependent kinases. Trend Cell Biol 6:393–397

41. Usui T, Yoshida M, Abe K, Osada H, Isono K, Beppu T (1991) Uncoupled cell cycle without mitosis induced by a protein kinase inhibitor, K-252a. J Cell Biol 115:1275–1282

42. Abe K, Yoshida M, Usui T, Horinouchi S, Beppu T (1991) Highly synchronous culture of fibroblasts from G2 block caused by staurosporine, a potent inhibitor of protein kinases. Exp Cell Res 192:122–127

43. Gadbois DM, Hamaguchi JR, Swank RA, Bradbury EM (1992) Staurosporine is a potent inhibitor of p34^{cdc2} and p34^{cdc2}-like kinases. Biochem Biophys Res Commun 184:80–85

44. Kitagawa M, Higashi H, Takahashi IS, Okabe T, Ogino H, Taya Y, Nishimura S, Okuyama A (1994) A cyclin-dependent kinase inhibitor, butyrolactone I, inhibits phosphorylation of RB protein and cell cycle progression. Oncogene 9:2549–2557

45. Deguchi A, Imoto M, Umezawa K (1996) Inhibition of G1 cyclin expression in normal rat kidney cells by inostamycin, a phosphatidylinositol synthesis inhibitor. J Biochem 120:1118–1122

46. Bollag DM, McQueney PA, Zhu J, Hensens O, Koupal L, Liesch J, Goetz M, Lazarides E, Woods CM (1995) Epothilones, a new class of microtubule-stabilizing agents with a taxol-like mechanism of action. Cancer Res 55:2325–2333

47. Hamaguchi T, Sudo T, Osada H (1995) RK-682, a potent inhibitor of tyrosine phosphatase, arrested the mammalian cell cycle progression at G1 phase. FEBS Lett 372:54–58

48. Usui T, Marriott G, Inagaki M, Schwarp G, Osada H (1999) Protein phosphatase 2A inhibitors, phoslactomycins. Effects on the cytoskeleton in NIH/3T3 cells. J Biochem 125:960–965

49. Kagamizono T, Hamaguchi T, Ando T, Sugawara K, Adachi T, Osada H (1999) Phosphatoquinones A and B, novel tyrosine phosphatase inhibitors produced by Streptomyces sp. J Antibiot 52:75–80

50. Greene LA, Tischler AS (1976) Establishment of a noradrenergic clonal line of rat adrenal pheochromocytoma cells which respond to nerve growth factor. Proc Natl Acad Sci, USA 73:2424–2428

51. Troppmair J, Bruder JT, App H, Cai H, Liptak L, Szeberenyi J, Cooper GM, Rapp UR (1992) Ras controls coupling of growth factor receptors and protein kinase C in the membrane to Raf-1 and B-Raf protein serine kinases in the cytosol. Oncogene 7:1867–73

52. Yao R, Cooper GM (1995) Requirement for phosphatidylinositol-3 kinase in the prevention of apoptosis by nerve growth factor. Science 267:2003–2006

53. Omura S, Fujimoto T, Otoguro K, Matsuzaki K, Moriguchi R, Tanaka H, Sasaki Y (1991) Lactacystin, a novel microbial metabolite, induces neuritogenesis of neuroblastoma cells. J Antibiot 44:113–116

54. Fenteany G, Standaert RF, Lane WS, Choi S, Corey EJ, Schreiber SL (1995) Inhibition of proteasome activities and subunit-specific amino-terminal threonine modification by lactacystin. Science 268:726–731

55. Koizumi S, Contreras ML, Matsuda Y, Hama T, Lazarovici P, Guroff G (1988) K-252a: a specific inhibitor of the action of nerve growth factor on PC 12 cells. J Neurosci 8:715–721

56. Berg MM, Sternberg DW, Parada LF, Chao MV (1992) K-252a inhibits nerve growth factor-induced trk proto-oncogene tyrosine phosphorylation and kinase activity. J Biol Chem 267:13–16

57. Kakeya H, Takahashi I, Okada G, Isono K, Osada H (1995) Epolactaene, a novel neuritogenic compound in human neuroblastoma cells, produced by a marine fungus. J Antibiot 48:733–735

58. Kakeya H, Onozawa C, Sato M, Arai K, Osada H (1997) Neuritogenic effect of epolactaene derivatives on human neuroblastoma cells which lack high affinity nerve growth factor receptors. J Med Chem 40:391–394

59. Green D, Kroemer G (1998) The central executioners of apoptosis : caspases or mitochondria ? Trend Cell Biol 8:267–271

60. Jacobson MD (1996) Reactive oxygen species and programmed cell death. Trend Biol Sci 21:83–86

61. Korsmeyer SJ, Yin XM, Oltvai ZN, Veis-Novack DJ, Linette GP (1995) Reactive oxygen species and the regulation of cell death by the Bcl-2 gene family. Biochim Biophys Acta 1271:63–66

62. Schlapbach R, Fontana A (1997) Differential activity of bcl-2 and ICE enzyme family protease inhibitors on Fas and puromycin-induced apoptosis of glioma cells. Biochim Biophys Acta 1359:174–180

63. Xia Z, Dickens M, Raingeaud J, Davis RJ, Greenberg ME (1995) Opposing effects of ERK and JNK-p38 MAP kinases on apoptosis. Science 270:1326–1331

64. Graves JD, Gotoh Y, Draves KE, Ambrose D, Han DKM, Wright M, Chernoff J, Clark EA, Krebs EG (1998) Caspase-mediated activation and induction of apoptosis by the mammalian Ste20-like kinase Mst1. EMBO J 17:2224–2234

65. Kakeya H, Onose R, Osada H (1998) Caspase-mediated activation of a 36-kDa myelin basic protein kinase during anticancer drug-induced apoptosis. Cancer Res 58:4888–4894

66. Watabe M, Kakeya H, Osada H (1999) Requirement of protein kinase (Krs/MST) activation for MT-21-induced apoptosis. Oncogene 18:5211–5220

67. Elliott JF, Lin Y, Mizel SB, Bleackley RC, Harnish DG, Paetkau V (1984) Induction of interleukin 2 messenger RNA inhibited by cyclosporin A. Science 226:1439–1441

68. Tocci MJ, Matkovich DA, Collier KA, Kwok P, Dumont F, Lin S, Degudicibus S, Siekierka JJ, Chin J, Huchinson N (1989) The immunosuppressant FK-506 selectively inhibits expression of early cell activation genes. J Immunol 143:718–726

69. Vezina C, Kudelski A, Sehgal SN (1975) Rapamycin (AY-22,989), a new antifungal antibiotic I. Taxonomy of the producing streptomycete and isolation of the active principle. J Antibiot 28:721–726

70. Bierer BE, Mattila PS, Standaert RF, Herzenberg LA, Burakoff SJ, Crabtree G, Schreiber SL (1990) Two distinct signal transmission pathways in T lymphocytes are inhibited by complexes formed between an immunophilin and either FK506 or rapamycin. Proc Natl Acad Sci USA 87:9231–9235

71. Masuda T, Mizutani S, Iijima M, Odai H, Suda H, Ishizuka M, Takeuchi T, Umezawa H (1987) Immunosuppressive activity of 15-deoxyspergualin and its effect on skin allografts in rats. J Antibiot 40:1612–1618

72. Nemoto K, Hayashi M, Abe F, Nakamura T, Ishizuka M, Umezawa H (1987) Immuno-suppressive activities of 15-deoxyspergualin in animals. J Antibiot 40:561–562
73. Nadler SG, Tepper MA, Schacter B, Mazzucco CE (1992) Interaction of the immuno-suppressant deoxyspergualin with a member of the Hsp70 family of heat shock pro-teins. Science 258:484–486
74. Endo A (1979) Monacolin K, a new hypo-cholesterolemic agent produced by a-Monascus species. J Antibiot 32:852–854
75. Matsuoka T, Miyakoshi S, Tanzawa K, Nakahara K, Hosobuchi M, Serizawa N (1989) Purification and characterization of cytochrome P-450sca from Streptomyces carbophilus. Eur J Biochem 184:707–713
76. Dawson MJ, Farthing JE, Marshall PS, Middleton RF, O'Neill MJ, Shuttleworth A, Stylli C, Tait RM, Taylor PM, Wildman HG, Buss AD, Langley D, Hayes MV (1992) The squalestatins, novel inhibitors of squalene synthase produced by a species of Phoma: I. Taxonomy, fermentation, isolation, physico-chemical properties and biological activ-ity. J Antibiot 45:639–647
77. Bergstrom JD, Kurtz MM, Rew DJ, Amend AM, Karkas JD, Bostedor RG, et al., & Alberts AW (1993) Zaragozic acids: a family of fungal metabolites that are picomolar competitive inhibitors of squalene synthase. Proc Natl Acad Sci, USA 90:80–84
78. Tomoda H, Tabata N, Yang D-J, Takayanagi H, Nishida H, Omura S (1995) Pyripyropenes, novel ACAT inhibitors produced by Aspergillus fumigatus. III. Struc-ture elucidation of pyripyropenes E to L. J Antibiot 48:495–503
79. Cui H, Matsui K, Omura S, Schauer SL, Matulka RA, Sonenshein GE, Ju S (1997) Proteasome regulation of activation-induced T cell death. Proc Natl Acad Sci USA 94:7515–7520
80. Hamamoto T, Gunji S, Tsuji H, Beppu T (1983) Leptomycins A and B, new antifungal antibiotics. I. Taxonomy of the producing strain and their fermentation, purification and characterization. J Antibiot 36:639–645
81. Kudo N, Khochbin S, Nishi K, Kitano K, Yanagida M, Yoshida M, Horinouchi S (1997) Molecular cloning and cell cycle-dependent expression of mammalian CRM1, a protein involved in nuclear export of proteins. J Biol Chem 272:29742–29751
82. Fukuda M, Asano S, Nakamura T, Adachi M, Yoshida M, Yanagida M, Nishida E (1997) CRM1 is responsible for intracellular transport mediated by the nuclear export signal. Nature 390:308–311
83. Fornerod M, Ohno M, Yoshida M, Mattaj IW (1997) CRM1 is an export receptor for leucine-rich nuclear export signals. Cell 90:1051–1060

2
Cell Proliferation: From Signal Transduction to Cell Cycle

Minoru Yoshida

1 Introduction

Eukaryotic cells have highly coordinated mechanisms to control cell proliferation. These include mitogenic signaling and cell cycle control. Cells receive a variety of positive and negative signals from external (growth factors, stresses, etc.) and internal (DNA damage, microtubule integrity, etc.) conditions and must decide to start or cease the cell cycle in response to these signals. Malignant cells arise as a result of a stepwise progression of genetic events that include short-circuited signal transduction and unregulated cell cycle progression. This review focuses on several aspects of cell proliferation control and target molecules for bioprobes which are highly useful in exploring the molecular mechanisms of cell proliferation control. Inhibitors of cell proliferation include a variety of classical cytotoxic compounds such as DNA-damaging agents, membrane-attacking agents, macromolecular synthesis inhibitors, and inhibitors of the respiratory system. Although some of these cytotoxic compounds are useful in analyzing cellular functions, this review focuses on relatively recent inhibitors of cell proliferation, of which targets are involved in the regulation of signal transduction and the cell cycle.

2 Membrane/Cytoplasmic Signal Transduction

2.1 Growth Factors and Receptors

Growth factors are the first signal that cells receive from the environment. Growth factors cause cells in the resting or G0 phase to enter and proceed through the cell cycle. Clues to understanding how cells receive the growth factor signal and transduce to the second messengers were obtained from studies on the action of oncogenes. A number of retroviral oncogene products are similar to the protein kinase encoded by v-*src* that specifically phosphorylates protein tyrosine residues. Some of them were found to be derived from normal growth factor receptor genes. For example, purification and sequencing of epidermal growth factor (EGF) receptor revealed that the v-*erbB* product, which has sequence similarity to v-*src*, was a truncated form of EGF receptor. Thus, oncogenes activated by a variety of mechanisms frequently have been shown to encode growth factors and receptor tyrosine kinases that participate in mitogenic signaling. Normal mitogenic signaling by growth factors except for the transforming growth factor (TGF)-β/bone morphogenetic protein (BMP) family, of which receptors encode serine/threonine kinases, is mediated by their receptors with intrinsic tyrosine kinase activity. These receptors have an extracellular ligand-binding domain and an intracellular tyrosine kinase domain responsible for transducing the mitogenic signal. Ligand binding induces formation of receptor dimers or oligomers and molecular interaction between adjacent cytoplasmic domains leads to activation of kinase function [1].

The receptors themselves are often tyrosine-phosphorylated (autophosphorylation) after they are activated by ligand binding. Tyrosine phosphorylation may modulate kinase activity, but certainly affects the ability of the kinase to interact with substrates containing the Src homology (SH) region 2. Certain enzymes become physically associated with and are phosphorylated by the platelet-derived growth factor (PDGF) receptor. These proteins include phospholipase C (PLC-γ), phosphatidylinositol 3'-kinase (PI-3K), adaptor molecules like Ash/Grb2, and Src and Src-like tyrosine kinases. These molecules contain noncatalytic domains called SH2 and SH3. SH2 domains bind preferentially to the phosphorylated tyrosine residues, while SH3 domains bind the proline-rich regions of the signaling proteins. The protein-protein interaction via SH2 or SH3 is one of the major mechanisms by which the mitogenic signals are transduced [2].

PLC-γ hydrolyzes phosphatidylinositol 4,5-bisphosphate and generates two second messengers, inositol triphosphate and diacylglycerol. The former causes release of stored intracellular calcium and the latter activates protein kinase C (PKC). The activated receptor binds and activates PLC-γ by tyrosine phosphorylation, thereby causing rapid increases in the second messengers. PI-3K that phosphorylates the inositol ring of phosphatidylinositol in the 3' position is composed of the 110-kD catalytic subunit and the 85-kD regulatory subunit. The 85-kD subunit contains two SH2 domains and an SH3 domain and is tyrosine phosphorylated. The association of the 85-kD subunit with the receptor and phosphorylation leads to activation of the catalytic 110-kD subunit, which increases 3'-phosphorylated phosphatidylinositols [3]. These lipids produced may also be second messengers for activation of other proteins such as the atypical PKC family [4]. PI-3K is postulated to be involved in several important downstream pathways [5], including protein kinase B (PKB)/Akt, a growth-factor-regulated serine/threonine kinase containing a pleckstrin homology (PH) domain. Binding of PI-3K products to the PH domain results in translocation of PKB/Akt to the plasma membrane where it is activated by phosphorylation by phosphoinositide-dependent kinase 1. Activated PKB/Akt provides a survival signal that protects cells from apoptosis and also mediates growth-factor signaling involving 70/85 kD S6 kinases (p70^{S6k}) [6]. Activation of p70^{S6k} by serine phosphorylation results in 40S ribosomal protein S6 phosphorylation and is important for G1 cell cycle transition in a variety of cells [7].

The mechanism by which EGF or PDGF activates Ras, a small G protein, has been extensively studied [8,9]. Ras is a plasma membrane-associated guanine nucleotide-binding protein that cycles between a GTP-bound and a GDP-bound form. Mutations in Ras that increase the proportion of the GTP-bound state activate the biological functions of Ras and cause malignant transformation in normal cells. Inhibition of Ras by microinjection of an anti-Ras antibody prevents cellular transformation by the oncogenes encoding tyrosine kinases such as Src, indicating that Ras plays a pivotal role in pathways that transfer signals from protein-tyrosine kinases at the cell surface to transcription factors in the nucleus. Two proteins appear to mediate Ras activation: Grb2 (also called Ash and Sem5) and Sos. Grb2 is a cytosolic protein which consists of two SH3 domains and one SH2 domain. Sos is another cytosolic protein which contains a CDC25-related GTP/GDP-exchanging factor domain and a proline-rich SH3-binding domain. Sos facilitates the replacement of GDP by GTP. These two proteins form a stable complex through the proline-rich domain of Sos and a SH3 domain of Grb2. The phosphorylated EGF or PDGF receptor binds the Grb2/Sos complex through the SH2 domain of Grb2, thereby recruiting the cytosolic Grb2/Sos complex to the plasma membrane where Ras is localized. As a consequence, Sos can interact with Ras to facilitate the conversion from the GDP-from to the active GTP-form.

Another class of receptors distinct from membrane spanning tyrosine kinases is known to stimulate cell proliferation. These include receptors for cytokines, such as interleukins (IL), granulocyte-macrophage colony-stimulating factor (GM-CSF), G-CSF, and erythropoietin (Epo). These receptors are membrane glycoproteins with a single hydrophobic transmembrane domain. In contrast to EGF or PDGF receptors, these receptors possess no tyrosine kinase domain. Although little is known of the biochemical pathways by which these receptors stimulate proliferation, their activation can lead to the appearance of tyrosine phosphorylated proteins and increased amounts of GTP-bound Ras. The Src family of non-receptor type tyrosine kinases participate in signal transduction of this class of receptors. For example, binding of IL-2 to its receptor activates the tyrosine kinase Lck and Lyn. Janus kinase (JAK) 1 and 2 are another class of non-receptor type tyrosine kinases which are activated by some of the receptors for cytokines such as interferons [10]. JAK contains two possible tyrosine kinase-like domains in the C-terminal region. The ligand-activated Epo receptor binds and activates JAK2. The most important substrates for JAK appear to be a STAT (signal transducer and activator of transcription) family of proteins. STATs contain SH2 and SH3 domains as well as the tyrosine residues that are phosphorylated by JAK. The tyrosine phosphorylation facilitates formation of homodimers, which enable STATs to be localized in the nucleus [10].

2.2 Ras/Raf and MAP Kinase Cascade

Upon stimulation by growth factors, tyrosine kinase receptors recruit the cytosolic Grb2/Sos complex to the plasma membrane as described above, thereby converting Ras to an active GTP-bound from. Raf-1, a serine/threonine kinase, is an important effector of Ras in mammalian cells. Raf-1 functions downstream of Ras, which in its active, GTP-bound state binds directly to the N-terminal regulatory domain of Raf-1

[11]. This interaction serves to recruit Raf-1 to the cell membrane, which is necessary for Raf-1 activation. Raf-1 shares three regions of conservation, termed CR1, CR2 and CR3, with other Raf isomers and homologs. CR1 and CR2 are located in the N-terminal half and CR3 corresponds to the C-terminal kinase domain. Raf-1 contains two Ras-binding domains [12]. The best-characterized one is Ras-binding domain (RBD) and is contained within residues 51–131[13]. The interaction between RBD and Ras appears to then allow for a second Ras binding domain (residues 139–184) referred to as the cysteine-rich region (CRR) in Raf-1 to contact Ras. CR3 is masked intramolecularly by the N-terminal half of Raf-1, thereby stabilizing the inactive state, but this conformation can be disrupted upon binding to Ras.

Raf-1 is part of a highly conserved kinase cascade that mediates signaling from extracellular growth factors to mitogen-activated protein kinases (MAPK) [14]. The MAPK cascade consists of three distinct members of the protein kinase family, including MAPK, MAPK/ERK kinase (MEK, also known as MAPKK), and MEK kinase (MEKK, also known as MAPKKK). Raf-1 is a member of MEKK that phosphorylates and thereby activates MEK, and the activated form of MEK in turn phosphorylates and activates MAPK. Activated MAPK may translocate to the cell nucleus and regulate the activities of transcription factors. The best characterized MAPK cascade consists of Raf, MEK1, 2, and ERK1,2. Mitogenic signals stimulate this pathway, and proliferation can be blocked by inhibiting it. At least two other MAPKs that mediate responses to cellular stress, namely c-Jun amino-terminal kinase/stress-activated protein kinase (JNK/SAPK) and p38, are known. The defining property of JNK/SAPK is that instead of the motif Thr-Glu-Tyr (TEY) on ERK, JNK/SAPK contains the sequence Thr-Pro-Tyr (TPY), which must be phosphorylated on T and Y for activation [15]. On the other hand, p38, a homolog of the yeast high-osmolarity glycerol response 1 (HOG1) MAPK, is defined by the sequence TGY in place of TEY in ERK [16]. A major advance in understanding signaling events in response to stress came from the discovery that stress such as heat shock, osmotic shock, cytokines, protein synthesis inhibitors, antioxidants, UV, and DNA-damaging agents will activate JNK/SAPK and p38. Growth inhibition may be a consequence of these MAPKs. These MAPKs are likely signaling participants in apoptosis. MEKK1-SEK1 (or MKK4) and ASK1-MKK3 (or MKK6) were identified as the upstream kinases for JNK/SAPK and p38, respectively [17].

2.3 Inhibitors of Signal Transduction

A large number of natural products and synthetic compounds were extensively screened for growth factor antagonists and inhibitors of receptor-associated protein-tyrosine kinases (Fig. 1). Suramin, a polyanionic compound, is known to inhibit growth factor binding to its receptor [18]. These growth factors include PDGF, hepatocyte growth factor (HGF), and vascular endothelial growth factor (VEGF), all of which are important for tumorigenesis and angiogenesis. Therefore, suramin is considered as an antitumor agent blocking receptor activation. Herbimycin A [19], genistein [20], erbstatin [21], and tyrphostins [22] were reported as protein-tyrosine kinase inhibitors. Herbimycin A is a benzoquinone ansamycin antibiotic that reverses the transformed morphology of *src*ts-transformed NRK cells to normal ones. This morphological change was explained by inactivation of Src protein kinase due to covalent binding of this antibiotic to the C-terminal conserved cysteine residue [23]. Recently, benzoquinone ansamycins such as herbimycin and geldanamycin were shown to interact with and inhibit Hsp90, a molecular chaperone [24]. Since Hsp90 is required for correct protein folding and membrane sorting of Src, strong inhibition of Src by herbimycin A may partly result from its inhibition of Hsp90. Genistein and erbstatin are the microbial metabolites that inhibit EGF receptor protein kinase. Genistein, an isoflavon derivative, inhibits protein-tyrosine kinase activity by competing with ATP. On the other hand, erbstatin has a structure similar to the substrate tyrosine. Therefore, erbstatin competes with peptide substrates. Tyrphostins were synthesized as more selective inhibitors for EGF receptor kinase and other tyrosine kinases based on erbstatin. Now, tyrphostins AG1478, AG1296, and AG490 are widely used as the specific tyrosine kinase inhibitors for EGF receptor, PDGF receptor, and JAK2 kinase, respectively. The tyrosine phosphorylated proteins must revert to their unphosphorylated form when signals are gone. Protein-tyrosine phosphatases play an important role in down-regulating signal transduction. Sodium orthovanadate is the most frequently used as an inhibitor of protein tyrosine phosphatases. Dephostatin [25] and RK-682 [26] are such inhibitors from a microbial origin.

A number of potent inhibitors of PKCs, serine/threonine kinases, have also been reported. Staurosporine was isolated from bacteria and identified as a potent inhibitor of PKC activity (Fig. 1). Its analogs UCN-01 and CGP 41251 [27], candidates for

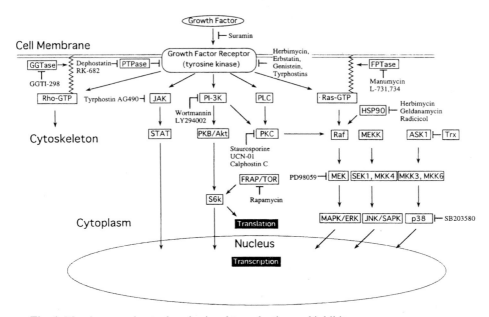

Fig. 1. Membrane and cytoplasmic signal transduction and inhibitors

antitumor drugs, inhibit conventional PKC isozymes more potently than "novel" and "atypical" ones. Although inhibitory activity of staurosporine and its analogs were very potent, they also inhibit a broad range of serine/threonine kinases such as CDK, probably because their mode of enzyme inhibition is in competition with ATP [28]. Calphostin C is a more specific inhibitor for PKC, which may interact with the regulatory domain of PKC [29]. 1-O-Octadecyl-2-O-methyl-rac-glycero-3-phosphocholine (ET-18-OCH3) is a synthetic diether phospholipid that is competitive with phosphatidylserine binding to the regulatory domain of PKC [30]. Auranofin, a lipophilic gold compound with antirheumatic activity, was shown to be a specific PKC inhibitor interacting with the catalytic domain [31].

Inostamycin (*see* p.81), which had been isolated from *Streptomyces* sp. as an inhibitor of phosphatidylinositol (PI) turnover, was shown to block CDP-diacyl glycerol (DG): inositol transferase that catalyzes the synthesis of PtdIns from CDP-DG and inositol [32]. Inostamycin inhibits cell cycle progression in normal rat kidney cells in G1 phase, by down-regulating expression of cyclins D1 and E, suggesting that inositol turnover is essential for signal transduction leading to G1 cyclin synthesis [33].

Rapamycin is a potent immunosuppressant which specifically inhibits G1- to-S-phase progression, leading to cell-cycle arrest in yeast and mammals (Fig. 1). One of the rapamycin-binding proteins was identified to be FK506-binding protein (FKBP12), with which FK506 was known to inhibit calcineurin. However, rapamycin does not inhibit calcineurin but blocks activation of p70^{S6k} that is involved in translational control [34,35]. PI kinase homologs, TOR1, TOR2 (target of rapamycin), and FKBP-rapamycin associated protein (FRAP), were identified as the target of the FKBP-rapamycin complex from yeast and mammals, respectively [36,37], which were shown to be rapamycin-sensitive regulators of p70^{S6k}. Both the kinase activity and an amino-terminal domain of FRAP are required for this regulation [38]. PI-3K also regulates p70^{S6k} activity, probably through the activation of PKB/Akt. Important roles of PI-3K in cell cycle control and differentiation have been shown by its specific inhibitors, wortmannin [39] and LY294002 . Wortmannin inhibits PI-3K by directly associating with the 110-kD catalytic subunit. Rapamycin and wortmannin cause similar downregulation of the translation rate of some mRNA species such as D-type cyclins, suggesting that p70^{S6k} is important for translation of a certain subset of mRNAs [40].

Since Ras plays a pivotal role in oncogenic signal transduction leading to aberrant cell proliferation in tumors, a variety of approaches have been applied for discovery of inhibitors of Ras-mediated signal transduction. To acquire transforming potential, the precursor of the Ras oncoprotein must undergo prenylation of the cysteine residue located in a carboxyl-terminal tetrapeptide. Protein prenylation is catalyzed by three prenyl transferases, farnesyl protein transferase (FPTase), geranylgeranyltransferase (GGTase) I and II. Ras is farnesylated, while the Rho family of small G proteins are geranylgeranylated. Inhibitors of the enzyme that catalyzes Ras farnesylation, FPTase, have therefore been suggested as anticancer agents for tumors in which Ras contributes to transformation and development (Fig. 1). Manumycin and its derivatives were isolated as FPTase inhibitors by a microbial screen using a yeast strain with conditional deficiency in the *GPA1* gene. Manumycin suppresses the lethality of *gpa1* disruption by inhibiting FPTase [41]. The analogs and peptidomimetics of the tetrapeptide that undergoes farnesylation were extensively synthesized [42,43]. L-731,735 is a potent and selective inhibitor of FPTase in vitro. A prodrug of this compound, L-731,734, inhibited Ras farnesylation and decreased the ability of v-*ras*-transformed cells to form colonies in soft agar but had no effect on the efficiency of colony formation of cells transformed by other oncogenes.

Mevalonate is an important precursor for cholesterol and protein prenylation such as farnesylation and geranylgeranylation. Hydroxymethylglutaryl (HMG)-CoA reductase is a rate-limiting enzyme to produce mevalonate and a target for drugs for hypercholestolemia. Inhibitors of HMG-CoA reductase, such as compactin, lovastatin, and pravastatin, block not only cholesterol synthesis but also protein prenylation. Lovastatin causes G1 cell cycle arrest, and this effect was canceled by the addition of mevalonate but not by cholesterol, suggesting that protein prenylation is required for cell cycle progression [44]. Although the most likely target essential for the cell cycle had been thought to be Ras farnesylation, this possibility was ruled out by the fact that lovastatin was able to block the cell cycle even in cells expressing myristoylated Ras which are capable of membrane sorting and functioning in the presence of lovastatin [45]. In fact, a specific inhibitor of FPTase did not inhibit cell cycle progression of NIH3T3 fibroblasts, whereas GGTI-298, a specific inhibitor of GGTase I, blocked the cell cycle [46] by inducing p21$^{Waf1/Cip1}$, a CDK inhibitor [47]. It is therefore likely that inhibition of protein geranylgeranylation, not farnesylation, required for the G1/S phase transition, is the primary reason for the cell cycle arrest by lovastatin.

Since activation of Raf-1 requires direct interaction with Ras, it is likely that inhibition of Ras/Raf binding would be another approach to blocking Ras-mediated signal transduction. Radicicol, known as a Src kinase inhibitor [48], was shown to block Ras/Raf interaction in a yeast two-hybrid screen system. In vivo Ras/Raf-1 binding in v-Ha-*ras*-transformed cells was blocked by low concentrations of radicicol, while in vitro binding of glutathion *S*-transferase-fused Ras to a maltose binding protein-fused RIP3 containing the Ras-binding domain (RBD) of Raf-1 was not inhibited by radicicol. It was suggested that radicicol interacts indirectly with the region except with RBD of Raf-1, thereby inhibiting a conformational change of Raf-1 prerequisite for binding to Ras [49]. Recently, radicicol as well as geldanamycin were shown to bind to the N-terminal ATP/ADP- binding domain of Hsp90, with radicicol displaying nanomolar affinity, and inhibiting the inherent ATPase activity of Hsp90 which is essential for its function in vivo (Fig. 1). Crystal structure determinations of Hsp90 N-terminal domain complexes with geldanamycin and radicicol identify key aspects of their nucleotide mimicry [50]. Raf-1 is known to interact with the molecular chaperone Hsp90, which has been suggested to function as a regulator of the Raf-1 structure and activity. Therefore, it seems likely that inhibition of Hsp90 function by radicicol is due to the decreased Ras/Raf-1 binding in vivo.

MAPKK and MAPK such as MEK1, 2 and ERK1,2 are the important down-stream targets of Ras/Raf. Thus, the inhibition of the MAPK pathway may also block the Ras-mediated signal. PD98059 has been reported as a MEK inhibitor [51], while SB203580 [52] is an inhibitor for p38 MAPK (Fig. 1).

3 Nuclear Signal Transduction

3.1 Nuclear Protein Import and Export

Signal transduction from plasma membrane-associated receptors requires not only catalytic activation of signaling molecules but a transient spatial redistribution within cells. Indeed, activation of both ERK1 and ERK2 is accompanied by their rapid relocalization from the cytoplasm to the nucleus, and phosphorylation of target transcription factors. Persistent localization of active ERK to the nucleus increases the frequency of cellular transformation in fibroblasts, suggesting that the duration of nuclear localization of active ERK is critical for ultimate cellular response [53].

The import of proteins into the nucleus occurs through the nuclear pore complexes (NPC) which allow diffusion of small molecules and can accommodate the active transport of particles as large as several million dalton in weight. The active import is energy-dependent and is conferred by virtue of a nuclear localization signal (NLS) that is rich in basic amino acids. The translocation of the NLS-bearing protein into the nucleoplasm is mediated by the saturable import machinery that recognizes basic NLSs [54]. The import receptor for NLS was identified to be a heterodimeric complex of importin α and importin β (also known as karyopherin α and β or PTACs). Importin α recognizes and binds basic NLSs, while importin β mediates docking of the cargo-importin complex with the NPC, in a temperature-independent manner. Translocation into the nucleoplasm is mediated by the Ran GTPase cycle, and requires energy in the form of GTP hydrolysis [55]. Importin β binds directly to the NPC and Ran-GTP. In turn, Ran-GTP binding to importin β dissociates the importin α-β heterodimer. Ran-GTP binds to the amino terminus of importin β, whereas importin α binds to the carboxy-terminal region of importin β. Release of importin α upon Ran-GTP binding to importin β is therefore probably explained by a conformational change in importin β.

A second pathway for import of proteins to the nucleus has also been identified. A subset of hnRNP proteins (heterogeneous nuclear RNA-binding proteins) is imported by a signal of 38 amino acids, termed M9, which is different in character from basic NLSs [56]. M9 is recognized by transportin, a specific M9 receptor. Transportin is distantly related to importin β, and translocation of the cargo-transportin complex is also mediated by Ran but is independent of GTP hydrolysis.

Ran is a Ras homolog in the nucleus, which switches between a GDP-bound and a GTP-bound state by nucleotide exchange and GTP hydrolysis, like all G proteins. RCC1 (regulator of chromosome condensation 1) is a major guanine nucleotide exchange factor (GEF) for Ran and generates Ran-GTP [57]. RanGAP1 is the GTPase-activating protein and causes conversion of Ran-GTP into Ran-GDP [58]. A characteristic feature of the Ran system is the asymmetric distribution of Ran-GTP and Ran-GDP. RCC1 is chromatin-bound and generates Ran-GTP in the nucleus. In contrast, RanGAP1 is cytoplasmic, and therefore depletes Ran-GTP from the cytoplasm. This asymmetric distribution of Ran explains why the Ran-GTP-mediated dissociation of importin heterodimer occurs specifically in the nucleus that follows cargo translocation into the nucleus [59].

RNAs must exit the nucleus through the nuclear pores to the cytoplasm, where they function. One of the earliest indications that proteins mediate RNA export came from HIV-1. The viral protein Rev recognizes a specific HIV RNA sequence, the Rev-responsive element, and mediates export of the unspliced viral RNA. A leucine-rich sequence in Rev was identified as a nuclear export signal (NES), which may be recognized by a saturable export receptor [60]. Similar NESs were found in many proteins including PKIα [61], which also act as the transport signal. These results suggest the presence of a cellular carrier protein responsible for the nuclear export of proteins. Recently, CRM1, which had been originally identified as a gene product required for chromosome region maintenance [62], was identified to be the NES receptor and renamed as exportin 1. This finding came from the molecular analysis of the mechanism by which leptomycin B, a small molecule cell cycle inhibitor produced by a *Streptomyces*, caused cell cycle arrest, as described below. CRM1/exportin 1 shares low but significant homology to importin β that recruits the complex of importin α and the NLS-containing proteins to the NPC by directly interacting with the NPC. CRM1/exportin 1 is shown to convey NES-bearing proteins through the NPC in a Ran-GTP-dependent manner. Thus, the NES-mediated intracellular transport system is a universal and conserved mechanism by which subcellular localization of proteins is controlled in cells. Increased evidence has suggested that many signaling molecules have NLS or NES or both, by which they relocalize in response to various stimuli. For instance, unstimulated MEK binds and anchors MAPK in the cytoplasm by its NES. The phosphorylated, activated MAPK can be released from MEK and enter the nucleus, although it is still unclear how MAPK dissociates from the complex [53]. On the other hand, MAPKAP kinase 2, an NLS-bearing nuclear substrate of MEK and p38, is exported from the nucleus by phosphorylation of a threonine residue (T317) that is overlapped with the potential NES [63].

3.2   Nuclear Signaling by Protein Phosphorylation

The nuclear import of the signaling molecules result in the initiation of specific gene transcription. Since the induction of signal-dependent transcription does not require de novo protein synthesis, it is likely that specific transcription factors would be activated by protein modifications such as phosphorylation in response to stimuli. Phosphorylation of transcription factors has been most extensively studied.

Recent extensive studies of the Ras-Raf-MAPK cascade have revealed that members of at least six subfamilies of Ets proteins (Ets, Yan, Elg, Pea3, Erf and ternary complex factor (TCF)) are nuclear targets of this pathway [64]. On the basis of their structural feature, these proteins are classified into two subfamilies, Ets and TCF. The Ets subfamily possess a highly conserved N-terminal regulatory domain and a C-terminal motif that comprises the DNA-binding domain (ETS-DBD). MAPK phosphorylates a single threonine residue within the N-terminal regulatory domain. By contrast, TCF subfamily members feature an ETS-DBD at the N-terminus. Unlike the Ets subfamily, the C-terminal transactivation domain of the TCFs contain multiple serine and threonine residues that can be phosphorylated by MAPK. Ets proteins regulate specific genes through interactions with other transcription factors at complex sites on DNA, termed Ras-responsive elements (RRE). Ets-1 and AP-1 proteins, together with Ras, are required for full functional activity of RRE. On other types of RRE, Ets proteins assemble complexes with the ubiquitous protein serum-response factor (SRF) that recognized serum-response elements (SRE), or with the dedicated pituitary-specific factor Pit-1. SREs are present in the promoters of many immediate early genes such as c-*fos* and function as RREs that mediate responses to many extracellular stimuli. The c-*fos* SRE is continuously occupied in vivo by SRF, and Ets proteins of the TCF subfamily. An SRF dimer binds with high affinity to the sequence CC(A/T)6GG within SRE. TCFs are inefficient in binding the SRE without SRF, but when they are recruited by the SRF dimer, a stable ternary complex is formed leading to unmasking the transactivation domain of SRF.

AP-1, a heterodimer composed of c-Jun and c-Fos, was identified as the protein that binds the TPA (12-O-tetradecanoylphorbol-13-acetate) responsive element (TRE) located upstream of a variety of TPA-inducible genes [65]. Transcriptional activity of AP-1 is enhanced by not only TPA but also growth factors, cytokines, and stresses such as UV irradiation. c-Jun can be phosphorylated in both N- and C-terminal regions. The DNA binding activity of AP-1 is negatively regulated by phosphorylation in the C-terminal region of c-Jun, which is constitutive and decreased by stimulation. At least two of these sites, Thr-231 and Ser-249, have been shown to be phosphorylated by casein kinase II (CKII). On the other hand, phosphorylation of the N-terminal region upon a variety of stimuli and oncogenic transformation by activated Ras is required for transcriptional activation of AP-1. The use of affinity columns containing the N-terminal activation domain of c-Jun led to identification of at least two kinase activities, 46 and 55 kDa in size, which bind to this region and phosphorylate it on serines 63 and 73. The kinases were named JNKs, and were also independently identified as cycloheximide activated protein kinases, SAPKs. As mentioned above, the JNK subfamily including JNK1, JNK2, and JNK3 (also known as SAPKα, SAPKβ, and SAPKγ) belongs to the MAPK superfamily protein kinases [17].

CRE binding protein (CREB) was identified as a protein recognizing the cAMP-responsive elements (CRE) [66]. Signals that increases cAMP cause activation of CREB. When cAMP binds inactive protein kinase A (PKA) consisting of the regulatory and catalytic subunits, the catalytic subunit dissociates from the complex and is translocated into the nucleus. CREB is thus activated by PKA through phosphorylating on the serine 133. The same serine residue is also phosphorylated by calmodulin-dependent kinase, suggesting that CREB is also activated by calcium-mediated signals. Recently, CREB formed a large protein complex containing CREB-binding protein (CBP). CBP was found to have histone acetyltransferase activity, which can remodel the chromatin structure (see below).

3.3   Nuclear Signaling by Protein Acetylation

The organization of chromatin is also crucial for the regulation of gene expression. In particular, both the positioning and properties of nucleosomes influence promoter-specific transcription in response to extracellular or intracellular signals. The nucleosome core contains DNA of 146 bp (base pairs) tightly wrapped around a central histone octamer comprising two molecules of each of core histones. The four core histones (H2A, H2B, H3 and H4) are subject to a variety of enzyme-catalyzed post-translational modifications, thereby modulating the chromatin functions. Of the modifications, acetylation has been the most extensively studied. The primary sites of histone acetylation are specific lysine residues in the positively charged N-terminal tails that protrude from the octamer, which are important for both histone-DNA and histone-nonhistone proteins interactions. The neutralization of the positive charge by acetylation has long been proposed to lead to loosening histone-DNA contacts, which facilitates the accessibility of a variety of factors to DNA [67]. The acetyl groups on histone molecules continuously turn over and the net level of acetylation is controlled by an equilibrium between the two specific enzyme activities, histone acetyltransferase (HAT) and deacetylase (HDAC). The genes encoding these enzymes have not been cloned until recently and the mechanism of regulation of the enzyme activities is still largely unknown. Recently, a human histone deacetylase (HDAC1) was isolated [68] as a protein that binds trapoxin [69]. HDAC1 was significantly similar to the yeast transcriptional regulator, Rpd3 [68]. On the other hand, several different human histone acetyltransferases (hGCN5, P/CAF, p300/CBP and TAF$_{II}$250) have been identified [70]. Recent studies have shown that HAT acts as the transcriptional coactivator whereas HDAC is associated with the corepressors such as Sin3 and NcoR/SMRT [71]. It was shown that the adenoviral oncoprotein E1A disturbed the normal cellular interaction of P/CAF with p300/CBP [72]. Furthermore, overexpression of P/CAF in cultured cells inhibited cell cycle progression and counteracted the activity of E1A [73]. A genetic defect in one of the two copies of human CBP gene is associated with Rubinstein-Taybi syndrome which is accompanied by high malignancy [74]. These findings suggest that control of the chromatin function by histone acetylation is one of the critical targets of oncoproteins and that its deregulation may lead to cellular transformation.

Recently, a number of nuclear proteins other than histones were reported to be acetylated by HAT. These include p53 [75] and GATA-1 [76]. Although little is known about the biological importance of acetylation of nonhistone nuclear proteins, acetylation apparently affects the DNA binding activity of p53 and GATA-1. These observations suggest that acetylation is a broad and general mechanism by which functions of transcription factors are modified.

3.4   Inhibitors of Nuclear Signal Transduction

Recent impressive speed in understanding the mechanisms of nucleo-cytoplasmic transport was accelerated by leptomycin B, a *Streptomyces* metabolite blocking nuclear export of NES-bearing proteins (Fig. 2). A study on leptomycin B-resistant mutants of *Schizosaccharomyces pombe* revealed that one class of mutants fell in the essential gene, *crm1+*, indicating that leptomycin B inhibits the CRM1 function [77]. Leptomycin B was recently rediscovered during the course of screening for the inhibitors of Rev nuclear export [78]. Furthermore, a human homolog of CRM1 was cloned and was found to be localized preferentially in the nuclear envelope and the nucleoplasm [79,80]. These results suggest that CRM1 is involved in the nuclear export of Rev and other proteins. Several groups provided strong evidence that CRM1 is a nuclear export receptor that recognizes Rev-like NESs [81–84]. CRM1 binds NES-bearing proteins in a Ran-GTP-dependent manner [81]. Thus, CRM1 has been renamed exportin 1. Leptomycin B binds covalently to a cysteine residue localized in the central conserved domain of CRM1, leading to inactivation of the protein function [85].

Several compounds have been reported to inhibit binding of transcription factors to their recognition DNA elements, although their specificity of inhibition is not well understood (Fig. 2). K1115A is an anthraquinone derivative that inhibits the binding of AP-1 to its recognition sites. Both collagenase production in IL-1α-stimulated cells and ornithine decarboxylase production in phorbol ester-stimulated cells were blocked by K1115A [86]. Recently, GE3, a novel anti-proliferative agent [87], was shown to inhibit transcription mediated by E2F, an important transcription factor involved in cell cycle control (see below). The gel shift assay showed that GE3 inhibited binding of E2F to its specific DNA element but not those of other transcription factors such as AP-1 and Sp1. Indeed, GE3 suppresses gene expression of *cyclin A*, *cdc2*, *polα*, and c-*myc*, of which transcription is controlled by E2F.

n-Butyrate has been known as an inhibitor of HDAC as well as an inducer of differentiation and cell cycle arrest. Although it has been considered to express its pleiotropic effects through the inhibition of HDAC, lack of specificity makes it difficult to draw a definitive conclusion [88]. Trichostatin A, an inducer of differentiation and morphological reversion of transformed cells with cell cycle arresting activity, was shown to be a potent inhibitor of the histone deacetylases (Fig. 2). Trichostatin A causes accumulation of hyperacetylated histones by its reversible inhibition of the deacetylation in the cells. The histone deacetylase from a trichostatin A-resistant cell line is resistant to trichostatin A, indicating that the enzyme is the primary target [89]. Trapoxin A is structurally unrelated to trichostatin A but causes inhibition of histone deacetylase through irreversible binding to the enzyme [69]. As described above, human HDAC1 was first isolated by using the affinity to a trapoxin analogue. Both of these HDAC inhibitors induce a variety of biological responses of cells such as induction of differentiation, arrest of the cell cycle, and inhibition of embryogenesis [88]. Trichostatin A selectively induces the expression of specific genes for gelsolin or p21[Waf1/Cip1]. These findings suggest that the increased acetylation of core histone induced by these HDAC inhibitors directly or indirectly affects the transcriptional activity of a subset of genes involved in cell cycle control and differentiation.

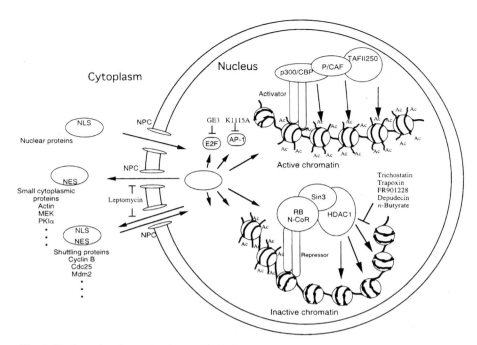

Fig. 2. Nuclear signal transduction and inhibitors

Thus, trichostatin A and trapoxin A are useful in analyzing the role of histone acetylation in chromatin structure and function as well as in determining the genes whose
activities are regulated by histone acetylation. Recently, a number of transcriptional
repressors such as Mad/Max, Ume6, unliganded nuclear receptors [71], and a methy-
lated CpG binding protein MeCP2 [90] were shown to recruit the HDAC complex to
the promoter regions. Trichostatin A and trapoxin A can suppress transcriptional
repression by these repressor-corepressor complexes. FR901228 [91], depudecin [92],
and oxamflatin [93], structurally unrelated compounds inducing morphological reversion of transformed cells, have recently been shown to be additional members of the
HDAC inhibitor group (Fig. 2).

4 Cell Cycle Control

The master checkpoint, originally proposed by Pardee and called the restriction (R)
point, defines a time in G1 at which cells are committed to enter S phase and no longer
respond to growth conditions [94]. Comprehensive studies on the R point have now
identified the D-type cyclins, Cdk4/Cdk6, pRb and certain CDK inhibitors as key
regulators of a common pathway controlling the commitment to enter S phase. On the
other hand, biochemical studies on the maturation-promoting factor (MPF) in oocytes and genetic studies on yeast cell division cycle (cdc) mutants converged to the

notion that Cdc2/cyclin B is a master regulator for the onset of M phase [95]. The kinase activity of these CDKs are regulated by a variety of checkpoint mechanisms following mitogen stimulation and damage to DNA and other cellular architectures.

4.1   Cyclin-Dependent Kinase

Cyclin-dependent kinases (CDKs) act as the engine that drives the cell cycle. The proper sequence of cell cycle events is governed by a control system consisting of CDKs and their regulatory subunits, cyclins. The first CDK called Cdc2 was identified as the product of a gene of *Schizosaccharomyces pombe* that complements the ts mutation of a fission yeast cell division cycle (cdc) mutant *cdc2*. The Cdc2 protein was required for both G1/S and G2/M transitions and was postulated to be the master regulator for the cell cycle. In *Saccharomyces cerevisiae*, a homologous gene termed *CDC28* performs similar functions. The Cdc2 human homologue was cloned because of its ability to complement an *S. pombe* temperature-sensitive *cdc2* mutation [96]. Cdc2 is also required for G2/M transition in mammalian cells. However, it was shown that in higher eukaryotes the G1/S transition is controlled by different Cdc2 homologues. While a single CDK triggers the major transitions of the yeast cell cycle, mammalian cells accommodate multiple CDK genes. A CDK requires association with a cyclin partner and concomitant phosphorylation/dephosphorylation of specific residues on CDK to become an active enzyme [97]. The first cyclin/CDK known to be activated after mammalian cells are released from a quiescent state is composed of a D-type cyclin and either Cdk4 or Cdk6 depending on the cell type. D-type cyclins are absent in quiescent cells and their expression is stimulated by growth factors. It is therefore likely that the expression of the D-type cyclins are the downstream target of signal transduction through the Ras-MAPK cascade. Cdk4/Cdk6 complexed with the D-type cyclins phosphorylate the retinoblastoma protein (pRb). The existence of a common pathway with an upstream function of cyclin D1 and pRb as the downstream target has been derived from cyclin D1 knockout experiments in pRb-positive and negative cells [98]. During the G0 to S transition, other G1 cyclin/CDK complexes such as cyclin E/CDK2 in mid- to late G1 and cyclin A/Cdk2 in late G1 to early S are sequentially activated following cyclin D/Cdk4 or Cdk6 activation [99]. Microinjection of antibodies to either cyclin E or cyclin A blocked G1/S transition, suggesting that both cyclins are required for the cell cycle progression. Cyclin protein levels are controlled strictly through the stage-specific expression and proteolytic destruction by the ubiquitin-dependent proteasome pathway. The G1 cyclins contain PEST sequences that are thought to confer instability, leading to a very short half-life. Deletion or mutations in the PEST sequence or in the destruction box block cyclin degradation, probably by preventing ubiquitination and subsequent proteolysis [100]. For example, cyclin E levels oscillate periodically: cyclin E is degraded after cells have entered S phase, and Cdk2 associates then with cyclin A.

4.2 Post-Translational Regulation of CDK

The kinase activity of CDKs is regulated tightly by not only transcription of cyclin genes and degradation of cyclins, but also modification of the kinase subunits by phosphorylation. Although bacterially produced CDKs form complexes with cyclins, they do not show detectable activity due to the absence of phosphorylation. The activity increases upon phosphorylation of a highly conserved threonine residue in the T loop by a CDK-activating kinase (CAK) [101]. CAK also belongs to the CDK family, which consists of cyclin H and Cdk7. In contrast to the phosphorylation in the T loop, Cdc2 is negatively regulated by phosphorylation at Tyr-15 and Thr-14 [102]. Wee1 is responsible for the phosphorylation at Tyr-15 in the nucleus, while Myt1 phosphorylates both Tyr-15 and Thr-14 in the cytoplasm. These kinases inactivated cyclin B/Cdc2 in the interphase. This effect is counteracted abruptly by Cdc25 phosphatase at the onset of M phase, which removes phosphates from both Tyr-15 and Thr-14. Cyclin B/Cdc2 phosphorylates a Ser/Thr-Pro motif, and mitotic phosphoproteins containing phosphorylated Ser/Thr-Pro motifs are recognized by a monoclonal antibody MPM-2 in a mitosis-specific manner. These MPM-2 antigens include the important mitotic regulators Cdc25, Myt1, Wee1, Plk1, and Cdc27. In fact, Cdc25 is activated by phosphorylation by cyclin B/Cdc2, generating a positive feedback loop, which is important for the activation of the kinase complex in mitosis [103]. The human rotamase or peptidyl-prolyl cis-trans isomerase (PPIase) Pin1 is a conserved mitotic regulator essential for the G2/M transition of the eukaryotic cell cycle. Pin1 interacts directly with a subset of mitotic phosphoproteins on phosphorylated Ser/Thr-Pro motifs, being required for proper progression through mitosis [104]. The interaction with phosphorylated Cdc25 inhibits entry into mitosis [105]. Recently, juglone, a naphthoquinone derivative isolated from fruit shells and leaves of walnut trees, was shown to be an irreversible inhibitor of Pin1 and other related PPIases (Fig. 3). Interestingly, juglone does not inhibit other classes of PPIases such as cyclophilin and FKBP12, while cyclosporine and FK506 do not affect enzymatic activity of Pin1 [106].

The subcellular localization of cyclin B/Cdc2 is also important for regulation of the cell cycle. The cyclin B/Cdc2 complex is present in the cytoplasm in the interphase, although its substrates such as histone H1 and lamin are present in the nucleus. This cytoplasmic localization of cyclin B/Cdc2 is due to the leptomycin B-sensitive nuclear export by the NES receptor, CRM1/exportin 1. Cyclin B/Cdc2 is relocalized in the nucleus because of decreased nuclear export upon the initiation of mitosis. Cyclin B1 possesses an NES, which is inactivated by phosphorylation of the amino-terminal region of the protein [107].

4.3 CDK Inhibitors

It has been shown that a family of cyclin-dependent kinase inhibitors (CKIs) plays a major role in the control of the cell cycle engine [108,109]. Animal cell CKIs described to date are classified into two families: the Cip/Kip family, composed of p21 [Waf1/Cip1], p27^{Kip1}, and p57^{Kip2}, and the INK4 family, including p15, p16, p18, and p19 [110]. The first mammalian CKI to be identified was p21$^{Waf1/Cip1}$, which was shown to be induced by p53 and senescence. Expression of p21$^{Waf1/Cip1}$ correlates with terminal differentiation in several lineages and that in 10T1/2 fibroblasts, in which MyoD is sufficient to induce the expression of p21$^{Waf1/Cip1}$ independently of p53. It is therefore likely that p21$^{Waf1/Cip1}$ is involved in cell cycle exit by inducing G1 arrest in response to a p53-independent differentiation signal. However, p21-/- mice do not develop any tumors, suggesting that p21 itself is not a tumor suppressor. p27^{Kip1} is another CKI that shares partial identity with p21$^{Waf1/Cip1}$. p27 inhibits Cdk2, Cdk3, Cdk4 and Cdk6 complexes in vitro. The greatest homology to p21 is observed in its amino terminal region that contains the CDK inhibitory domain. p27 can block CDK activation by preventing its phosphorylation by CAK, in addition to the direct inhibition of active CDK. Although the mRNA levels of p27 are constant throughout the cell cycle, the protein levels change due to ubiquitin-mediated degradation. p27 is proposed as an essential component of the pathway that connects mitogenic signals to the cell cycle at the R point, correlating the accumulation of this protein with inactivation of G1 cyclin/CDK complexes following growth factor depletion. p27 may play important roles in contact inhibition and inhibition of cell growth in the absence of anchorage [111]. In contrast to p21-/- mice, p27-/- animals showed an increase in size, although no malignant tumors developed. A novel member of this family, p57^{Kip2}, has been isolated on the basis of homology with p21 and p27. p57 binds and inhibits several CDK complexes, and the level of p57 mRNA seems to be more tissue-specific than p27.

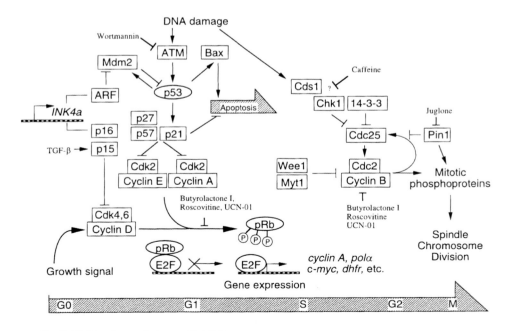

Fig. 3. Cell cycle regulation and inhibitors

p16[INK4a/MTS1], a member of INK4 family, was identified as being associated with Cdk4 by a two-hybrid screening. p16 can block Cdk4 and Cdk6 function by sequestering the catalytic subunit [112]. In contrast to the Cip/Kip family CKIs, the INK4 family CKIs fail to inhibit cyclin E/Cdk2 or cyclin A/Cdk2. Thus, the D-type cyclin-dependent kinase activity is the target for p16. Although p16 overexpression induces G1 arrest in normal fibroblasts, the ability of p16 to induce cell-cycle arrest is lost in cells lacking functional pRb [113]. Thus, loss of p16, overexpression of D-cyclins and loss of pRb have similar effects on G1 progression, and may represent a common pathway to tumorigenesis. The p16 gene was mapped onto 9q21, a locus frequently mutated or repressed by DNA methylation in cancers. In fact, p16-/- mice showed high incidence of tumors and increased sensitivity to carcinogenesis. Recently, the p16INK4a locus was shown to produce a second transcript coding for p19ARF (alternative reading frame) [114]. p19ARF stabilizes p53 by physically interacting with Mdm2. p19ARF binding to Mdm2 blocks Mdm2-induced p53 degradation [115,116]. Therefore, INK4a encodes two different growth inhibitory proteins that act upstream of both pRb and p53. p15INK4b/MTS2, another member of this family, is expressed 30-fold after treatment of human keratinocytes with TGF-β, and this correlates with increased association with its targets CDK4 or CDK6.

4.4 pRb

The retinoblastoma susceptibility gene is a tumor suppressor and its product (pRb) is a repressor of progression towards S phase. During G1 phase, pRb in its hypophosphorylated form binds to several transcription factors of the E2F family, constraining their activity on some promoters and actively repressing transcription [117]. Phosphorylation of pRb by CDKs in G1 untethers pRb from E2Fs. The cyclin D-dependent kinases Cdk4 and Cdk6 trigger pRb phosphorylation, which is likely completed by cyclin E/Cdk2 as cells approach the G1/S boundary. The mechanism by which pRb acts as a tumor suppressor and its function in the normal cell cycle control originates from investigations unraveling the mechanisms of how oncoviral proteins function to deregulate cell cycle control. The SV40 T antigen, the adenovirus E1A protein, and the human papilloma virus-16 E7 protein interact with the hypophosphorylated form of pRb, thereby releasing E2Fs. The finding that the same sequence motif in the viral proteins (LXCXE) is responsible for the oncogenic activity of the viral proteins has led to the model that the viral oncoproteins promote growth by sequestering pRb that normally suppresses E2F-directed transcription of genes necessary for cell cycle progression into S phase. Mapping of the regions of pRb necessary for interactions with the oncoproteins has identified two distinct domains A and B, termed A/B pocket. This pocket is also involved in binding E2F.

Hypophosphorylated pRb binds to the activation domain of E2F and silences specific genes that are active in S phase and are regulated by E2F. Recently, the mechanism by which pRb actively represses the promoter was shown to be transcriptional repression by histone deacetylation [118,119]. Hypophosphorylated pRb binds HDAC1 through the Rb A/B pocket domain. HDAC1 contains an IXCXE sequence, similar to the LXCXE motif used by viral transforming proteins to contact pRb. Association with HDAC1 is reduced by mutations in the pocket or by binding of the human papilloma virus E7. pRb can recruit HDAC1 to E2F to repress the E2F-regulated promoter of the genes such as cyclin E. Trichostatin A, a specific HDAC inhibitor, rescued the pRb-mediated repression of a chromosomally integrated E2F-regulated promoter.

4.5   Small Molecules that Induce pRb Hypophosphorylation

Small molecules that directly affect activity of cell cycle regulators were illustrated in Fig. 3. CDKs are serine/threonine kinases that are inhibited by general protein kinase inhibitors such as staurosporine and UCN-01. Butyrolactone I is the first Cdc2 and Cdk2 specific inhibitor isolated from a fungal broth [120]. These kinase inhibitors induced a G1 cell cycle block probably due to hypophosphorylation of pRb [121,122]. Purine analogs that behave as competitive inhibitors for ATP binding to Cdc2 have been extensively studied, and roscovitine was shown to display high efficiency and high selectivity [123]. The crystal structure of a complex between Cdk2 and roscovitine revealed that the purine portion of the inhibitor binds to the adenine binding pocket of Cdk2.

A large number of natural and synthetic compounds that inhibit signal transduction or nuclear functions indirectly reduce pRb phosphorylation by a variety of mechanisms, thereby inhibiting G1/S transition. For instance, phosmidosine, a nucleotide antibiotic, and inostamycin, an inhibitor of PI turnover, reduce the expression of cyclin D, which results in hypophosphorylation of pRb [33,124]. PI-3K inhibitors such as wortmannin and LY294002 also inhibit increases in both mRNA and protein levels of cyclin D1 and phosphorylation of pRb in response to mitogens [125]. Herbimycin A, a drug that binds to Hsp90 and induces the destruction of tyrosine kinases, causes the down-regulation of cyclin D and an Rb-dependent growth arrest in the G1 phase of the cell cycle [40]. Rapamycin, an inhibitor of FRAP/TOR, causes a decline in the level of D-cyclins probably by reducing the protein synthesis rate through repressing p70^{S6k} activity. On the other hand, a GGTase I inhibitor such as GGTI-298 [47], prostaglandin A2 [126], and HDAC inhibitors such as n-butyrate and trichostatin A induce up-regulation of p21$^{Waf1/Cip1}$, leading to G1 cell cycle arrest.

Many important cell cycle regulators such as cyclins and CKIs are rapidly degraded when they become unnecessary in the cell. Most of these proteins are ubiquitinated, through which they are recognized by the specific destruction machinery, proteasome [100]. The ubiquitinated proteins undergo ATP-dependent degradation by 26S proteasome, consisting of the 20S core containing 28 subunits and two 19S regulatory particles. Proteasome is involved in the degradation of many intracellular short-lived proteins such as c-Jun and p53 in addition to cyclins and CKIs. Lactacystin is a *Streptomyces* metabolite that induces neuronal differentiation of a mouse neuroblastoma cell line, Neuro-2a. A cellular lactacystin-binding protein was isolated and identified to be a β-subunit of 20S proteasome. Lactacystin potently inhibited proteasome activity while other proteases were not inhibited at all [127]. X-ray crystallography showed that lactacystin bound to the N-terminal threonine residue, which was predicted to be the catalytic amino acid residue of proteasome by in vitro mutagenesis. Lactacystin causes G1 cell cycle arrest probably due to the inhibition of degradation of CKIs.

4.6   p53 and Checkpoints

Cell cycle checkpoints enhance genetic fidelity by causing arrest at specific stages of the cell cycle when previous events have not been completed. The G1 checkpoint controls arrest the cell cycle after DNA damage, allowing repair to take place before mutations can be perpetuated by DNA replication. In multicellular organisms, DNA damage can also induce apoptotic cell death, protecting the organism at the expense of the individual cell. p53, another important tumor suppressor protein, is a transcription factor that can inhibit cell cycle progression or induce apoptosis in response to stress or DNA damage [128]. Inactivation of p53 attenuates both of these responses. Loss of functional p53 through various mechanisms is the most common event in human cancer, occurring in over half of all tumors. p53 is short-lived and expressed at very low levels in normal cells, but it is stabilized and accumulates in cells that undergo DNA damage. Among the gene products induced by p53 is p21$^{Waf1/Cip1}$, which can arrest the cell cycle. Other key targets are Bax and Mdm2, the former of which induces apoptosis by counteracting Bcl-2, while the latter acts in a feedback loop to limit the action of p53 by inhibiting transcriptional activity and by facilitating its destruction. Recently, it was shown that phosphorylation of p53 at serine 15 occurs after DNA damage and that this leads to reduced interaction of p53 with Mdm2 [129]. Phosphorylation at serine-15 is catalyzed by ATM kinase, which was predicted to be an activator of p53 in the DNA-damaging agents signal. ATM kinase is the Ataxia telangiectasia gene product that shares a phosphoinositide 3-kinase-related domain with several proteins, some of which are protein kinases. Wortmannin, an inhibitor of PI-3K, can also inhibit ATM kinase [130]. Mdm2 has an NES that is recognized by an export receptor CRM1/exportin 1. Using leptomycin B, an export inhibitor, nuclear export was shown to be required for Mdm2-dependent degradation of p53. This finding suggests that p53 is escorted by Mdm2 to the cytoplasm, where p53 is specifically degraded by a ubiquitin-proteasome pathway. Most recently, p53 itself was shown to have an intrinsic NES in its oligomerization domain, which is masked by tetramer formation upon activation, thereby inhibiting its nuclear export [131].

Mutation of p53 results in genomic instability, which increases the probability of evolution toward malignancy. Most of the mutations in human cancer occur in one of the two copies of p53 genes, which act as the dominant negative regulator of wild-type p53. The SV40 large T antigen, the adenovirus E1B protein, and the human papilloma virus E6 protein target p53 function. Although loss of function of pRb by large T, E1A, and E7 can bypass p53-mediated G1 arrest, it induces E2F and p53-dependent apoptosis. Thus, the DNA tumor viruses not only cancel the pRb function to drive cells into S phase but also inhibit p53 to prevent host cell suicide. p53-induced apoptosis is also involved in the sensitivity to anticancer drugs. It is therefore reasonable that introduction of the wild-type p53 gene into tumor cells lacking functional p53 results in restoration of normal cell cycle regulation and high sensitivity to DNA-damaging agents. Thus, gene therapy using wild-type p53 is undergoing clinical trials.

The G2 checkpoints also regulate the cell cycle by monitoring DNA replication and damage. Cdc2, the kinase that induces mitosis, is regulated by the checkpoints that couple mitosis to the completion of DNA replication and repair. The repair checkpoint kinase Chk1 regulates Cdc25, a phosphatase that activates Cdc2. Phosphorylation of Cdc25 by Chk1 creates a binding site in Cdc25 for 14-3-3 proteins. Rad24, a 14-3-3 protein that is important in the DNA-damage checkpoint in fission yeast, functions as an attachable nuclear export signal that enhances the nuclear export of Cdc25 in response to DNA damage [132]. Thus, Cdc25 cannot be accumulated in the nucleus where its substrate Cdc2 is present. On the other hand, *Xenopus* Cdc25 bound to 14-3-3 in the interphase oocytes appears to decrease in the nuclear import activity by masking an NLS rather than the enhanced nuclear export in Cdc25 [133].

Caffeine is well known to induce cells to override the G2 checkpoints. Genotoxic agents such as ionizing radiation, inhibitors of topoisomerase I and II induce G2 arrest through the checkpoint control. Caffeine completely abrogates this arrest and induces cell progress into M phase. The Chk1 kinase phosphorylates Cdc25 to promote binding to 14-3-3 proteins, preventing it from activating Cdc2. This phosphorylation is abolished by caffeine, suggesting that caffeine inhibits the Chk1 kinase or its upstream regulators [134]. Moreover, caffeine causes uncoupling of mitosis from the completion of DNA replication in BHK cells [135]. 2-Aminopurine [136] and staurosporine [137], known inhibitors of protein kinases, also induce premature mitosis by abrogating the G2 checkpoints. UCN-01 abrogates the S-phase arrest or delay induced by camptothecin, a topoisomerase I inhibitor [138].

5 Future Prospects

As described above, many of the protein apparatuses that control critical events of signal transduction and the cell cycle have been identified. Their properties now make it possible to integrate the knowledge of a number of distinct research areas, such as the cell cycle, differentiation and oncology. The classical genetic approach to understanding the protein function in cells involves making mutations in genes that alter the function of the encoded protein. An alternative approach is to alter the function of the protein directly by using a cell permeable small molecule that binds to the protein in its intracellular environment. This approach using a chemical as a bioprobe allows conditional alteration of function, causing either a loss or gain of function of the intracellular protein. Many small molecules inhibit progression of the cell cycle by binding to a protein required for cell division, thus helping to determine the function of the protein. Conversely, an understanding of signal transduction and cell cycle events helps in understanding the mechanisms of action for many growth inhibitors. Thus, in the field of cell cycle regulation, the marriage of cell biology and chemistry of small molecules has already proven fruitful, with continued promise for the future. Some such small molecule ligands have played remarkably important roles in identifying critical proteins. Examples of this are FKBP, FRAP, HDAC, and exportin. However, there are a large number of compounds whose target molecules are still unknown. It is clearly important for the future biology of cell proliferation to elucidate the molecular mode of action of these compounds.

References

1. Cantley LC, Auger KR, Carpenter C, Duckworth B, Graziani A, Kapeller R, Soltoff S (1991) Oncogenes and signal transduction. Cell 64:281–302
2. Pawson T (1995) Protein modules and signalling networks. Nature 373:573–580
3. Vanhaesebroeck B, Leevers SJ, Panayotou G, Waterfield MD (1997) Phosphoinositide 3-kinases: a conserved family of signal transducers. Trends Biochem Sci 22:267–272
4. Akimoto K, Takahashi R, Moriya S, Nishioka N, Takayanagi J, Kimura K, Fukui Y, Osada S, Mizuno K, Hirai S, Kazlauskas A, Ohno S (1996) EGF or PDGF receptors activate atypical PKCλ through phosphatidylinositol 3-kinase. EMBO J 15:788–798
5. Fukui Y, Ihara S, Nagata S (1998) Downstream of phosphatidylinositol–3 kinase, a multifunctional signaling molecule, and its regulation in cell responses. J Biochem 124:1–7
6. Downward J (1998) Mechanisms and consequences of activation of protein kinase B/ Akt. Curr Opin Cell Biol 10:262–267
7. Chung J, Grammer TC, Lemon KP, Kazlauskas A, Blenis J (1994) PDGF- and insulin-dependent p70^{S6k} activation mediated by phosphatidylinositol-3-OH kinase. Nature 370:71–75
8. Egan SE, Giddings BW, Brooks MW, Buday L, Sizeland AM, Weinberg RA (1993) Association of Sos Ras exchange protein with Grb2 is implicated in tyrosine kinase signal transduction and transformation. Nature 363:45–51
9. Buday L, Downward J (1993) Epidermal growth factor regulates p21^{ras} through the formation of a complex of receptor, Grb2 adapter protein, and Sos nucleotide exchange factor. Cell 73:611–620
10. Darnell JJ, Kerr IM, Stark GR (1994) Jak-STAT pathways and transcriptional activation in response to IFNs and other extracellular signaling proteins. Science 264:1415–1421
11. Avruch J, Zhang X, Kyriakis JM (1994) Raf meets Ras: completing the framework of a signal transduction pathway. Trends Biochem Sci 19:279–283
12. Morrison DK, Cutler Jr R (1997) The complexity of Raf-1 regulation. Curr Opin Cell Biol 9:174–179
13. Vojtek AB, Hollenberg SM, Cooper JA (1993) Mammalian Ras interacts directly with the serine/threonine kinase Raf. Cell 74:205–214
14. Robinson MJ, Cobb MH (1997) Mitogen-activated protein kinase pathways. Curr Opin Cell Biol 9:180–186
15. Kyriakis JM, Banerjee P, Nikolakaki E, Dai T, Rubie EA, Ahmad MF, Avruch J, Woodgett JR (1994) The stress-activated protein kinase subfamily of c-Jun kinases. Nature 369:156–160
16. Han J, Lee JD, Bibbs L, Ulevitch RJ (1994) A MAP kinase targeted by endotoxin and hyperosmolarity in mammalian cells. Science 265:808–811
17. Minden A, Karin M (1997) Regulation and function of the JNK subgroup of MAP kinases. Biochim Biophys Acta 1333:F85–F104

18. Keating MT, Escobedo JA, Williams LT (1988) Ligand activation causes a phosphory-lation-dependent change in platelet-derived growth factor receptor conformation. J Biol Chem 263:12805–12808

19. Uehara Y, Fukazawa H (1991) Use and selectivity of herbimycin A as inhibitor of protein-tyrosine kinases. Methods Enzymol 201:370–379

20. Akiyama T, Ishida J, Nakagawa S, Ogawara H, Watanabe S-i, Itoh N, Shibuya M, Fukami Y (1987) Genistein, a specific inhibitor of tyrosine-specific protein kinases. J Biol Chem 262:5592–5595

21. Umezawa K, Hori T, Tajima H, Imoto M, Isshiki K, Takeuchi T (1990) Inhibition of epidermal growth factor-induced DNA synthesis by tyrosine kinase inhibitors. FEBS Lett 260:198–200

22. Gazit A, Yaish P, Gilon C, Levitzki A (1989) Tyrphostins I: synthesis and biological activity of protein tyrosine kinase inhibitors. J Med Chem 32:2344–2352

23. Fukazawa H, Mizuno S, Uehara Y (1990) Effects of herbimycin A and various SH-reagents on p60v-src kinase activity in vitro. Biochem Biophys Res Commun 173:276–282

24. Whitesell L, Mimnaugh EG, De CB, Myers CE, Neckers LM (1994) Inhibition of heat shock protein Hsp90-pp60v-src heteroprotein complex formation by benzoquinone ansamycins: essential role for stress proteins in oncogenic transformation. Proc Natl Acad Sci USA 91:8324–8328

25. Imoto M, Kakeya H, Sawa T, Hayashi C, Hamada M, Takeuchi T, Umezawa K (1993) Dephostatin, a novel protein tyrosine phosphatase inhibitor produced by Streptomy-ces. I. Taxonomy, isolation, and characterization. J Antibiot (Tokyo) 46:1342–1346

26. Hamaguchi T, Sudo T, Osada H (1995) RK-682, a potent inhibitor of tyrosine phos-phatase, arrested the mammalian cell cycle progression at G1 phase. FEBS Lett 372:54–58

27. Gescher A (1998) Analogs of staurosporine: potential anticancer drugs? Gen Pharmacol 31:721–728

28. Kawakami K, Futami H, Takahara J, Yamaguchi K (1996) UCN-01, 7-hydroxyl-staurosporine, inhibits kinase activity of cyclin- dependent kinases and reduces the phosphorylation of the retinoblastoma susceptibility gene product in A549 human lung cancer cell line. Biochem Biophys Res Commun 219:778–783

29. Kobayashi E, Nakano H, Morimoto M, Tamaoki T (1989) Calphostin C (UCN-1028C), a novel microbial compound, is a highly potent and specific inhibitor of protein kinase C. Biochem Biophys Res Commun 159:548–553

30. Daniel LW, Civoli F, Rogers MA, Smitherman PK, Raju PA, Roederer M (1995) ET-18-OCH3 inhibits nuclear factor-κB activation by 12-O- tetradecanoylphorbol-13-acetate but not by tumor necrosis factor-alpha or interleukin 1α. Cancer Res 55:4844–4849

31. Froscio M, Murray AW, Hurst NP (1989) Inhibition of protein kinase C activity by the antirheumatic drug auranofin. Biochem Pharmacol 38:2087–2089

32. Imoto M, Taniguchi Y, Umezawa K (1992) Inhibition of CDP-DG: inositol transferase by inostamycin. J Biochem (Tokyo) 112:299–302

33. Deguchi A, Imoto M, Umezawa K (1996) Inhibition of G1 cyclin expression in normal rat kidney cells by inostamycin, a phosphatidylinositol synthesis inhibitor. J Biochem (Tokyo) 120:1118–1122

34. Kuo CJ, Chung J, Fiorentino DF, Flanagan WM, Blenis J, Crabtree GR (1992) Rapamycin selectively inhibits interleukin-2 activation of p70 S6 kinase. Nature 358:70–73

35. Chung J, Kuo CJ, Crabtree GR, Blenis J (1992) Rapamycin-FKBP specifically blocks growth-dependent activation of and signaling by the 70 kd S6 protein kinase. Cell 69:1227–1236

36. Kunz J, Henriquez R, Schneider U, Deuter-Reinhard M, Movva NR, Hall MN (1993) Target of rapamycin in yeast, TOR2, is an essential phosphatidylinositol kinase homolog required for G1 progression. Cell 73:585–596
37. Brown EJ, Albers MW, Shin TB, Ichikawa K, Keith CT, Lane WS, Schreiber SL (1994) A mammalian protein targeted by G1-arresting rapamycin-receptor complex. Nature 369:756–758
38. Brown EJ, Beal PA, Keith CT, Chen J, Shin TB, Schreiber SL (1995) Control of p70 s6 kinase by kinase activity of FRAP in vivo. Nature 377:441–446
39. Yano H, Nakanishi S, Kimura K, Hanai N, Saitoh Y, Fukui Y, Nonomura Y, Matsuda Y (1993) Inhibition of histamine secretion by wortmannin through the blockade of phosphatidylinositol 3-kinase in RBL-2H3 cells. J Biol Chem 268:25846–25856
40. Muise-Helmericks RC, Grimes HL, Bellacosa A, Malstrom SE, Tsichlis PN, Rosen N (1998) Cyclin D expression is controlled post-transcriptionally via a phosphatidylinositol 3-kinase/Akt-dependent pathway. J Biol Chem 273:29864–29872
41. Hara M, Akasaka K, Akinaga S, Okabe M, Nakano H, Gomez R, Wood D, Uh M, Tamanoi F (1993) Identification of Ras farnesyltransferase inhibitors by microbial screening. Proc Natl Acad Sci USA 90:2281–2285
42. Kohl NE, Mosser SD, deSolms SJ, Giuliani EA, Pompliano DL, Graham SL, Smith RL, Scolnick EM, Oliff A, Gibbs JB (1993) Selective inhibition of ras-dependent transformation by a farnesyltransferase inhibitor. Science 260:1934–1937
43. James GL, Goldstein JL, Brown MS, Rawson TE, Somers TC, McDowell RS, Crowley CW, Lucas BK, Levinson AD, Marsters JJC (1993) Benzodiazepine peptidomimetics: potent inhibitors of Ras farnesylation in animal cells. Science 260:1937–1942
44. Keyomarsi K, Sandoval L, Band V, Pardee AB (1991) Synchronization of tumor and normal cells from G1 to multiple cell cycles by lovastatin. Cancer Res 51:3602–3609
45. DeClue JE, Vass WC, Papageorge AG, Lowy DR, Willumsen BM (1991) Inhibition of cell growth by lovastatin is independent of ras function. Cancer Res 51:712–717
46. Vogt A, Qian Y, McGuire TF, Hamilton AD, Sebti SM (1996) Protein geranylgeranylation, not farnesylation, is required for the G1 to S phase transition in mouse fibroblasts. Oncogene 13:1991–1999
47. Adnane J, Bizouarn FA, Qian Y, Hamilton AD, Sebti SM (1998) p21([WAFI/CIPI]) is upregulated by the geranylgeranyltransferase I inhibitor GGTI-298 through a transforming growth factor beta- and Sp1- responsive element: involvement of the small GTPase rhoA. Mol Cell Biol 18:6962–6970
48. Kwon HJ, Yoshida M, Fukui Y, Horinouchi S, Beppu T (1992) Potent and specific inhibition of p60[v-src] protein kinase both in vivo and in vitro by radicicol. Cancer Res 52:6926–6930
49. Ki SW, Kasahara K, Kwon HJ, Eishima J, Takesako K, Cooper JA, Yoshida M, Horinouchi S (1998) Identification of radicicol as an inhibitor of in vivo Ras/Raf interaction by the yeast two-hybrid screen system. J Antibiot 51:936–944
50. Roe SM, Prodromou C, O'Brien R, Ladbury JE, Piper PW, Pearl LH (1999) Structural basis for inhibition of the Hsp90 molecular chaperone by the antitumor antibiotics radicicol and geldanamycin. J Med Chem 42:260–266
51. Alessi DR, Cuenda A, Cohen P, Dudley DT, Saltiel AR (1995) PD 098059 is a specific inhibitor of the activation of mitogen- activated protein kinase kinase in vitro and in vivo. J Biol Chem 270:27489–94
52. Cuenda A, Rouse J, Doza YN, Meier R, Cohen P, Gallagher TF, Young PR, Lee JC (1995) SB203580 is a specific inhibitor of a MAP kinase homologue which is stimulated by cellular stresses and interleukin-1. FEBS Lett 364:229–233

53. Fukuda M, Gotoh I, Adachi M, Gotoh Y, Nishida E (1997) A novel regulatory mechanism in the mitogen-activated protein (MAP) kinase cascade. Role of nuclear export signal of MAP kinase kinase. J Biol Chem 272:32642–32648
54. Dingwall C, Laskey RA (1991) Nuclear targeting sequences: a consensus? Trends Biochem Sci 16:478–481
55. Görlich D, Vogel F, Mills AD, Hartmann E, Laskey RA (1995) Distinct functions for the two importin subunits in nuclear protein import. Nature 377:246–248
56. Siomi H, Dreyfuss G (1995) A nuclear localization domain in the hnRNP A1 protein. J Cell Biol 129:551–560
57. Bischoff FR, Ponstingl H (1991) Catalysis of guanine nucleotide exchange on Ran by the mitotic regulator RCC1. Nature 354:80–82
58. Bischoff FR, Krebber H, Kemph T, Hermes I, Ponstingl H (1995) Human RanGTPase activating protein RanGAP1 is a homolog of yeast RNA1p involved in messenger RNA processing and transport. Proc Natl Acad Sci USA 92:1749–1753
59. Izaurralde E, Kutay U, von Kobbe C, Mattaj IW, Görlick D (1997) The asymmetric distribution of the constituents of the Ran system is essential for transport into and out of the nucleus. EMBO J 16:6535–6547
60. Fischer U, Huber J, Boelens WC, Mattaj IW, Lührmann R (1995) The HIV-1 Rev activation domain is a nuclear export signal that accesses an export pathway used by specific cellular RNAs. Cell 82:475–483
61. Wen W, Meinkoth JL, Tsien RY, Taylor SS (1995) Identification of a signal of rapid export of proteins from the nucleus. Cell 82:463–473
62. Adachi Y, Yanagida M (1989) Higher order chromosome structure is affected by cold-sensitive mutations in a *Schizosaccharomyces pombe* gene *crm1*[+] which encodes a 115-kD protein preferentially localized in the nucleus and at its periphery. J Cell Biol 108:1159–1207
63. Engel K, Kotlyarov A, Gaestel M (1998) Leptomycin B-sensitive nuclear export of MAPKAP kinase 2. EMBO J 17:3363–3371
64. Wasylyk B, Hagman J, Gutierrez-Hartmann A (1998) Ets transcription factors: nuclear effectors of the Ras-MAP-kinase signaling pathway. Trends Biochem Sci 23:213–216
65. Karin M, Liu Zg, Zandi E (1997) AP-1 function and regulation. Curr Opin Cell Biol 9:240–246
66. Montminy M (1997) Transcriptional regulation by cyclic AMP. Annu Rev Biochem 66:807–822
67. Turner BM (1991) Histone acetylation and control of gene expression. J Cell Sci 99:13–20
68. Taunton J, Hassig CA, Schreiber SL (1996) A mammalian histone deacetylase related to the yeast transcriptional regulator Rpd3p. Science 272:408–411
69. Kijima M, Yoshida M, Sugita K, Horinouchi S, Beppu T (1993) Trapoxin, an antitumor cyclic tetrapeptide, is an irreversible inhibitor of mammalian histone deacetylase. J Biol Chem 268:22429–22435
70. Wade PA, Pruss D, Wolffe AP (1997) Histone acetylation: chromatin in action. Trends Biochem Sci 22:128–132
71. Torchia J, Glass C, Rosenfeld MG (1998) Co-activators and co-repressors in the integration of transcriptional responses. Curr Opin Cell Biol 10:373–383
72. Janknecht R, Hunter T (1996) A growing coactivator network. Nature 383:22–23
73. Yang XJ, Ogryzko VV, Nishikawa J, Howard BH, Nakatani Y (1996) A p300/CBP-associated factor that competes with the adenoviral oncoprotein E1A. Nature 382:319–324

74. Petrij F, Giles RH, Dauwerse HG, Saris JJ, Hennekam RCM, Masuno M, Tommerup N, van Ommen G-JB, Goodman RH, Peters DJM, Breuning MH (1995) Rubinstein-Taybi syndrome caused by mutations in the transcriptional co-activator CBP. Nature 376:348–351

75. Gu W, Roeder RG (1997) Activation of p53 sequence-specific DNA binding by acetylation of the p53 C-terminal domain. Cell 90:595–606

76. Boyes J, Byfield P, Nakatani Y, Ogryzko V (1998) Regulation of activity of the transcription factor GATA-1 by acetylation. Nature 396:594–598

77. Nishi K, Yoshida M, Fujiwara D, Nishikawa M, Horinouchi S, Beppu T (1994) Leptomycin B targets a regulatory cascade of crm1, a fission yeast nuclear protein, involved in control of higher order chromosome structure and gene expression. J Biol Chem 269:6320–6324

78. Wolff B, Sanglier J-J, Wang Y (1997) Leptomycin B is an inhibitor of nuclear export: inhibition of nucleo-cytoplasmic translocation of the human immunodeficiency virus type 1 (HIV-1) Rev protein and Rev-dependent mRNA. Chem Biol 4:139–147

79. Fornerod M, Deursen Jv, Baal Sv, Reynolds A, Davis D, Murti KG, Fransen J, Grosveld G (1997) The human homologue of yeast CRM1 is in a dynamic subcomplex with CAN/Nup214 and a novel nuclear pore component Nup88. EMBO J 16:807–816

80. Kudo N, Khochbin S, Nishi K, Kitano K, Yanagida M, Yoshida M, Horinouchi S (1997) Molecular cloning and cell cycle-dependent expression of mammalian CRM1, a protein involved in nuclear export of proteins. J Biol Chem 272:29742–29751

81. Fornerod M, Ohno M, Yoshida M, Mattaj IW (1997) CRM1 is an export receptor for leucine-rich nuclear export signals. Cell 90:1051–1060

82. Stade K, Ford CS, Guthrie C, Weis K (1997) Exportin 1 (Crm1p) is an essential nuclear export factor. Cell 90:1041–1050

83. Fukuda M, Asano S, Nakamura T, Adachi M, Yoshida M, Yanagida M, Nishida E (1997) CRM1 is responsible for intracellular transport mediated by the nuclear export signal. Nature 390:308–311

84. Kudo N, Wolff B, Sekimoto T, Schreiner EP, Yoneda Y, Yanagida M, Horinouchi S, Yoshida M (1998) Leptomycin B inhibition of signal-mediated nuclear export by direct binding to CRM1. Exp Cell Res 242:540–547

85. Kudo N, Matsumori N, Taoka H, Fujiwara D, Schreiner EP, Wolff B, Yoshida M, Horinouchi S (1999) Leptomycin B inactivates CRM1/exportin 1 by covalent modification at a cysteine residue in the central conserved region. Proc Natl Acad Sci USA 96:9112–9117

86. Goto M, Masegi M, Yamauchi T, Chiba K, Kuboi Y, Harada K, Naruse N (1998) K1115 A, a new anthraquinone derivative that inhibits the binding of activator protein-1 (AP-1) to its recognition sites. I. Biological activities. J Antibiot 51:539–544

87. Sakai Y, Yoshida T, Tsujita T, Ochiai K, Agatsuma T, Saitoh Y, Tanaka F, Akiyama T, Akinaga S, Mizukami T (1997) GE3, a novel hexadepsipeptide antitumor antibiotic, produced by Streptomyces sp. I. Taxonomy, production, isolation, physico-chemical properties, and biological activities. J Antibiot 50:659–664

88. Yoshida M, Horinouchi S, Beppu T (1995) Trichostatin A and trapoxin: novel chemical probes for the role of histone acetylation in chromatin structure and function. BioEssays 17:423–430

89. Yoshida M, Kijima M, Akita M, Beppu T (1990) Potent and specific inhibition of mammalian histone deacetylase both in vivo and in vitro by trichostatin A. J Biol Chem 265:17174–17179

90. Nan X, Ng HH, Johnson CA, Laherty CD, Turner BM, Eisenman RN, Bird A (1998) Transcriptional repression by the methyl-CpG-binding protein MeCP2 involves a histone deacetylase complex. Nature 393:386–389

91. Nakajima H, Kim YB, Terano H, Yoshida M, Horinouchi S (1998) FR901228, a potent antitumor antibiotic, is a novel histone deacetylase inhibitor. Exp Cell Res 241:126–133

92. Kwon HJ, Owa T, Hassig CA, Shimada J, Schreiber SL (1998) Depudecin induces morphological reversion of transformed fibroblasts via the inhibition of histone deacetylase. Proc Natl Acad Sci USA 95:3356–3361

93. Kim YB, Lee K-H, Sugita K, Yoshida M, Horinouchi S (1999) Oxamflatin is a novel antitumor compound that inhibits mammalian histone deacetylase. Oncogene 18:2461–2470

94. Pardee AB (1989) G1 events and regulation of cell proliferation. Science 246:603–608

95. Norbury C, Nurse P (1992) Animal cell cycles and their control. Annu Rev Biochem 61:441–470

96. Lee MG, Nurse P (1987) Complementation used to clone a human homologue of the fission yeast cell cycle control gene cdc2. Nature 327:31–35

97. Morgan DO (1995) Principles of CDK regulation. Nature 374:131–134

98. Lukas J, Bartkova J, Rohde M, Strauss M, Bartek J (1995) Cyclin D1 is dispensable for G1 control in retinoblastoma gene-deficient cells independently of cdk4 activity. Mol Cell Biol 15:2600–2611

99. Sherr CJ (1993) Mammalian G1 cyclins. Cell 73:1059–1065

100. Hilt W, Wolf DH (1996) Proteasomes: destruction as a programme. Trends Biochem Sci 21:96–102

101. Solomon MJ (1993) Activation of the various cyclin/cdc2 protein kinases. Curr Opin Cell Biol 5:180–186

102. Nurse P (1990) Universal control mechanism regulating onset of M-phase. Nature 344:503–508

103. Nigg EA (1995) Cyclin-dependent protein kinases: key regulators of the eukaryotic cell cycle. Bioessays 17:471–480

104. Yaffe MB, Schutkowski M, Shen M, Zhou XZ, Stukenberg PT, Rahfeld JU, Xu J, Kuang J, Kirschner MW, Fischer G, Cantley LC, Lu KP (1997) Sequence-specific and phosphorylation-dependent proline isomerization: a potential mitotic regulatory mechanism. Science 278:1957–1960

105. Shen M, Stukenberg PT, Kirschner MW, Lu KP (1998) The essential mitotic peptidyl-prolyl isomerase Pin1 binds and regulates mitosis-specific phosphoproteins. Genes Dev 12:706–720

106. Hennig L, Christner C, Kipping M, Schelbert B, Rucknagel KP, Grabley S, Kullertz G, Fischer G (1998) Selective inactivation of parvulin-like peptidyl-prolyl cis/trans isomerases by juglone. Biochemistry 37:5953–5960

107. Hagting A, Karlsson C, Clute P, Jackman M, Pines J (1998) MPF localization is controlled by nuclear export. EMBO J 17:4127–4138

108. Hunter T, Pines J (1994) Cyclins and cancer. II: Cyclin D and CDK inhibitors come of age. Cell 79:573–582

109. Elledge SJ, Harper JW (1994) Cdk inhibitors: on the threshold of checkpoints and development. Curr Opin Cell Biol 6:847–852

110. Grana X, Reddy EP (1995) Cell cycle control in mammalian cells: role of cyclins, cyclin dependent kinases (CDKs), growth suppressor genes and cyclin-dependent kinase inhibitors (CKIs). Oncogene 11:211–219

111. Kawada M, Yamagoe S, Murakami Y, Suzuki K, Mizuno S, Uehara Y (1997) Induction of p27[Kip1] degradation and anchorage independence by Ras through the MAP kinase signaling pathway. Oncogene 15:629–637

112. Serrano M, Hannon GJ, Beach D (1993) A new regulatory motif in cell-cycle control causing specific inhibition of cyclin D/CDK4. Nature 366:704–707

113. Lukas J, Parry D, Aagaard L, Mann DJ, Bartkova J, Strauss M, Peters G, Bartek J (1995) Retinoblastoma-protein-dependent cell-cycle inhibition by the tumour suppressor p16. Nature 375:503–506

114. Kamijo T, Zindy F, Roussel MF, Quelle DE, Downing JR, Ashmun RA, Grosveld G, Sherr CJ (1997) Tumor suppression at the mouse INK4a locus mediated by the alternative reading frame product p19ARF. Cell 91:649–659

115. Bates S, Phillips AC, Clark PA, Stott F, Peters G, Ludwig RL, Vousden KH (1998) p14ARF links the tumour suppressors RB and p53. Nature 395:124–125

116. Zhang Y, Xiong Y, Yarbrough WG (1998) ARF promotes MDM2 degradation and stabilizes p53: ARF-INK4a locus deletion impairs both the Rb and p53 tumor suppression pathways. Cell 92:725–734

117. Bartek J, Bartkova J, Lukas J (1996) The retinoblastoma protein pathway and the restriction point. Curr Opin Cell Biol 8:805–814

118. Brehm A, Miska EA, McCance DJ, Reid JL, Bannister AJ, Kouzarides T (1998) Retinoblastoma protein recruits histone deacetylase to repress transcription. Nature 391:597–601

119. Magnaghi JL, Groisman R, Naguibneva I, Robin P, Lorain S, Le VJ, Troalen F, Trouche D, Harel BA (1998) Retinoblastoma protein represses transcription by recruiting a histone deacetylase. Nature 391:601–605

120. Kitagawa M, Okabe T, Ogino H, Matsumoto H, Suzuki-Takahashi I, Kokubo T, Higashi H, Saitoh S, Taya Y, Yasuda H, Ohba Y, Nishimura S, Tanaka N, Okuyama A (1993) Butyrolactone I, a selective inhibitor of cdk2 and cdc2 kinase. Oncogene 8:2425–2432

121. Abe K, Yoshida M, Usui T, Horinouchi S, Beppu T (1991) Highly synchronous culture of fibroblasts from G2 block caused by staurosporine, a potent inhibitor of protein kinases. Exp Cell Res 192:122–127

122. Schnier JB, Nishi K, Goodrich DW, Bradbury EM (1996) G1 arrest and down-regulation of cyclin E/cyclin-dependent kinase 2 by the protein kinase inhibitor staurosporine are dependent on the retinoblastoma protein in the bladder carcinoma cell line 5637. Proc Natl Acad Sci USA 93:5941–5946

123. Meijer L, Borgne A, Mulner O, Chong JP, Blow JJ, Inagaki N, Inagaki M, Delcros JG, Moulinoux JP (1997) Biochemical and cellular effects of roscovitine, a potent and selective inhibitor of the cyclin-dependent kinases cdc2, cdk2 and cdk5. Eur J Biochem 243:527–536

124. Kakeya H, Onose R, Liu PC, Onozawa C, Matsumura F, Osada H (1998) Inhibition of cyclin D1 expression and phosphorylation of retinoblastoma protein by phosmidosine, a nucleotide antibiotic. Cancer Res 58:704–710

125. Takuwa N, Fukui Y, Takuwa Y (1999) Cyclin D1 expression mediated by phosphatidylinositol 3-kinase through mTOR-p70(S6K)-independent signaling in growth factor-stimulated NIH 3T3 fibroblasts. Mol Cell Biol 19:1346–1358

126. Gorospe M, Liu Y, Xu Q, Chrest FJ, Holbrook NJ (1996) Inhibition of G1 cyclin-dependent kinase activity during growth arrest of human breast carcinoma cells by prostaglandin A2. Mol Cell Biol 16:762–770

127. Fenteany G, Standaert RF, Lane WS, Choi S, Corey EJ, Schreiber SL (1995) Inhibition of proteasome activities and subunit-specific amino-terminal threonine modification by lactacystin. Science 268:726–731

128. Lane DP (1992) Cancer. p53, guardian of the genome. Nature 358:15–16.

129. Shieh SY, Ikeda M, Taya Y, Prives C (1997) DNA damage-induced phosphorylation of p53 alleviates inhibition by MDM2. Cell 91:325–334
130. Banin S, Moyal L, Shieh S, Taya Y, Anderson CW, Chessa L, Smorodinsky NI, Prives C, Reiss Y, Shiloh Y, Ziv Y (1998) Enhanced phosphorylation of p53 by ATM in response to DNA damage. Science 281:1674–1677
131. Stommel JM, Marchenko ND, Jimenez GS, Moll UM, Hope TJ, Wahl GM (1999) A leucine-rich nuclear export signal in the p53 tetramerization domain: regulation of subcellular localization and p53 activity by NES masking. EMBO J 18:1660–1672
132. Lopez-Girona A, Furnari B, Mondesert O, Russell P (1999) Nuclear localization of Cdc25 is regulated by DNA damage and a 14-3-3 protein. Nature 397:172–175
133. Yang J, Winkler K, Yoshida M, Kornbluth S (1999) Maintenance of G2 arrest in the Xenopus oocyte: A role for 14-3-3-mediated inhibition of Cdc25 nuclear import. EMBO J 18:2174–2183
134. Kumagai A, Guo Z, Emami KH, Wang SX, Dunphy WG (1998) The Xenopus Chk1 protein kinase mediates a caffeine-sensitive pathway of checkpoint control in cell-free extracts. J Cell Biol 142:1559–1569
135. Schlegel R, Pardee AB (1986) Caffeine-induced uncoupling of mitosis from the completion of DNA replication in mammalian cells. Science 232:1264–1266
136. Andreassen PR, Margolis RL (1992) 2-Aminopurine overrides multiple cell cycle checkpoints in BHK cells. Proc Natl Acad Sci USA 89:2272–2726
137. Yoshida M, Usui T, Tsujimura K, Inagaki M, Beppu T, Horinouchi S (1997) Biochemical differences between staurosporine-induced apoptosis and premature mitosis. Exp Cell Res 232:225–239
138. Shao RG, Cao CX, Shimizu T, O'Connor PM, Kohn KW, Pommier Y (1997) Abrogation of an S-phase checkpoint and potentiation of camptothecin cytotoxicity by 7-hydroxystaurosporine (UCN-01) in human cancer cell lines, possibly influenced by p53 function. Cancer Res 57:4029–4035

3
Differentiation

HIROYUKI OSADA

1 Introduction

The human body consists of about 6×10^{13} cells derived from a fertilized egg. A balance of proliferation and differentiation is coordinated during development. The cell should have a control mechanism that decides whether the cell should differentiate or proliferate under each environmental condition, such as nutritional state.

The molecular control of development has been studied most extensively in two organisms with short life times and relatively small genomes: the fruit fly *Drosophila melanogaster* and the nematode *Caernorhabditis elegans*. In mammals, cell cultures have served as surrogate models for whole organisms in identifying molecules that effect cell commitment to particular types of differentiation. A simple approach in this exponentially growing field follows that commonly used for discussing oncogenes, and classifies the relevant molecules according to a nuclear, cytoplasmic or extracellular locus of action.

During embryonic development, precursor cells differentiate to functional matured cells under the control of cytokines and hormones. As tumor cells, including leukemia and neuroblastoma, are thought to disrupt the differentiation programs, it is worth while exploiting differentiation inducers of cells. Cytokines, neurotrophins, and so on are important differentiation-inducing proteins for blood cells, neurons, bones, muscle cells, liver cells, and epithelial cells differentiated from precursor cells. These proteins are important in the maintenance of homeostasis of human or animal bodies; however, it is still difficult to use them as animal or human therapeutics. Small molecular weight compounds that mimic differentiation-inducing proteins are more useful as therapeutic agents as well as bioprobes.

A cell based screening system using precursor cells which can be differentiated to mature cells in vitro is very effective in isolating small molecular weight differentiation-inducing compounds from microbial metabolites. Sometimes, specific cell lines are constructed to express a specific cytokine receptor or signal transducing molecule which is involved in the differentiation. The screening strategies used to discover novel differentiation inducers from microbial metabolites and their biological activities are described in this chapter.

2 Differentiation of Hematopoietic Cells

Hematopoietic cells, including erythrocytes and lymphocytes, are differentiated from stem cells originally derived from bone marrow (Fig. 1). The hematopoietic stem cells with the multipotency proliferate and differentiate under the control of various cytokines. Abnormalities in this program result in hematological diseases including leukemia. Many cytokines are found as differentiation regulators. There is a cytokine network that has positive regulators such as colony-stimulating factors (CSFs) and interleukins (ILs) and negative regulators such as transforming growth factor (TGF) β and tumor necrosis factor (TNF).

The analysis of differentiation induced by cytokines led to important information for understanding physiological differentiation. When proliferating cells differentiate to mature cells, the cells lose proliferating ability. Some human and mouse leukemia cells can be induced to terminally differentiate to mature macrophages or granulocytes by some differentiation-inducing cytokines [1]. It is suggested that induction of terminal differentiation of leukemic cells is potentially valuable in the therapy of leukemia, thus putting forward the idea of differentiation therapy. A part of leukemic cells could be induced to differentiate with cytokines, but another part could not [2]. It is expected that insensitive clones to differentiation-inducing cytokines could be induced to differentiate by other non-peptide compounds [3].

2.1 Erythrocytes

The development of mature blood cells from hematopoietic stem/progenitor cells is tightly controlled by multiple protein factors that are required for both limited growth

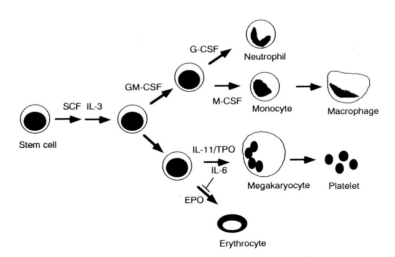

Fig. 1. Differentiation of blood cells

and terminal differentiation. The two most prominent cytokines that regulate erythropoiesis are stem cell factor (SCF) [4] and erythropoietin (EPO) [5]. SCF is required for the initial stage of the hematopoietic differentiation. SCF and IL3 synergistically promote the growth and differentiation of progenitor cells, such as erythroid and myeloid progenitors. EPO is a 34-kD glycoprotein that promotes survival, proliferation, and differentiation of erythroid progenitor cells [5].

The binding of EPO to its receptor (EPO-R) causes receptor dimerization and triggers a phosphorylation cascade in target cells [6]. Although the EPO-R lacks intrinsic enzymatic activity, it couples with the protein tyrosine kinase c-Kit and the cytosolic protein tyrosine kinase JAK2. JAK2 phosphorylates multiple tyrosine residues on the cytoplasmic domain of the EPO-R, making a docking site for the SH2 domain of signaling molecules such as protein tyrosine phosphatases (SHP-1 and SHP-2), phosphatidylinositol-3-kinase, and STAT5. Subsequently MAP-kinase cascade and AP-1 transcription-factor are activated. During this period, the cyclin-dependent kinase (cdk)-inhibitors p21 and p27 are highly induced, and then the cells stop proliferation [7].

Differentiation of this lineage also needs hematopoietic specific trans-acting factors including GATA-1 and nuclear factor-erythroid 2 (NF-E2). GATA-1 is a major activator of hematopoietic gene expression and is absolutely required for erythropoiesis [8]. NF-E2 is a hematopoietic-specific basic-leucine zipper protein required for the β-globin expression [9].

2.2 Megakaryocytes

The hematopoietic pathway for megakaryocytic and erythroid differentiation are closely related. These lineages share common hematopoietic specific transcription factors such as GATA-1 and NF-E2. Therefore the separation between megakaryocytes and erythrocytes occurs relatively late in hematopoietic hierarchy. In the early stage of megakaryocytic development, IL-6 and IL-3 work synergistically and support development and proliferation of progenitors [10]. IL-6 is shown to inhibit the gene expression of erythroid specific gene such as β and γ globin and increase megakaryocyte specific genes such as glycoprotein IIb/IIIa [11]. For terminal maturation of megakaryocytes, thrombopoietin and IL-11 are required [12, 13].

2.3 Granulocytes (Neutrophils)

Granulocytes are one of the major populations in the circulating cells of the blood and play an important role in the mammalian defense system. IL-3, GM-CSF, and G-CSF, are involved in the generation of granulocytes [14]. IL-3 and GM-CSF stimulate the survival, proliferation and differentiation of bipotent granulocyte-macrophage progenitors. GM-CSF acts through a heterodimeric receptor consisting of a cytokine-specific α-chain and a common β-chain. IL-3 and IL-5 share this receptor G-CSF acts specifically on granulocytic progenitor cells.

The binding of G-CSF to its receptor specifically activates JAK1 and JAK2 kinases, as well as STAT3 [15]. In the process of functional differentiation, the neutrophilic progenitor cells pass through several stages of maturation during which they acquire a number of functional properties that are necessary for mature neutrophils to eradicate microbes. The capacity for microbial killing requires oxidative processes that can be measured by the ability to reduce nitroblue tetrazolium (NBT). The biochemical basis of NBT reduction in neutrophils is superoxide generation, and cytochemical NBT reduction is an expression of the respiratory burst. The expression of both Mac-1 antigen (CD11b/CD18) and Fc γ II/III receptor (CD32/CD16) on the cell surface enables neutrophils to recognize microbes as foreign through opsonins, serum components that coat microorganisms. Interaction of Mac-1 antigen or Fc γ II/III receptor with opsonins promotes both phagocytosis and superoxide generation.

2.4 Monocytes/Macrophages

Monocytes/macrophages play a variety of important roles, including phagocytosis, release of proinflammetory cytokines, antigen presentation, and incorporation of lipids. IL-3 and GM-CSF exert their effects during the early stages of hematopoiesis, acting on both macrophagic and granulocytic progenitor cells [14]. In late stages, macrophage colony-stimulating factor (M-CSF) stimulates survival, proliferation and maturation of committed monocytic progenitors. M-CSF binds to its receptor known as the proto-oncogene c-*fms* [16]. This cell lineage also differentiates to monocyte-derived lineages such as osteoclasts and dendritic cells [17]. It is known that GM-CSF spurs differentiation of peripheral blood monocytes (PBMs) into dendritic cells when combined with IL-4 and/or TNF-α. In contrast, M-CSF and IL-4 promote differentiation of PBMs into osteoclasts.

2.5 Differentiation Inducible Cell Lines

Murine erythroleukemia (MEL) cells are well characterized as a model system for erythroid differentiation. Friend virus-infected mouse leukemia cell line is a typical MEL line that is blocked in its differentiation to erythrocytes. MEL cells are differentiated to the erythrocytes by non-peptide differentiation-inducing compounds. The handling of MEL cells is much easier than human cell lines and sensitive to differentiation-inducing compounds. Dimethyl sulfoxide (DMSO), hexametylene bisacetamide and sodium butyrate are known as potent erythroid differentiation inducers [18, 19].

K-562, a human erythroleukemia cell line that express both erythroid and megakaryocytic-specific genes in the uninduced state and has the capacity to differentiate along the erythrocytes or megakaryocytes. Treatment of this cell with 1-β-D-arabinofuranosylcytosine (ara-C), sodium butyrate, hemin, or 5-azacytidine results in growth inhibition and erythroid differentiation [20, 21]. On the other side, 12-O-tetradecanoyl phorbol-13 acetate (TPA) and staurosporine induced significant phenotypic differentiation evidenced by cellular expression of the CD41 differentiation-specific megakaryocytic cell surface marker. Like normal progenitor cells committed for erythrocytes/megakaryocytes, treatments with EPO or IL-6 induce erythroid or megakaryocytic differentiation, respectively [22, 23].

Treatment of human promyelocytic leukemia HL-60 cells with TPA ultimately induces the differentiation of these cells along the monocytes/macrophage lineage, whereas treatment with retinoic acid or DMSO induces granulocytic/neutrophilic differentiation [24–26].

Treatment of the human monoblastoid cell line U937 with various differentiation inducers, such as TPA, 1,25-dihydroxyvitamin D3, and all-trans retinoic acid, induces differentiation of this cell to the monocytes/macrophage lineage [27–29].

2.6 Differentiation-Inducing Compounds

Trichostatin A was discovered as a fungistatic antibiotic from the culture broth of *Streptomyces platensis* and later it was found to be a potent inducer of differentiation in MEL (Friend leukemia) cells [30]. Trichostatin A also inhibited the cell cycle progression of normal rat fibroblasts in the G1 and G2 phases at very low concentrations. These biological activities of trichostatin A are due to the inhibition of histone deacetylase of mammalian cell lines [31].

Herbimycin A, a benzoquinonoid ansamycin antibiotic, was originally discovered as a herbicidal compound from a fermentation broth of *Streptomyces* sp. [32] and later found to inhibit tyrosine phosphorylation by $p60^{v-src}$ kinase [33]. Herbimycin A induced the reduction of tyrosine phosphorylation level in the human leukemia K562, mouse erythroleukemia (MEL) and embryonic carcinoma (F9) cells, and induced differentiation [34, 35].

Daunomycin is one of the most characterized antileukemic compounds, which induces the terminal differentiation of erythroleukemia cells such as K-562 [36]. Respinomycins are a novel group of anthracycline antibiotics produced by *Streptomyces xanthocidicus* [37, 38]. Although four components (A1, A2, B, C and D) of respinomycins were isolated, respinomycins A1 and A2 induced the differentiation of human leukemia K-562 cells. The differentiation-inducing mechanism of respinomycin is different from that of daunomycin (Fig. 2). Daunomycin suppressed cell growth immediately after the treatment of K-562 cells, and induced almost 100% of cells to become erythrocytes, however, respinomycin did not suppress cell growth and induced differentiation in a half of the total number of cells.

3 Differentiation of Neuron Cells

3.1 Neuron Cells from Precursor Cells

Neural precursor cells have been of interest historically as the building blocks of the embryonic central nerve system (CNS). The majority of in vitro studies of the regulation of neural-cell proliferation by polypeptide growth factors, and in vivo studies of neural lineage, argue for the presence of precursors with limited proliferation or

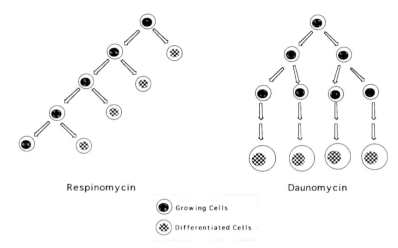

Respinomycin   Daunomycin

⬤ Growing Cells

⊛ Differentiated Cells

Fig. 2. Terminal differentiation of erythroid leukemia K-562 cells induced by two different anthracyclines

lineage potential in the mammalian CNS. This is in contrast to renewable tissues, such as the blood or immune system, skin epithelium and the epithelium of the small intestinal crypts, which contain specialized, self-renewing cells known as stem cells. However, recent in vitro and in vivo studies lead to the conclusion that neural stem cells, with self-renewal and multilineage potential, are present in the embryonic through to adult mammalian forebrain [39].

Neurons and glia are generated primarily during different developmental periods (neurons prenatally and glia postnatally), from different precursors. However, in vitro lineage analysis supports the hypothesis that neuronal and glia cells can be derived from multipotent precursor cells. For example, it was demonstrated that a small number of single septal precursor cells from a rat at embryonic day 13 maintain multipotency in culture, giving rise to neurons and astrocytes [40]. Moreover, it was demonstrated that nearly 20% of embryonic cortical precursors give rise to neurons and oligodendrocytes [41]. The signals that regulate unipotent and multipotent progenitor cells are beginning to be understood. Basic fibroblast growth factor (bFGF) has been shown to induce or enhance the limited proliferation of CNS neuronal precursors.

While the majority of CNS progenitor cells exhibit a limited proliferative potential in culture, under specific conditions, certain mammalian cells can be induced to exhibit an extended proliferative potential and hence generate a large number of progeny-one of the defining characteristics of a stem cell.

3.2 Control of Differentiation

It is becoming increasingly evident that significant changes in gene expression occur during the course of neuronal differentiation. Thus, it should be possible to gain information about the biochemical events by identifying differentially expressed genes in neuronal differentiation. The PC12 cell line is a useful model system to investigate the molecular mechanism underlying neuronal differentiation and has been used extensively for the study of the molecular events that underlie the biological actions of nerve growth factor (NGF). It is reported that NGF stimulates the expression of several types of genes, such as c-*fos* [42], CDK inhibitor p21$^{WAF1/CIP1}$ [43, 44], sodium channels [45], nitric oxide synthase [46] and clusterin [47]. However, it remains unclear which gene product(s) are responsible for the neuronal differentiation.

Neurotrophic molecules have a profound influence on developmental events in the nervous system such as naturally occurring cell death, differentiation and process outgrowth. NGF is largely known as a target-derived factor responsible for the survival and maintenance of the phenotype of specific subsets of peripheral neurons and basal forebrain cholinergic nuclei during development and maturation. Many sympathetic neurons depend on NGF for survival into adulthood, whereas postnatal sensory neurons cease to rely on this factor. Although the trophic actions of NGF are restricted to a few types of neurons, other classes of neurons are likely to depend on specific neurotrophic molecules in a similar manner [48].

Precursors of neuron and glia responsive to EGF/TGF-α were isolated from late embryonic forebrain and retina, respectively [49]. The EGF-responsive precursor gave rise to neurons whose neurotransmitter phenotype, for example GABA and substance P, was represented in the adult forebrain (striatum). In addition, as the mRNAs for both the EGF and the EGF receptor were detected in forebrain, this EGF-responsive precursor might play an in vivo in forebrain histogenesis.

Success in the studies of neurotrophic factors will serve as a foundation for utilizing a combination of genetic and epigenetic regulation of central nerve system stem cells and progeny, in vivo and ex vivo, as novel approaches to cell replacement in the treatment of brain injury or disease.

3.3 Differentiation Inducible Cell Lines

Human neuroblastoma, SH-SY5Y cells, which were established from a patient are typical malignant neuroblastoma cells because the cells lack the high affinity NGF receptor, Trk. The cells express low-affinity NGF receptors (p75), TrkB [50], a signal transducing receptor unit for brain-derived neurotrophic factor (BDNF) and p53 [51]. This cell line does not respond to NGF because it does not have Trk, but the cells can be differentiated to extend neurites by several compounds, including phorbol ester [52], retinoic acid [53], insulin [54], staurosporine [55], epolactaene [56], MT-5 and MT-21 [57]

Neuro2A cells are derived from mouse neuroblastoma. Omura et al. discovered lactacystin [58] by cell based screening using this cell line. Retinoic acid and cAMP analogs also induced neurite-outgrowth in Neuro2A cells [59].

PC12 cells, rat pheochromocytoma cells, are the most characterized cells in respect to NGF signal transduction (Fig. 3). PC12 cells, which express TrkA and p53, are differentiated by NGF, cAMP analogs, staurosporine [60], MT-19, MT-20 [61], and erbstatin analogs [62].

3.4 Differentiation-Inducing Compounds

Lactacystin is a *Streptomyces* metabolite that induces differentiation of Neuro2A cells, a mouse neuroblastoma cell line. Lactacystin causes a transient increase of the intracellular cAMP level and morphological changes including neuritogenesis in Neuro2A cells [58]. Tritium-labeled lactacystin was used to identify the 20S proteasome as its specific cellular target. Lactacystin appears to modify covalently the highly conserved amino-terminal threonine of the mammalian proteasome subunit X (also called MB1) [63].

Epolactaene was isolated from the culture broth of *Penicillium* sp. as a novel non-peptide neurotrophic compound. Epolactaene arrested the cell cycle progression at the G0/G1 phase and induced the neurite outgrowth of SH-SY5Y cells. The differentiation markers, phosphorylated neurofilaments and acetylcholine esterase activity, were enhanced 2 days after treatment with epolactaene [56, 57].

Staurosporine, a potent protein kinase inhibitor isolated from microbial metabolites, is known to mimic the effect of NGF in promoting neurite outgrowth in PC12 cells [64, 65]. Staurosporine induced the activation of a kinase with an apparent molecular mass of 57 kD, which was thought to be JNK3. The kinase phosphorylates transcription factors including c-Jun, Elk-1 and ATF2, as well as MBP, suggesting that it plays a role in gene induction. Furthermore, staurosporine induced immediate-early genes including *Nur*77 and c-*fos*, but not c-*jun* [60].

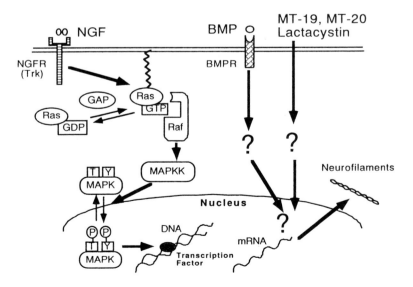

Fig. 3. Signal transduction of NGF and neuritogenic compounds in PC12 cells

4 Differentiation of Skeletal Cells (Bone and Muscle Cells)

Bone tissue is continuously remodeled by precisely coordinated mechanisms of the balance of two cell types, osteoblasts and osteoclasts (Fig. 4). Osteoblasts are derived from mesenchymal cells and osteoclasts, multinucleated giant cells from hematopoietic precursor cells [66, 67]. These cells are responsible for bone deposition and resorption, respectively.

4.1 Osteoblasts

HOS cells, which are established from human carcinoma tissue, are immature osteoblasts. The alkaline phosphatase activity, an osteoblastic marker enzyme, of the cells increases in response to vitamin D [68] or estrogen [69]. Most biologically active form, 1,25-dihydroxyvitamin D3 and its analogs have functions and therapeutic potential that extend beyond those of regulating bone mineralization and intestinal calcium transport. Meantime estrogens have been shown to be essential for maintaining a sufficiently high bone mineral density and ER- expression has been demonstrated in bone cells.

Bone morphogenetic protein-2 (BMP-2) inhibits terminal differentiation of C2C12 myoblasts and converts them into osteoblast lineage cells [70]. Smad proteins are vertebrate homologues of Mad protein in *Drosophila* and are involved in the signal transduction of TGF-β families. C2C12 cells expressed Smad1, Smad2, Smad4, and Smad5 mRNAs, and expression levels were not altered by treatment with BMP-2 or TGF-β. When Smads were transiently transfected into C2C12 cells, both Smad1 and Smad5 induced alkaline phosphatase activity and decreased the activity of myogenin

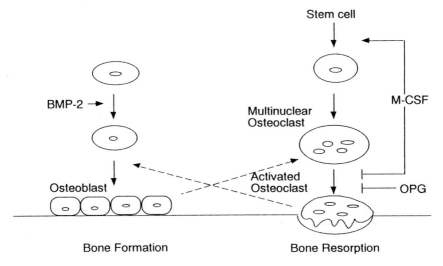

Fig. 4. Remodeling of bone cells

promoter. Thus both Smad1 and Smad5 are specifically involved in the intracellular BMP signals which inhibit myogenic differentiation and induce osteoblast differentiation in C2C12 cells [71].

With respect to the molecular aspect of osteoblastic differentiation in the animal, our knowledge largely depends on the analysis of two osteoblast specific *cis*-elements named OSE1 and OSE2 [72]. These elements were first identified in the promoter region of the osteoblast specific gene, osteocalcin. Further genetic studies have revealed that the transcription factor, CBFA1/OSF2 which binds to the OSE2 functions as a key effector in osteoblast differentiation [73–79].

4.2 Osteoclasts

The osteoclast is responsible for bone resorption. Compared to osteoblast differentiation, the molecular basis of the osteoclast differentiation is little understood. Major reasons are the lack of the knowledge about osteoclast specific genes and of established osteoclast cell lines. Most of the information obtained so far is based on a genetic study using transgenic mice. Transcription factors PU.1, c-Fos and NFκB deficient mice can't generate osteoclasts [80–82]. The mice deficient in macrophage colony stimulating factor (M-CSF) fail to produce osteoclasts [83]. On the other hand, osteoprotegerin (OPG) deficient mice showed a severe decrease in bone density suggesting the involvement of the gene in negative regulation for the osteoclast differentiation [84].

4.3 Myoblasts

Myoblast differentiation and fusion to multinucleated muscle cells can be studied in myoblasts grown in culture. The existence of a master gene for myogenic lineage was predicted by the evidence that C3H10T1/2 can be converted to myoblast phenotype by exposure to 5-azacytidine [85]. Continuous efforts have been made to elucidate the molecular mechanisms of myogenic differentiation using animals and cultured cells, including C3H10T1/2 and C2C12 [86, 87]. Recent progress in myogenic differentiation have been reviewed [88, 89].

5 Differentiation of Liver: Hepatocytes

5.1 Differentiation of Liver Precursor Cells

The liver is an epithelial organ that contains two major differentiated cell types: the hepatocytes and the bile ductule cells [90]. Hepatocytes are present throughout the liver and perform several metabolic functions, such as bile and blood protein production. Bile ductule cells are limited to the periportal region, which is where the blood enters the liver via the portal vein and hepatic artery, and form a bile ductule that collects the bile and lets bile flow into the common bile duct. Although the phenotypes of differentiated hepatocytes and bile ductule cells are different, it is thought that both cells are derived from a common precursor cell known as the oval cell. This derivation is made for the following reasons. 1) According to the histochemical characteristics, embryonic livers consist of hepatocytes and bile ductule cells, both cells produce α-fetoprotein (AFP) and albumin [91, 92]. 2) Some carcinogenic protocols result in the appearance of cells of unknown origin with oval-shaped nuclei in the periportal region. These so-called oval cells have characteristics of both hepatocytes and bile ductule cells [93]. 3) Oval cells can differentiate into either hepatocytes or bile ductule cells in vitro, depending on the culture medium [94].

5.2 Gene Expression and Growth

Present knowledge of transcriptional regulation in the liver has been derived primarily from the analysis of promoter and enhancer elements of genes selectively expressed in hepatocytes. These studies led to the identification of four families of liver-enriched transcription factors that participate in the restricted expression of liver genes in the hepatocytes. These families are characterized by structurally related DNA binding domains, include the variant homeodomain containing family of HNF1 (hepatocyte nuclear factor 1) [95], the leucine zipper C/EBP (CCAAT/enhancer binding protein) family [96, 97], the HNF3 winged helix family [98], and members of the nuclear receptor superfamily. HNF4 [99], CoupTF1 (chicken ovalbumin upstream promoter transcription factor)/Ear3, and Arp-1/CoupTF II [100] are the nuclear receptors (Fig. 5).

Hepatocyte differentiation, starting with the commitment of the definitive endoderm, followed by induction of the definitive endoderm into hepatic precursor cells and ending with liver organogenesis, has been correlated with the induction of different liver-enriched regulatory factors and a tentative hierarchy has been proposed

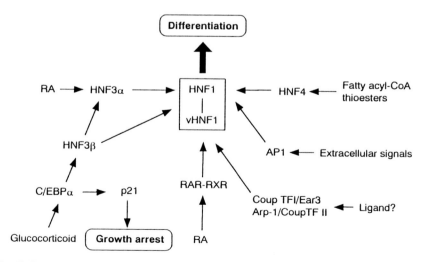

Fig. 5. Schematic representation of the regulatory network of liver-enriched transcription factor, showing cross-regulation

[101]. HNF3 proteins are placed at an early position in the transcriptional hierarchy that define regionalization within the definitive endoderm being expressed very early in the definitive endoderm lineage. Other regulators induced at the onset of liver differentiation include HNF4 and HNF1. HNF1, however, appears only at the time of liver organogeneis. Recent evidence suggest that a regulatory cascade of C/EBP proteins is involved in the development of the hepatocyte lineage [102]. The relatively late expression of these factors together with the ability to promote cell cycle arrest make it unlikely that a similar cascade is involved in hepatocyte differentiation; instead, they may be critical in terminal differentiation and in the maintenance of a differentiated nonproliferative state.

During normal hepatic differentiation from precursor cells to matured cells, several cytokines appear to affect the proliferation and differentiation of hepatic cells. There are three "primary" cytokines associated with normal hepatocyte proliferation and differentiation in vivo: TGF-α, hepatocyte growth factor (HGF), and acidic fibroblast growth factor (aFGF) [103]. In addition, TGF-β is also expressed during hepatic regeneration, and it has been proposed that TGF-β may provide at least part of the negative growth signals controlling liver size. The cellular distribution of the cytokines differs: TGF-α and aFGF transcripts are found in both Ito cells (non parenchymal cells) and hepatic precursor cells, whereas the HGF transcripts are only found in Ito cells. The TGF-β1 transcripts are located mainly in Ito cells, but the early population of hepatic precursor cells also contain the TGF-β1 transcripts. Recently, a novel ligand/receptor system, the stem cell factor (SCF)/c-*kit* system, which may be uniquely involved in the earliest stages of hepatic precursor cell activation, was discovered [104]. The SCF/c-*kit* signal transduction system is believed to play a fundamental role in the survival, and migration of precursor cells in hepatic differentiation.

5.3 Phenotype of Established Cell Lines

Models of expression of liver specific genes characterize the phases of hepatocyte differentiation, from primitive endodermal cell to adult hepatocytes. Hepatocellular carcinomas, many of which have been converted to cell lines, also express some liver specific genes. Though neoplastic transformation and growth in tissue culture are both thought to distort normal in vivo gene expression, cell lines are the most suitable systems for experimental manipulation of gene controls.

Two major human hepatocarcinoma cell lines (HepG2 and Hep3B) have been isolated from the liver biopsies of two children with primary hepatocarcinoma. Seventeen kinds of the major human plasma proteins which contain AFP, albumin, α1-antitrypsin, transferrin and plasminogen are synthesized and secreted by these cells [105]. The Hep3B also produces the two major polypeptides of the hepatitis B virus surface antigens (HBsAg); the HepG2 line does not synthesize HBsAg. Hepatitis B virus (HBV) has been epidemiologically implicated as a probable causative agent of the majority of hepatocarcinomas. It is reported that Hep3B forms tumors when injected into nude mice. The cell line thus provides an excellent model for the investigation of human plasma protein synthesis and hepatitis B virus-host cell relationship.

Other human hepatoma (HuH-7, PLC/PRF/5, huH-1, huH-4, KIM-1) and hepatoblastoma (HuH-6) cell lines also exhibit various liver-specific functions [106]. HuH-7 produces a large amount of AFP and a moderate level of albumin. On the other hand, huH-1 produces a high level of albumin but only a detectable amount of AFP. These cell lines provide useful model systems for investigating the molecular basis of the AFP and albumin gene-expression in hepatomas.

5.4 Differentiation Inducers

Retinoic acid (RA) is a natural acidic derivative of vitamin A, which acts as a morphogen or a teratogen and has effects the proliferation and differentiation of various cell types. RA also affects gene expression in various cells, for example, transferrin and albumin genes are transcriptionally influenced in the human hepatoma cell line Hep3B [107]. Recent reports suggest that RA induces the expression of retinoic acid receptor (RAR) α, β and γ mRNA in Hep3B cells. RA regulates the promoter activity of some liver specific genes through direct interaction of the RAR with its response element in each gene.

Glucocorticoid is a derivative of steroid hormones. Its receptor in the nucleus binds as a homodimer to hexanucleotide motifs (AGAACA or AGGTCA), arranged in an inverted repeat configuration. Transcriptional regulation by steroid hormones is thought to be mediated by specific interaction of the hormone-receptor complex with a DNA element, termed the "glucocorticoid responsive element" (GRE). In the case of HuH-7 and five other human hepatoma cells, dexamethasone treatment has invariably resulted in increased secretion of AFP [106]. Other researchers revealed that glucocorticoid stimulated the expression of C/EBPα resulting in a steroid-induced G1 cell cycle arrest, and induced phosphoenolpyruvate carboxykinase, a rate controlling gluconeogenic enzyme, via C/EBPβ protein in some hepatoma cells [108, 109].

6 Epidermal Differentiation

The epidermis is an excellent example of dynamic tissue in which highly regulated mechanisms exist to control cell proliferation and differentiation. According to in vivo studies, the proliferative compartment of the epidermis, termed stem cells, reside in the basal layer and the progeny cells migrate suprabasally. These cells change morphologically and biochemically, lose their proliferative potential and terminally differentiate to form the stratum corneum. Nonetheless, the precise biological mechanisms remained unclear until the establishment of the methods of culturing normal human keratinocytes in serum-free medium [110, 111]. Keratinocytes have now been widely used to study the balance between proliferation and differentiation.

6.1 Program of Epithelial Differentiation

The epidermis is a stratified squamous epithelium that forms the protective covering of the skin. Only the basal layer of epidermal cells has the capacity for DNA synthesis and proliferation. Under an unidentified trigger of terminal differentiation, basal cells begin the journey to the skin surface. In transit, it undergoes a series of morphological and biochemical changes that culminate in the production of dead, flattened, enucleated squamous cells, which are sloughed from the surface, and continually replaced by inner cells differentiating outward.

Epidermal differentiation is characterized by procession through a defined program of changes in cellular architecture, gene expression and enzyme activation [112]. Differentiation of the basal keratinocytes results in a permanent loss of growth potential and the subsequent expression of differentiation markers; first expressed are the two suprabasal keratins K1 and K10 followed by other proteins, such as loricrin, involucrin, filaggrin and epidermal transglutaminase [113]. Cornified envelope formation is specific to later stages of differentiation.

6.2 Molecular Controls of Epidermal Gene Expression

Based on the evidence from other differentiation systems (e.g. muscle differentiation and the MyoD family), expression of a few transcriptional regulatory genes is central to the manifestation and maintenance of the differentiated state [114–116]. It seems that a knowledge of the major transcription factors controlling keratinocyte-specific gene expression will be of central importance in the quest for elucidating the molecular mechanisms underlying epidermal differentiation. There are a number of keratinocyte genes that have already been isolated and characterized. These include genes encoding: (a) the basal keratins K5 and K14 [117–119]; (b) the suprabasal keratins K1, K10, K6, and K16 [120, 121]; (c) the cornified envelope protein, involucrin [122]; and (d) filaggrin [123]. Sequence comparisons have provided a few clues as to possible common regulatory elements involved in controlling keratinocyte-specific gene expression. One of these is the CK 8-mer sequence 5' AAGCCAAA3', found upstream from a number of epidermal genes. A different sequence, 5' GCCTGCAGGC3' located 5' from the TATA box of the human K14 gene, was shown to be the target for the binding of an AP2-like factor, KER1, which is especially abundant in keratinocytes, and present in lower amounts in simple epithelial cells [124, 125].

6.3 Role of Growth Factors and Calcium

It is thought that at least three steps are involved in terminal differentiation of keratinocyte: reversible growth arrest; irreversible loss of proliferative potential; and the expression of terminally differentiated phenotype. The search for an optimal culture system uncovered a number of extracellular regulators controlling the balance between growth and terminal differentiation. Stimulators of keratinocyte growth are EGF, TGF-α, low (10^{-7}–10^{-10} M) concentrations of retinoic acid (RA), keratinocyte growth factor (KGF) and two cytokines, IL-6 and IL-1. KGF is an epithelial cell specific mitogen that is secreted by stromal cells derived from major epithelial organs. KGF is as potent as EGF in stimulating proliferation of keratinocytes [117, 126]. Exposure of KGF- or EGF-stimulated keratinocytes to 10^{-3} M calcium, an inducer of differentiation, led to cessation of cell growth. However, immunological analysis of early and late markers of terminal differentiation, K1 and filaggrin, respectively, revealed striking differences in keratinocytes propagated in the presence of these growth factors. With KGF, the differentiation response was associated with expression of both markers whereas their appearance was retarded or blocked by EGF [127]. TGF-α, which also interacts with the EGF receptor, gave a similar response to that observed with EGF. These findings functionally distinguish KGF from the EGF family and support the role of KGF in the normal proliferation and differentiation of epithelial cells [128].

Withdrawal from the cell cycle seems to be a prerequisite for irreversible commitment of a keratinocyte to terminally differentiate. The most extensively studied negative regulators are the TGF-β, which act at picomolar concentrations to inhibit DNA synthesis and cell division. While TGF-β inhibit growth of basal cells, their natural expression in epidermis seems to be predominantly in suprabasal, differentiating layers [129]. Indeed during mammalian development, expression of TGF-β mRNA coincides with epidermal stratification and keratinization. Because the effects of TGF-β on basal cell growth appear to be largely reversible, it has been assumed that TGF-β alone is not enough to induce terminal differentiation, a process thought to be irreversible.

It was shown that 10^{-6} M RA inhibited proliferation and concomitantly induced active TGF-β in mouse keratinocytes. Conversely, when exposed to 10^{-7}-10^{-6} M RA, human keratinocytes showed an increase in proliferation and cell migration, with no detectable TGF-β induction [130]. This discovery of an autocrine regulatory loop between retinoids and TGF-β in some keratinocytes under some conditions is exciting, and suggests that both factors may be involved in some common regulatory pathways.

When cultured in medium containing 5×10^{-5} M calcium, mouse keratinocytes grow as a monolayer, because desmosome assembly is inhibited [131]. Upon a switch to high (1.2×10^{-3} M) calcium, desmosomes form and cells stratify. After three to seven additional days in culture, squamous containing cornified envelopes appear in the medium. These data indicate that at least some features of terminal differentiation can be induced by calcium in vitro. Calcium also seems to play a role in mediating keratinization in vivo: certain calcium ionophores have recently been shown to enhance action of TPA in promoting epidermal differentiation [132].

While calcium is necessary for certain features of terminal differentiation (e.g., desmosome formation, stratification and transglutaminase activation) in human keratinocytes, its effects often seem more pronounced in regulating differentiation specific changes in cellular architecture than alterations in gene expression. For example, a high calcium medium does not induce appreciable levels of K1/K10 and filaggrin expression. In addition, some data showed that calcium induces a marked increase in TGF-β2 mRNA expression, thereby implicating calcium in both early and late stages of terminal differentiation. Hence, like other known regulators of epidermal differentiation, calcium seems to have pleiotropic effects, and further-more, it differs from other known regulators in having a distinct positive influence on many aspects of terminal differentiation. While additional experiments will be nec-essary to elucidate the precise relation between calcium, TGF-β and retinoids, it seems clear that calcium is intimately involved in the complex regulatory mecha-nisms underlying the balance between growth and keratinization.

6.4 Differentiation-Inducing Compounds

Reveromycin A was isolated as an inhibitor of mitogenic activity induced by EGF in mouse epidermal keratinocyte, Balb/MK cells [133]. Furthermore reveromycin A exhibited morphological reversion of src^{ts}-NRK cells, antiproliferative activity against human tumor cell lines and antifungal activity. Reveromycin A is a selective inhibi-tor of protein synthesis in eukaryotic cells [134].

References

1. Paran M, Sachs L, Barak Y, Resnitzky P (1970) In vitro induction of granulocyte differ-entiation in hematopoietic cells from leukemic and non-leukemic patients. Proc Natl Acad Sci USA 67:1542–1549
2. Fibach E, Hayashi M, Sachs L (1973) Control of normal differentiation of myeloid leukemic cells to macrophages and granulocytes. Proc Natl Acad Sci USA 70:343–346
3. Sachs L (1987) The molecular control of blood cell development. Science 238:1374–1379
4. Broudy VC (1997) Stem cell factor and hematopoiesis. Blood 90:1345–1364
5. Krantz SB (1991) Erythropoietin. Blood 77:419–434
6. Klingmuller U (1997) The role of tyrosine phosphorylation in proliferation and matura-tion of erythroid progenitor cells; Signals emanating from the erythropoietin receptor. Eur J Biochem 249:637–647
7. Panzenbock B, Bartunek P, Mapara MY, Zenke M (1998) Growth and differentiation of human stem cell factor/erythropoietin-dependent erythroid progenitor cells in vitro. Blood 92:3658–3668
8. Pevny L, Simon MC, Robertson E, Klein WH, Tsai S-F, D'Agati V, Orkin SH, Constantini F (1991) Erythroid differentiation in chimaeric mice blocked by a targeted mutation in the gene for transcription factor GATA-1. Nature 349:257–260
9. Andrews NC, Erdjument-Bromage H, Davidson MB, Tempst P, Orkin SH (1993) Eryth-roid transcription factor NF-E2 is a haematopoietic-specific basic-leucine zipper pro-tein. Nature 362:722–728
10. Akira S, Hirano T, Taga T, Kishimoto T (1990) Biology of multifunctional cytokines: IL-6 and related molecules (IL-1 and TNF). FASEB J 4:2860–2867
11. Ferry AE, Baliga SB, Monteiro C, Pace BS (1997) Globin gene silencing in primary erythroid cultures. J Biol Chem 272:20030–20037
12. Broudy VC, Lin NL, Kaushansky K (1995) Thrombopoietin (c-Mpl ligand) acts syner-gistically with erythropoietin, stem cell factor, and interleukin-11 to enhance murine megakaryocyte colony growth and increases megakaryocyte ploidy in vitro. Blood 85:1719–1726

13. Weich NS, Wang A, Fitzgerald M, Neben TY, Donaldson D, Giannotti J, Yez-Aldape J, Leven RM, Turner KJ (1997) Recombinant human interleukin-11 directly promotes megakaryocytopoiesis in vitro. Blood 90:3893–3902

14. Metcalf D (1995) The granulocyte-macrophage regulators: reappraisal by gene inactivation. Exp Hematol 23:569–572

15. Ihle JN, Kerr IM (1995) JAKs and STATs in signaling by the cytokine receptor superfamily. Trends Genet 11:69–74

16. Sherr CJ, Rettenmier CW, Sacca R, Roussel MF, Look AT, Stanley ER (1985) The c-fms proto-oncogene product is related to the receptor for the mononuclear phagocyte growth factor, CSF-1. Cell 41:665–676

17. Akagawa K, Takasuka N, Nozaki Y, Komuro I, Azuma M, Ueda M, Naito M, Takahashi K (1996) Generation of CD1+RelB+ dendritic cells and tartrate-resistant acid phosphatase-positive osteoclast-like multinucleated giant cells from human monocytes. Blood 88:4029–4039

18. Friend C, Scher W, Holland JG, Sato T (1971) Hemoglobin synthesis in murine virus-induced leukemic cells in vitro: stimulation of erythroid differentiation by dimethyl sulfoxide. Proc Natl Acad Sci USA 68:378–382

19. Noguchi T, Fukumoto H, Mishima Y, Obinata M (1987) Factors controlling induction of commitment of murine erythroleukemia (TSA8) cells to CFU-E (colony forming unit-erythroid). Development 101:169–174

20. Rutherford TR, Clegg JB, Weatherall DJ (1979) K562 human leukaemic cells synthesise embryonic haemoglobin in response to haemin. Nature 280:164–165

21. Lozzio CB, Lozzio BB, Machado EA, Fuhr JE, Lair SV, Bamberger EG (1979) Effects of sodium butyrate on human chronic myelogenous leukaemia cell line K562. Nature 281:709–710

22. Hoffman R, Murnane MJ, Jr BE, Prohaska R, Floyd V, Dainiak N, Forget BG, Furthmayr H (1979) Induction of erythropoietic colonies in a human chronic myelogenous leukemia cell line. Blood 54:1182–1187

23. Nakayama K (1998) Expression of IL-6, IL-6 receptor and its signal transducer gp130 mRNAs in megakaryocytic cell lines. Leuk Lymphoma 29:399–405

24. Breitman TR, Selonick SE, Collins SJ (1980) Induction of differentiation of the human promyelocytic leukemia cell line (HL-60) by retinoic acid. Proc Natl Acad Sci USA 77:2936–2940

25. Stendahl O, Dahlgren C, Hed J (1982) Physicochemical and functional changes in human leukemic cell line HL-60. J Cell Physiol 112:217–221

26. Tanaka H, Abe E, Miyaura C, Shiina Y, Suda T (1983) 1 α, 25-dihydroxyvitamin D3 induces differentiation of human promyelocytic leukemia cells (HL-60) into monocyte-macrophages, but not into granulocytes. Biochem Biophys Res Commun 117:86–92

27. Olsson IL, Breitman TR (1982) Induction of differentiation of the human histiocytic lymphoma cell line U937 by retinoic acid and cyclic adenosin 3':5'-monophosphate-including agents. Cancer Res 42:3924–3927

28. Gullberg U, Nilssen E, Einhorn S, Olsson I (1985) Combinations of interferon-γ and retinoic acid or 1 α, 25-dihydroxycholecalciferol induce differentiation of the human monoblast leukemia cell line U-937. Exp Hematol 13:675–679

29. Tsuchita S, Kobayashi Y, Goto Y, Okumura H, Nakae S, Konno T, Tada K (1982) Induction of maturation in cultured human monocytic leukemia cells by a phorbol diester. Cancer Res 42:1530–1536

30. Yoshida M, Nomura S, Beppu T (1987) Effects of trichostatins on differentiation of murine erythroleukemia cells. Cancer Res 47:3688–3691

31. Yoshida M, Kijima M, Akita M, Beppu T (1990) Potent and specific inhibition of mammalian histone deacetylase both in vivo and in vitro by trichostatin A. J Biol Chem 265:17174–17179

32. Omura S, Iwai Y, Takahashi Y, Sadakane N, Nakagawa A, Oiwa H, Hasegawa Y, Ikai T (1979) Herbimycin, a new antibiotic produced by a strain of Streptomyces. J Antibiot 32:255–261

33. Uehara Y, Murakami Y, Sugimoto Y, Mizuno S (1989) Mechanism of reversion of Rous sarcoma virus transformation by herbimycin A: reduction of total phosphotyrosine levels due to reduced kinase activity and increased turnover of $p60^{v\text{-}src1}$. Cancer Res 49:780–785

34. Honma Y, Okabe-Kado J, Hozumi M, Uehara Y, Mizuno S (1989) Induction of erythroid differentiation of K562 human leukemic cells by herbimycin A, an inhibitor of tyrosine kinase activity. Cancer Res 49:331–334

35. Kondo K, Watanabe T, Sasaki H, Uehara Y, Oishi M (1989) Induction of in vitro differentiation of mouse embryonal carcinoma (F9) and erythroleukemia (MEL) cells by herbimycin A, an inhibitor of protein phosphorylation. J Cell Biol 109:285–293

36. Tonini GP, Radzioch D, Gronberg A, Clayton M, Blasi E, Benetton G, Varesio L (1987) Erythroid differentiation and modulation of c-myc expression induced by antineoplastic drugs in the human leukemic cell line K562. Cancer Res 47:4544–4547

37. Ubukata M, Tanaka C, Osada H, Isono K (1991) Respinomycin A_1, a new anthracycline antibiotic. J Antibiot 44:1274–1276

38. Ubukata M, Osada H, Kudo T, Isono K (1993) Respinomycins, A_1, A_2, B, C and D, a novel group of anthracycline antibiotics. I. Taxonomy, fermentation, isolation and biological activities. J Antibiotics 46:936–941

39. Weiss S, Reynolds BA, Vescovi AL, Morshead C, Craig CG, van der Kooy D (1996) Is there a neural stem cell in the mammalian forebrain? Trends Neurosci 19:387–393

40. Temple S (1989) Division and differentiation of isolated CNS blast cells in microculture. Nature 340:471–473

41. Williams BP, Read J, Price J (1991) The generation of neurons and oligodendrocytes from a common precursor cell. Neuron 7:685–693

42. Milbrandt J (1986) Nerve growth factor rapidly induces c-fos mRNA in PC12 rat pheochromocytoma cells. Proc Natl Acad Sci USA 83:4789–4793

43. Billon N, van Grunsven LA, Rudkin BB (1996) The CDK inhibitor p21[WAFI/Cip1] is induced through a p300-dependent mechanism during NGF-mediated neuronal differentiation of PC12 cells. Oncogene 13:2047–2054

44. Yan GZ, Ziff EB (1997) Nerve growth factor induces transcription of the p21 [WAFI/CIPI] and cyclin D1 genes in PC12 cells. J Neurosci 17:6122–6132

45. Kalman D, Wong B, Horvai AE, Cline MJ, O'Lague PH (1990) Nerve growth factor acts through cAMP-dependent protein kinase to increase the number of sodium channels in PC12 cells. Neuron 4:355–366

46. Poluha W, Schonhoff CM, Harrington KS, Lachyankar MB, Crosbie NE, Bulseco DA, Ross AH (1997) A novel, nerve growth factor-activated pathway involving nitric oxide, p53, and p21[WAFI] regulates neuronal differentiation of PC12 cells. J Biol Chem 272:24002–24007

47. Gutacker C, Klock G, Diel P, Koch-Brandt C (1999) Nerve growth factor and epidermal growth factor stimulate clusterin gene expression in PC12 cells. Biochem J 339:759–766

48. Levi-Montalcini R, Skaper SD, Toso RD, Petrelli L, Leon A (1996) Nerve growth factor: from neutrophin to neurokine. Trends Neurosci 19:514–520

49. Reynolds BA, Tetzlaff W, Weiss S (1992) A multipotent EGF-responsive striatal embryonic progenitor cell produces neurons and astrocytes. J Neurosci 12:4565–4574

50. Ehrhard PB, Ganter U, Schmutz B, Bauer J, Otten U (1993) Expression of low-affinity NGF receptor and trkB mRNA in human SH-SY5Y neuroblastoma cells. FEBS Lett 330:287–292

51. Ronca F, Chan SL, Yu VC (1997) 1-(5-isoquinolinesulfonyl)-2-methylpiperazine induces apoptosis in human neuroblastoma cells, SH-SY5Y, through a p53-dependent pathway. J Biol Chem 272:4252–4260

52. Heikkila J, Jalava A, Eriksson K (1993) The selective protein kinase C inhibitor GF 109203X inhibits phorbol ester-induced morphological and functional differentiation of SH-SY5Y human neuroblastoma cells. Biochem Biophys Res Commun 197:1185–1193

53. Nicolini G, Miloso M, Zoia C, Di Silvestro A, Cavaletti G, Tredici G (1998) Retinoic acid differentiated SH-SY5Y human neuroblastoma cells: an in vitro model to assess drug neurotoxicity. Anticancer Res 18:2477–2481

54. Pecio-Pinto E, Ishii DN (1984) Effects of insulin, insulin-like growth factor-II and nerve growth factor on neurite outgrowth in cultured human neuroblastoma cells. Brain Res 302:323–334

55. Jalava A, Akerman K, Heikkila J (1993) Protein kinase inhibitor, staurosporine, induces a mature neuronal phenotype in SH-SY5Y human neuroblastoma cells through an α-, β-, and ζ-protein kinase C-independent pathway. J Cell Physiol 155:301–312

56. Kakeya H, Takahashi I, Okada G, Isono K, Osada H (1995) Epolactaene, a novel neuritogenic compound in human neuroblastoma cells, produced by a marine fungus. J Antibiot 48:733–735

57. Kakeya H, Onozawa C, Sato M, Arai K, Osada H (1997) Neuritogenic effect of epolactaene derivatives on human neuroblastoma cells which lack high-affinity nerve growth factor receptors. J Med Chem 40:391–394

58. Omura S, Fujimoto T, Otoguro K, Matsuzaki K, Moriguchi R, Tanaka H, Sasaki Y (1991) Lactacystin, a novel microbial metabolite, induces neuritogenesis of neuroblastoma cells. J Antibiot 44:113–116

59. Riboni L, Prinetti A, Bassi R, Caminiti A, Tettamanti G (1995) A mediator role of ceramide in the regulation of neuroblastoma Neuro2A cell differentiation. J Biol Chem 270:26868–26875

60. Yao R, Yoshihara M, Osada H (1997) Specific activation of a c-Jun NH_2-terminal kinase isoform and induction of neurite outgrowth in PC-12 cells by staurosporine. J Biol Chem 272:18261–18266

61. Yao R, Osada H (1997) Induction of neurite outgrowth in PC-12 cells by γ-lactam-related compounds via Ras-MAP kinase signaling pathway independent mechanism. Exp Cell Res 234:223–229

62. Watanabe Y, Kakeya H, Ikoma E, Umezawa K (1993) Induction of morphological and enzymic differentiation in rat pheochromocytoma PC12h cells by stable erbstatin analogues. Drugs Exp Clin Res 19:1–6

63. Fenteany G, Standaert RF, Lane WS, Choi S, Corey EJ, Schreiber SL (1995) Inhibition of proteasome activities and subunit-specific amino-terminal threonine modification by lactacystin. Science 268:726–731

64. Hashimoto S, Hagino A (1989) Staurosporine-induced neurite outgrowth in PC12h cells. Exp Cell Res 184:351–359

65. Rasouly D, Rahamim E, Lester D, Matsuda Y, Lazarovici P (1992) Staurosporine-induced neurite outgrowth in PC12 cells is independent of protein kinase C inhibition. Mol Pharmacol 42:35–43

66. Erlebacher A, Filvaroff EH, Gitelman SE, Derynck R (1995) Toward a molecular understanding of skeletal development. Cell 80:371–378
67. Reddi AH (1997) Bone morphogenesis and modeling: soluble signals sculpt osteosomes in solid state. Cell 89:159–161
68. Mulkins M, Manolagas SC, Deftos LJ, Sussman HH (1983) 1,25-Dihydroxyvitamin D3 increases bone alkaline phosphatase isoenzyme levels in human osteogenic sarcoma cells. J Biol Chem 258:6219–6225
69. Komm BS (1988) Estrogen binding, receptor mRNA, and biologic response in osteoblast-like osteosarcoma cells. Science 241:81–84
70. Hogan BLM (1996) Bone morphogenetic proteins: multifunctional regulators of vertebrate development. Genes Dev 10:1580–1594
71. Yamamoto N, Akiyama S, Katagiri T, Namiki M, Kurokawa T, Suda T (1997) Smad1 and smad5 act downstream of intracellular signalings of BMP-2 that inhibits myogenic differentiation and induces osteoblast differentiation in C2C12 myoblasts. Biochem Biophys Res Commun 238:574–580
72. Ducy P, Karsenty G (1995) Two distinct osteoblast-specific cis acting elements control expression of a mouse osteocalcin gene. Mol Cell Biol 15:1858–1869
73. Ducy P, Zhang R, Geoffroy V, Ridall A, Karsenty G (1997) Osf2/Cbfa1: a transcriptional activator of osteoblast differentiation. Cell 89:747–754
74. Geoffroy V, Ducy P, Karsenty G (1995) A PEBP2/AML-1-related factor increases osteocalcin promoter activity through its binding to osteoblast-*cis*-acting element. J Biol Chem 270:30973–30979
75. Mundlos S, Otto F, Mundlos C, Mulliken JB, Aylsworth AS, Albright S, Lindhout D, Cole WG, Henn H, Knoll JHM, Owen MJ, Mertelsmann R, Zabel BU, Olsen BR (1997) Mutations involving the transcription factor CBFA1 cause Cleidocranial Dysplasia. Cell 89:773–779
76. Komori T, Yagi H, Nomura S, Yamaguchi A, Sasaki K, Deguchi KY, Shimizu Y, Bronson RT, Gao YH, Inada M, Sato M, Okamoto R, Kitamura Y, Yoshiki S, Kishimoto T (1997) Targeted disruption of CBFA1 results in a complete lack of bone formation owing to maturational arrest of osteoblasts. Cell 89:755–764
77. Lee B, Thirunavukkarasu K, Zhou L, Pastore L, Baldini A, Geoffroy V, Ducy P, Karsenty G (1997) Missense mutations abolishing DNA binding OSF2/CBFA1 inpatients affected with cleidocranial dysplasia. Nature Genet 16:307–311
78. Merriman HL, van Wijnen AJ, Hiebert S, Bidwell JP, Fey E, Lian J, Stein J, Stein GS (1995) The tissue-specific nuclear matrix protein, NMP-2, is a member of the AML/CBF/PEBP2/RUNT domain transcription factor family: interactions with the osteocalcin gene for runt. Biochemistry 34:13125–13132
79. Otto F, Thornell AP, Crompton T, Denzel A, Gilmour AC, Rosewell IR, Stamp GWH, Beddington RSP, Mundlos S, Olsen BR, Selby PB, Owen MJ (1997) CBFA1, a candidate gene for Cleidocranial Dysplasia syndrome, is essential for osteoblast differentiation and bone development. Cell 89:765–771
80. Tondravi MM, McKercher SR, Anderson K, Erdmann JM, Quiroz M, Maki R, Teitelbaum SL (1997) Osteopetrosis in mice lacking haematopoietic transcription factor PU.1. Nature 386:81–84
81. Johnson RS, Spiegelman BM, Papaioannou V (1992) Pleiotropic effects of a null mutation in the c-fos proto-oncogene. Cell 71:577–586
82. Franzoso G, Carlson L, Xing L, Poljiak L, Shores EW, Brown KD, Leonardi A, Tran T, Boyce BF, Siebenlist U (1997) NF-kB in osteoclast and B-cell development. Genes Dev 11:3482–3496

83. Yoshida H, Hayashi S, Kunisada T, Ogawa M, Nishikawa S, Okamura H, Sudo T, Shultz LD, Nishikawa S (1990) The murine mutation osteopetrosis is in the coding region of the macrophage colony stimulating factor gene. Nature 345:442–444

84. Bucay N, Sarosi I, Dunstan CR, Morony S, Tarpley J, Capparelli C, Scully S, Tan HL, Xu W, Lancey DL, Boyle WJ, Simonet WS (1998) Osteoprotegerin-deficient mice develop early onset osteoporosis and arterial calcification. Gene Dev 12:1260–1268

85. Taylor SM, Jones PA (1979) Multiple new phenotypes induced in 10T1/2 and 3T3 cells treated with 5-azacytidine. Cell 17:771–779

86. Davis RL, Weintraub H, Lassar AB (1987) Expression of a single transfected cDNA converts fibroblasts to myoblasts. Cell 51:987–1000

87. Rudnicki MA, Schnegelsberg PN, Stead RH, Braun T, Arnold HH, Jaenisch R (1993) MyoD or Myf-5 is required for the formation of skeletal muscle. Cell 75:1351–1359

88. Borycki AG, Emerson CP (1997) Muscle determination: Another key player in myogenesis? Curr Biol 7:R620–R623

89. Taylor MV (1998) Muscle development: A transcriptional pathway in myogenesis. Curr Biol 8:R356–R358

90. Thorgeirsson SS (1993) Hepatic stem cells. Am J Pathol 142:1331–1333

91. Shiojiri N, Lemire JM, Fausto N (1991) Cell lineages and oval cell progenitors in rat liver development. Cancer Res 51:2611–2620

92. Sirica AE, Mathis GA, Sano N, Elmore LW (1990) Isolation, culture, and transplantation of intrahepatic biliary epithelial cells and oval cells. Pathobiology 58:44–64

93. Brill S, Holst P, Sigal S, Zvibel I, Fiorino A, Ochs A, Somasundaran U, Reid LM (1993) Hepatic progenitor populations in embryonic, neonatal, and adult liver. Proc Soc Exp Biol 204:261–269

94. Germain L, Blouin MJ, Marceau N (1988) Biliary epithelial and hepatocytic cell lineage relationships in embryonic rat liver as determined by the differential expression of cytokeratins, α-fetoprotein, albumin, and cell surface-exposed components. Cancer Res 48:4909–4918

95. Courtois G, Morgan JG, Campbell LA, Fourel G, Crabtree GR (1987) Interaction of a liver-specific nuclear factor with the fibrinogen and α 1-antitrypsin promoters. Science 238:688–692

96. Umek RM, Friedman AD, McKnight SL (1991) CCAAT-enhancer binding protein: a component of a differentiation switch. Science 251:288–292

97. Descombes P, Chojkier M, Lichtsteiner S, Falvey E, Schibler U (1990) LAP, a novel member of the C/EBP gene family, encodes a liver-enriched transcriptional activator protein. Genes Dev 4:1541–1551

98. Costa RH, Grayson DR, Darnell JEJ (1989) Multiple hepatocyte-enriched nuclear factors function in the regulation of transthyretin and α 1-antitrypsin genes. Mol Cell Biol 9:1415–1425

99. Sladek FM, Zhong WM, Lai E, Darnell Jr JE (1990) Liver-enriched transcription factor HNF-4 is a novel member of the steroid hormone receptor superfamily. Genes Dev 4:2353–2365

100. Mietus-Snyder M, Sladek FM, Ginsburg GS, Kuo CF, Ladias JA, Darnell JEJ, Karathanasis SK (1992) Antagonism between apolipoprotein AI regulatory protein 1, Ear3/COUP-TF, and hepatocyte nuclear factor 4 modulates apolipoprotein CIII gene expression in liver and intestinal cells. Mol Cell Biol 12:1708–1718

101. Ang SL, Wierda A, Wong D, Stevens KA, Cascio S, Rossant J, Zaret KS (1993) The formation and maintenance of the definitive endoderm lineage in the mouse: involvement of HNF3/forkhead proteins. Development 119:1301–1315

102. Hamamoto R, Kamihira M, Iijima S (1999) Growth and differentiation of cultured fetal hepatocytes isolated various developmental stages. Biosci Biotechnol Biochem 63:395–401

103. Michalopoulos GK (1990) Liver regeneration: molecular mechanisms of growth control. FASEB J 4:176–187

104. Fujio K, Hu Z, Evarts RP, Marsden ER, Niu CH, Thorgeirsson SS (1996) Coexpression of stem cell factor and c-kit in embryonic and adult liver. Exp Cell Res 224:243–250

105. Knowles BB, Howe CC, Aden DP (1980) Human hepatocellular carcinoma cell lines secrete the major plasma proteins and hepatitis B surface antigen. Science 209:497–499

106. Nakabayashi H, Taketa K, Yamane T, Oda M, Sato J (1985) Hormonal control of α-fetoprotein secretion in human hepatoma cell lines proliferating in chemically defined medium. Cancer Res 45:6379–6383

107. Hsu SL, Lin YF, Chou CK (1992) Transcriptional regulation of transferrin and albumin genes by retinoic acid in human hepatoma cell line Hep3B. Biochem J 283:611–615

108. Ramos RA, Nishio Y, Maiyar AC, Simon KE, Ridder CC, Ge Y, Firestone GL (1996) Glucocorticoid-stimulated CCAAT/enhancer-binding protein alpha expression is required for steroid-induced G1 cell cycle arrest of minimal-deviation rat hepatoma cells. Mol Cell Biol 16:5288–5301

109. Yamada K, Duong DT, Scott DK, Wang JC, Granner DK (1998) CCAAT/enhancer-binding protein beta is an accessory factor for the glucocorticoid response from the cAMP response element in the rat phosphoenolpyruvate carboxykinase gene promoter. J Biol Chem 274:5880–5887

110. Wille-Jr. J, Pittelkow MR, Shipley GD, Scott RE (1984) Integrated control of growth and differentiation of normal human prokeratinocytes cultured in serum-free medium: clonal analyses, growth kinetics, and cell cycle studies. J Cell Physiol 121:31–44

111. Boyce ST, Ham RG (1983) Calcium-regulated differentiation of normal human epidermal keratinocytes in chemically defined clonal culture and serum-free serial culture. J Invest Dermatol (Suppl) 81:335–405

112. Fuchs E (1990) Epidermal Differentiation: The bare essentials. J Cell Biol 111:2807–2814

113. Mdema JP, Sark-Jr. M, Backendorf C, Bos JL (1994) Calcium inhibits epidermal growth factor-induced activation of $p21^{ras}$ in human primary keratinocytes. Mol Cell Biol 14:7078–7085

114. Mangalam HJ, Albert VR, Ingraham HA, Kapiloff M (1989) A pituitary POU domain protein, Pit-1, activates both growth hormone and prolactin promoters transcriptionally. Genes Dev 3:946–958

115. Kemler I, Schaffner W (1990) Octamer transcription factors and the cell type-specificity of immunoglobulin gene expression. FASEB J 4:1444–1449

116. Davis RL, Cheng PF, Lassar AB, Weintraub H (1990) The MyoD DNA binding domain contains a recognition code for muscle-spicific gene activation. Cell 60:733–746

117. Rubin JS, Osada H, Finch PW, Taylor WG, Rudikoff S, Aaronson SA (1989) Purification and characterization of a newly identified growth factor specific for epithelial cells. Proc Natl Acad Sci USA 86:802–806

118. Lersch R, Stellmach V, Stocks C, Giudice G, Fuchs E (1989) Isolation, sequence, and expression of a human keratin K5 gene: transcriptional regulation of keratins and insights into pairwise control. Mol Cell Biol 9:3685–3697

119. Marchuk D, McCrohon S, Fuchs E (1984) Remarkable conservation of structure among intermediate filament genes. Cell 39:491–498

120. Johnson L, Idler W, Zhou XM, Roop D, Steinert P (1985) Structure of a gene for the human epidermal 67-kDa keratin. Proc Natl Acad Sci USA 82:1896–1900

121. Rosenberg M, Ray Chaudhury A, Shows TB, Beau MML, Fuchs E (1988) A group of type I keratin genes on human chromosome 17: characterization and expression. Mol Cell Biol 8:722–736

122. Eckert RL, Green H (1986) Structure and evolution of the human involucrin gene. Cell 46:583–589

123. Rothnagel JA, Steinert PM (1990) The structure of the gene for mouse filaggrin and a comparison of the repeating units. J Biol Chem 265:1862–1865

124. Leask A, Rosenberg M, Vassar R, Fuchs E (1990) Regulation of a human epidermal keratin gene : sequence and nuclear factors involved in keratinocyte-specific transcription. Genes Dev 4:1985–1998

125. Leask A, Byrne C, Fychs E (1991) Transcription factor AP2 and its role in epidermal-specific gene expression. Proc Natl Acad Sci, USA 88:7948–7952

126. Suzuki M, Itoh T, Osada H, Rubin JS, Aaronson SA, Suzuki T, Koga N, Saito T, Mitsui Y (1993) Spleen derived growth factor, SDGF-3, is identified as keratinocyte growth factor (KGF). FEBS Lett 328:17–20

127. Hines MD, Allen-Hoffman BL (1996) Keratinocyte growth factor inhibits cross-linked envelope formation and nucleosomal fragmentation in cultured human keratinocytes. J Biol Chem 271:6245–6251

128. Peehl DM, Wong ST, Rubin JS (1996) KGF and EGF differentially regulate the phenotype of prostatic epithelial cells. Growth Regul 6:22–31

129. Zheng C, Hoffman MP, McMillan T, Kleinman HK, O'Connell BC (1998) Growth factor regulation of the amylase promoter in a differentiating salivary acinar cell line. J Cell Physiol 177:628–635

130. Kuijpers AL, Van Pelt JP, Bergers M, Boegheim PJ, Den Bakker JE, Siegenthaler G, Van de Kerkhof PC, Schalkwijk J (1998) The effects of oral liarozole on epidermal proliferation and differentiation in severe plaque psoriasis are comparable with those of acitretin. Br J Dermatol 139:380–389

131. Hennings H, Michael D, Cheng C (1980) Calcium regulation for growth and differentiation of mouse epidermal cells in culture. Cell 29:245–254

132. Jaken S, Yuspa H (1988) Early signals for keratinocyte differentiation: role of calcium mediated inositol lipid metabolism in normal and neoplastic epidermal cells. Carcinogenesis 9:1033–1038

133. Osada H, Koshino H, Isono K, Takahashi H, Kawanishi G (1991) Reveromycin A, a new antibiotic which inhibits the mitogenic activity of epidermal growth factor. J Antibiot 44:259–261

134. Takahashi H, Yamashita Y, Takaoka H, Nakamura J, Yoshihama M, Osada H (1997) Inhibitory action of reveromycin A on TGF-α-dependent growth of ovarian carcinoma BG-1 in vitro and in vivo. Oncol Res 9:7–11

4
Apoptosis

Masaya Imoto

1 Introduction

Apoptosis is a morphologically distinct form of programmed cell death that plays a major role during development, homeostasis, and in many diseases including cancer, acquired immunodeficiency syndrome, and neurodegenerative disorders. Apoptosis can be triggered by a variety of stimuli including viral infection [1,2], growth factor withdrawal [3–8], and DNA damage resulting from radiation [9–11] and chemotherapeutic drugs [12]. The hallmark signs of apoptosis include chromatin condensation, internucleosomal cleavage of DNA, cell shrinkage and the formation of apoptotic bodies, which are ultimately engulfed by phagocytic cells. Apoptosis occurs through the activation of a cell-intrinsic suicide program. The basic machinery to carry out apoptosis appears to be present in essentially all mammalian cells at all times, but the activation of the suicide program is regulated by many different signals that originate from both the intracellular and the extracellular milieu. A study of the molecular events underlying apoptotic intracellular signalling pathways has, in recent years, resulted in the identification of a large number of molecules that are involved either in promoting or in suppressing the apoptotic response.

Much of our knowledge of the basal mechanism underlying programmed cell death derived from studies on developmental cell death in the developmentally deterministic nematode worm *Caenorhabditis elegans*.

Three genes, *ced-3*, *ced-4*, and *ced-9* encode the general apoptotic program in *C. elegans* [13]. *ced-9* negatively regulates apoptosis while *ced-4* and *ced-3* are required to execute the apoptotic program [14,15]. The *ced-9* gene is homologous to the human *bcl-2* gene, which is overexpressed in follicular lymphomas and contributes to a heightened state of resistance to cell death induced by a variety of agents including glucocorticoids and irradiation [16–19]. The Bcl-2 protein physically interacts with several homologous proteins that contribute the Bcl-2 protein family. Some of the members of this protein family are blockers of cell death such as Bcl-2 [20,21], Bcl-X_L [22], Bcl-w [23], Mcl-1 [24] whereas others such as Bax [25], Bak [26,27], Bad [28], Bik [29,30], Bid [31], and Bcl-X_s [22, 32] promote apoptosis and antagonize the function of Bcl-2. Bax can form a homodimer and also can heterodimerize with Bcl-2 or Bcl-X_L. The ratio of Bcl-2 or Bcl-X_L to Bax dictates the susceptibility of cells to apoptotic stimuli [33]. When Bax is in excess, multiple death stimuli result in apoptosis. Determining precisely how Bcl-2(-like) proteins function has been surprisingly difficult. In various experimental systems, Bcl-2 has been found to interact with calcineurin, p53 binding protein, lamin, NIP-1,2, and -3, R-Ras, Raf-1, Bag-1, and galectin-3, as well as interacting with other Bcl-2 family members [34]. Recently it has been proposed that Bcl-2 functions by forming pores to allow ions or small molecules to cross the outer mitochondrial membrane [35]. As release of cytochrome c from the intermitochondrial space has been associated with apoptosis in a number of systems, and as this release is prevented by Bcl-2, it has been hypothesized that the key function of Bcl-2(-like) proteins is to somehow retain cytochrome c in the mitochondria [36,37].

The key pro-apoptotic gene, *ced-3*, encodes a putative cystein protease showing sequence similarity to the family of mammalian interleukin-1β-converting enzyme (ICE)-related proteases (now called caspases) [38, 39]. The caspase family (caspase-1–10) represents a new class of intracellular cystein proteases with known or suspected roles in cytokine maturation and apoptosis. These enzymes display a preference for Asp in the P1 position of substrates [40, 41]. To better understand the roles of proteases in apoptosis, peptide substrate specificities for the caspases were determined [42, 43]. The results divided caspases into three distinct groups. Members of Group I (caspases 1,4, and 5) all prefer the tetrapeptide sequence WEHD. In contrast, the optimal peptide recognition motif for Group II caspases (2,3, and 7) is DEXD. The similarity between caspase-3 and caspase-7 is particularly striking; their specificity profiles are virtually indistinguishable. The caspases in Group III (6, 8, and 9) prefer the sequence (L/V)EXD. A comparison of the specificities of these enzymes reveals the following. First, aside from their stringent requirement for Asp in P1, P4 is the most critical determinant of specificity. Members of Group I can accommodate large aromatic/hydrophobic amino acids in this position, whereas those in Group II require Asp for efficient catalysis. The Group III caspases tolerate many different amino acids in P4 but prefer those with larger aliphatic side chain. Second, all of these enzymes prefer, to varying degrees, Glu in P3. Finally, in general, liberal substitutions are tolerated in P2. One notable exception is caspase-9, which has a stringent specificity for His in P2. The optimal recognition motif for Group III caspases resembles activation sites within several effector caspase proenzymes, specifically caspase-3 and caspase-7, implicating these enzymes as upstream components in a proteolytic cascade that serves to amplify the death signal. Regarding the mechanism of activation of family members other than caspase-3 and caspase-7, it is interesting to note that all of those with relatively long N-terminal peptides have specificities that are similar to their own activation sequences, suggesting that these enzymes may employ an autocatalytic mechanism of activation. This observation implies that the N-terminal peptide plays an essential role in autocatalysis, perhaps in mediating dimerization between two proenzyme molecules. In addition, the results secure a functional relationship between CED-3 and its closely related Group II human homologs. The optimal recognition motif for these enzymes (DEXD) is similar or identical to the cleavage sites in several cell maintenance and/or repair proteins that are proteolytically cleaved during apoptosis, including poly (ADP-ribose) polymerase (PARP) [44], Gelsolin [45], DNA-dependent protein kinase [46], D4-GDI [47], huntingtin [48], PKC δ [49] and θ [50], retinoblastoma protein [51], PAK2 [52] and DFF/ICAD [53–55]. Several of these are known endogenous substrates for caspase-3. It now appears that all members of this group function to cripple or destroy essential homeostatic pathways during the effector phase of apoptosis.

Ced-4 has been determined genetically to function downstream of ced-9 but upstream of ced-3 [56,57]. Ced-4 is homologous to the recently identified human protein, Apaf-1, which participates in the activation of caspase-3 [58]. Likewise, Apaf-1 functions downstream of Bcl-2 but upstream of caspase-3. Bcl-2 may function upstream of Apaf-1 by regulating the release of cytochrome c from mitochondria. Cytochrome c is a required cofactor for Apaf-1 function [58, 59]. Apaf-1 binds caspase-9 only in the presence of cytochrome c and dATP. Binding leads to the cleavage of caspase-9, converting it to an active protease [60]. Active caspase-9 then cleaves and activates caspase-3, thereby setting in motion the events that lead to DNA fragmentation and cell death.

On the other hand, recent studies have demonstrated that the apoptosis is controlled via a number of signaling events. These signals, some of which are mediated via membrane receptors, are either protective or inductive of cell death. Fas and tumor necrosis factor receptor belong to a family of membrane receptors that are thought to mediate death signals [61, 62], and the growth factor receptors are believe to transduce protective signals [63–65]. However, characterization of the signaling pathways involved in the receptor-mediated intracellular regulation of apoptosis is an area under intense investigation. A better understanding of the molecular events regulating the process of apoptosis is therefore required to develop novel and more efficacious agents modulating apoptotic signaling pathways.

2   Tyrosine Kinase

Tyrosine kinase activity was reported to regulate the process of apoptosis. Abl protein inhibited cytokine withdrawal-mediated [66] or Fas-mediated apoptosis [67]. Inhibition of a number of the Janus family of tyrosine kinases (Jak2) induced apoptosis in acute lymphoblastic leukemia [68]. Tyrosine kinase activity thus seems to be necessary for the prevention of apoptosis. However, the intracellular events that lead to apoptosis following tyrosine kinase inhibition have been obscure. Erbstatin was originally isolated from Streptomyces as an inhibitor of epidermal growth factor (EGF)-receptor tyrosine kinase [69]. It also inhibits nonreceptor-type tyrosine kinase, such as Src, Fyn and Lck, whereas it does not inhibit serine/threonine protein kinases, such as protein kinase C and protein kinase A [70, 71]. Erbstatin induced morphological apoptosis and DNA-fragmentation in human small cell lung carcinoma (SCLC) cells [72]. Erbstatin-induced apoptosis was inhibited by H_2O_2 scavengers such as N-acetylcysteine (NAC) and reduced form glutathione (GSH), without affecting erbstatin-inhibiting tyrosine phosphorylation. Furthermore, erbstatin was shown by means of flow cytometry to induce H_2O_2 generation. Erbstatin-induced H_2O_2 generation and apoptosis were suppressed by inhibition of protein synthesis. Thus, erbstatin-in-

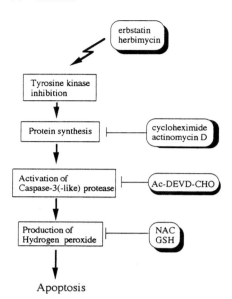

Fig. 1. Apoptotic pathway induced by tyrosine kinaseinhibitor

duced apoptosis was due to H_2O_2 generation via newly synthesized protein [72]. On the other hand, induction of apoptosis by erbstatin resulted in activation of caspase-3(-like) proteases, and inhibition of caspase-3(-like) protease activity by Ac-DEVD-CHO, a specific inhibitor of caspase-3(-like) proteases, reduced the extent of cell death and H_2O_2 generation (73). Also, expression of apoptotic protein Bax was induced by erbstatin, and this Bax expression was not inhibited by an H_2O_2 scavenger, but by Ac-DEVD-CHO. Thus, generation of intracellular H_2O_2 and Bax expression in erbstatin-induced apoptosis were modulated by the activation of caspase-3(-like) proteases in SCLC cells [73]. Similar results were observed by using another type of thyrosine kinase inhibitor, herbimycin A [74]. Tyrosine kinase inhibitor-induced apoptotic death signaling is illustrated in Fig. 1.

On the other hand, persistent invasion of malignant glioma tumor cells into the adjacent normal brain parenchyma renders surgical restriction incomplete and necessitates adjuvant treatment such as radiation and chemotherapy [75]. However, most gliomas eventually become drug-resistant, limiting the effectiveness of chemotherapy. A number of mechanisms may contribute to cellular drug resistance, including reduced intracellular drug concentrations, rapid inactivation of the drug, and increased rate of DNA repair [76]. Inhibition of apoptosis, a genetically controlled form of cell death, may also be important for drug resistance because the primary mechanism by which most chemotherapeutic agents having disparate modes of action and cellular targets induce cell death appears to be apoptosis. The malignant progression of gliomas involves accumulation of genetic alterations that inactivate tumor suppressor genes such as p53, p16, RB or activate oncogenes including the EGF receptor [77, 78]. The most common alterations of the EGF receptor that occurred in human malignant glioma were deletion of exons 2–7, resulting in truncation of the extracellular

Fig. 2. Structure of tyrphostin AG1478

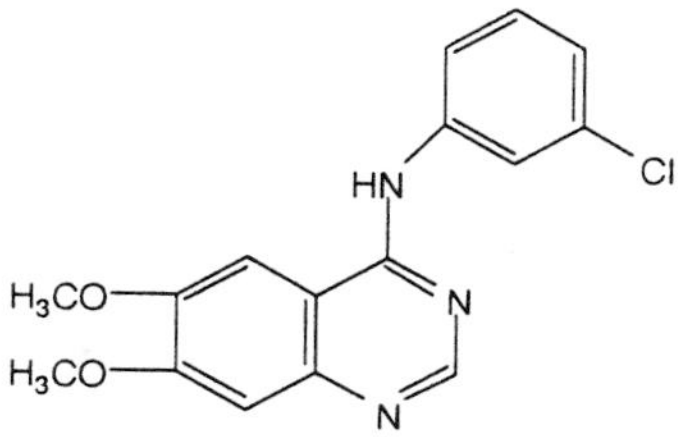

domain (DEGFR) [79]. Introduction of DEGFR into the U87MG human glioma cell line resulted in cell surface expression of a truncated receptor having a ligand-independent, weak but constitutively active, and unattenuated tyrosine kinase and enhanced tumorigenecity in nude mice [80], which was mediated by both an increase in proliferation and a decrease in apoptosis of tumor cells. In contrast, overexpression of wild-type (wt) EGFR did not confer a similar growth advantage [81, 82]. Furthermore, DEGFR expression in glioma cells confers resistance to some commonly utilized chemotherapeutic agents, including CDDP, taxol, and vincristine. The resistance was associated with suppression of drug-induced apoptosis, which was largely mediated by increased expression of Bcl-X_L and subsequent inhibition of caspas-3-like protease activation. These effects required constitutive signaling by DEGFR, because overexpression of tyrosine kinase-deficient DEGFR(DK) or wt EGFR had no such effects. To confirm that DEGFR confers chemotherapeutic agent resistance through its signal transduction, the tyrosine kinase inhibitor, AG1478 was used. AG1478 (Fig. 2) has been shown to have a more than 10-fold greater specificity for inhibition of DEGFR relative to wt EGFR [83]. Treatment of DEGFR expressing cells with 15 µM AG1478 resulted in marked reduction of Bcl-X_L expression. Their apoptosis index increased slightly after a single continuous treatment with 15–20 µM AG1478 for 2 days compared with that by the vehicle control, dimethyl sulfoxide. Thus, AG1478 is unlikely to be a direct apoptotic stimulus. Combination treatment of DEGFR expressing cells with AG1478 and CDDP, a DNA-damaging agent known to induce apoptosis in tumor cells and that often has been used in glioma therapy, caused significant apoptosis in a dose-dependent and synergistic manner. Similar results were obtained when the more potent DEGFR-selective tyrophostin, AG1517, was used at a lower concentration (7.5 µM). In contrast, the nonspecific and less potent tyrophostins, AG1479 and AG1536, had no effect on apoptosis induction in these cells, indicating that the observed synergistic effects likely were specific to the tyrosine kinase activity of DEGFR. CDDP-induced caspase activation in DEGFR expressing cells also was enhanced by cotreatment with AG1478 or AG1517, and the colony-forming efficiency of DEGFR expressing cells treated with the CDDP/AG1478 combination was significantly less than when treated with CDDP alone. These results suggest that constitutively active DEGFR signaling is responsible for Bcl-X_L up-regulation and inhibition of drug-induced apoptosis. This contention is supported by studies that have shown that HER-2/erbB-2-expressing lung cancer cells have synergistic responses to the erbB-2-specific tyrophostin, AG825, and the chemotherapeutic agents CDDP and etoposide [84].

3 Ceramide Synthesis

Sphingolipid metabolisms (Fig. 3) participate in key events of signal transduction and cell regulation. In the sphingomyelin cycle, a number of extracellular agents and insults cause the activation of sphingomyelinases, which act on membrane sphingomyelin and release ceramide [85–87]. The inducers of sphingomyelin hydrolysis and ceramide accumulation are 1,25-dihydroxyvitamin D3, tumor necrosis factor-α, endotoxin, interferon-gamma, interleukin-1, Fas ligands, CD28, dexamethasone, retinoic acid, progesterone, ionizing irradiation, chemotherapeutic agents, heat, and nerve growth factor (NGF) [88]. A growing number of cell-surface receptors are now being shown to generate signals that trigger the hydrolysis of sphingomyelin to release difusible ceramides. Ceramides have been implicated as key mediators in signaling pathways, with outcomes as diverse as cell proliferation and growth arrest. Furthermore, ceramide has been associated with apoptosis induction because synthetic, cell permeable ceramide (C2-ceramide) can induce caspase activation and apoptosis in various cell types. In several malignant and nonmalignant cell lines, ceramide rapidly and specifically induces apoptosis while closely related lipids remain inactive [89]. Ceramide activates proteases of caspases [90]. Activation of pro-caspase by ceramide and induction of apoptosis are inhibited by overexpression of Bcl-2 [91, 92] which suggests that Bcl-2 functions downstream of ceramide. On the other hand, overexpression of Bcl-2 does not reduce the amounts of ceramide produced in response to extracellular agents. The effects of ceramide on pro-caspases can be dissociated from activation of Rb, and vice versa; similarly, activation of PKC antagonizes

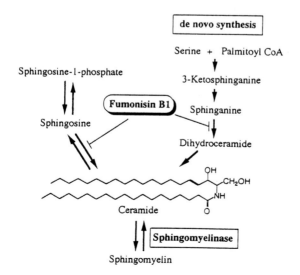

Fig. 3. Spihingolipid metabolism

the effects of ceramide on apoptosis but not on cell cycle arrest. Therefore, ceramide may relay a stress signal, whereby the specific cellular outcome appears to be determined by additional downstream modulators. Indeed, in some cell lines, ceramide may actually protect from apoptosis, possibly by preferentially steering cells into cell cycle arrest. Biologically relevant and direct targets of ceramide should be activated by ceramide in vitro, and they should mediate the most proximal effects of ceramide in cells. Ceramide-activated protein phosphatase (CAPP) is one of the candidates for direct targets of ceramide action [93, 94]. In vitro, ceramide activates a serine-threonine protein phosphatase. This phosphatase is related to the PP2A family of phosphatases because it copurifies with the heterotrimeric form of PP2A and is inhibited potently by okadaic acid in vitro. It is also activated by ceramides with various N-linked acyl groups, but it not activated by other sphingolipids or neutral lipids. Several lines of evidence suggest a role for CAPP in mediating at least some of the cellular activities of ceramide. First, some of ceramide's effects on cells such as apoptosis are inhibited by low concentrations of okadaic acid. Second, CAPP is activated in vitro by ceramide but not by dihydroceramide, which is inactive in eliciting ceramide effects on cells. Further studies will be required to determine the relevant physiologic substrates for CAPP. Ceramide-activated protein kinase (CAPK) is a another candidate for direct targets of ceramide action [95, 96]. CAPK is a membrane-associated kinase with a substrate specificity for serine or threonine in proximity to proline. Treatment of cells with sphingosine (but not C2-ceramide) results in a unique phosphorylation of Thr^{669} on the epidermal growth factor (EGF) receptor, and C8-ceramide mimics the effects of sphingosine and EGF [97, 98]. In vitro, ceramide does not activate this kinase in a partially purified preparation, which raises the possibility that this kinase is not directly regulated by ceramide [99].

Fumonisins are mycotoxins produced by *Fusarium moniliforme* . Recently, fumonesin B1 was identified as a specific inhibitor of ceramide synthesis [100, 101]. The chemotherapeutic agent CPT-11 induced apoptotic cell death in mouse fibroblast 4B1 cells and CPT-11-intiated cytolytic activity was prevented by both caspase inhibitors YVAD-CHO and DEVD-CHO, or fumonisin B1, and accelerated by sphingomyelin, suggesting the direct involvement of ceramide synthesis and the caspase cascade [102]. In addition, apoptosis was induced by both native and C2-ceramide and prevented by YVAD-CHO and DEVD-CHO, suggesting the possible involvement of ceramide in caspase cascade operation. On the other hand, caspase-1(-like) protease activity was prevented by fumonisin B1 and YVAD-CHO, but not by DEVD-CHO. In contrast, fumonisin B1, YVAD-CHO, and DEVD-CHO all prevented caspase-3(-like) protease activity. These results suggest that ceramide synthesis acts as a dominant regulator in CPT-11-intiated death signaling and sequentially operates the caspase cascade.

4 Fas Signaling

Fas(APO-1/CD95) and its ligand have been identified as important signal-mediators of apoptosis [103]. The elimination of tumor cells by cytotoxic T-lymphocytes (CTLs) is, in part, executed by inducing apoptosis through the cell surface antigen Fas on the target cells. The structural organization of Fas indicates that it is a member of the tumor necrosis factor receptor superfamily. Several Fas-interacting signal transducing molecules, have been identified using yeast two-hybrid and biochemical approaches, including Fas-associated phosphatase-1 (FAP-1) [104], FADD/MORT1/ CAP-1/CAP-2 [105–107] and RIP [108] (Fig. 4). All but FAP-1 associate with the functional cell death domain of Fas and overexpression of FADD/MORT1, or RIP

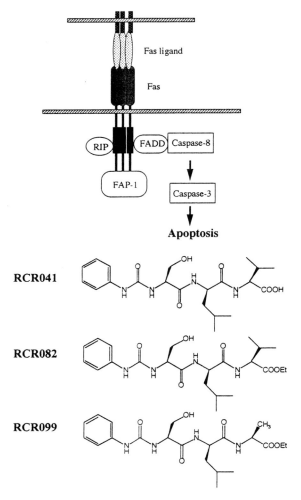

Fig. 4. FAS signaling and FAP-1 inhibitor

induces apoptosis in cells transfected with these proteins. In contrast, FAP-1 binds to the tSXV motif of the Fas C-terminus via its PDZ(GLGF) domain and negatively regulates Fas-mediated apoptosis [109]. The expression of FAP-1 was detected in many colon cancer cell lines. On the other hand, no FAP-1 expression was detected in the normal colon tissue. To examine the contribution of the FAP-1 PDZ domain binding to the Fas tSXV motif on Fas signal transduction, Kataoka et al. designed and synthesized a low molecular weight chemical compound, RCR041(Ph-NHCO-SLV), based on the tripeptide sequence of the Fas C-terminus(-SLV) (Fig. 4). RCR041 exhibited a stronger ability to inhibit binding than native Fas C-terminal peptide in vitro. RCR082(Ph-NHCO-SLV-OEt), an esterized form of RCR041, penetrated the cell surface membrane and inhibited the binding between Fas and the PDZ domain of FAP-1 in Fas transfected mouse L929 cells. RCR082 induced apoptosis in FAP-1 positive human colon cancer DLD-1 cells when anti-Fas antibody was added, whereas RCR099(Ph-NHCO-SLA-OEt), a compound with a valine to alanine substitution, had little effect on Fas-mediated apoptosis. These results suggest that the interaction of Fas and Fap-1 has a very important role in Fas-mediated apoptosis signal transduction. The expression level of FAP-1 is closely correlated with relative resistance to Fas-induced apoptosis in a variety of human tumor cell lines that express Fas. Therefore, the Fas C-terminal analog may be useful in the design of treatments for FAP-1 positive tumors that are resistance to Fas-mediated CTL cytotoxicity.

5 c-Jun NH$_2$-Terminal Kinase

JNK(c-Jun NH$_2$-terminal kinase)/SAPK(stress-activated protein kinase), characterized as a member of MAPK (mitogen-activated protein kinase) family, is activated by various forms of environmental stresses, such as UV irradiation or exposure to toxic agents [110–111]. Like other members of the MAPK family, JNK/SAPK requires phosphorylation at Thr-183 and Tyr-185 for its enzymatic activation [111]. This phosphorylation state is controlled by, at least, a kinase and a phosphatase, such as SEK1/ JNKK and MKP1, respectively [112–114]. The activated JNK/SAPK, in turn, phosphorylates the c-Jun transcription factor at Ser-63 and Ser-73 within its N-terminal transactivation domain, which induces the expression of c-Jun-responsive genes, such as c-jun itself [115]. These stresses are triggers for apoptosis in certain cells and tissues. In that sense, the JNK/SAPK signaling pathway, as one of stress responses, might be functionally involved in cellular survival or apoptosis.

Etoposide (VP-16) or camptothecin (CPT), which induced apoptosis in myeloid leukemia U937 cells, activated JNK1, transient c-jun expression, and caspase-3(-like) proteases. The phorbol ester-resistant U937 variant, UT16 cells, displayed a decreased susceptibility to apoptosis induced by these drugs. The drugs did not cause JNK1 activation, c-jun expression, or activation of caspase-3 (-like) proteases in UT16 cells [116]. Benzyloxycarbonyl-AspCH2OC(O)-2,6-dichlorobenzene (Z-Asp), a preferential inhibitor of caspase-3(-like) proteases, blocked the apoptosis of U937 cells. However, Z-Asp did not inhibit JNK1 activation in either VP-16- or CPT-treated U937 cells.

The JNK1 antisense oligonucleotides diminished protein expression of JNK1 and inhibited drug-induced apoptosis of U937 cells, whereas sense control oligonucleotides did not. Consistent with this observation, the antisense oligonucleotide-treated cells did not respond to VP-16 or CPT with Z-Asp-sensitive proteases. Thus, JNK1 triggers the DNA damaging drug-induced apoptosis of U937 cells by activating Z-Asp-sensitive caspase-3(-like) proteases (Fig. 5). Tumor necrosis factor-α (TNF-α) is a multifunctional cytokine produced by many cell types. It elicits a wide range of biological responses including cell proliferation, differentiation, and apoptosis, depending on the cell type and its state of differentiation [117]. Rat mesangial cells are normally resistant to TNF-α-induced apoptosis. They are made susceptible to the apoptotic effect of TNF-α when pretreated with actinomycin D, cycloheximide or vanadate. TNF-α alone stimulated a single transient JNK activity peak. However, when the cells were pretreated with actinomycin D, cycloheximide or vanadate, TNF-α stimulated a second sustained JNK activity peak. Therefore, a TNF-α-inducible phosphatase was responsible for preventing a sustained activation of JNK and consequent apoptosis in these cells [118]. Ro31-8220, although originally identified as a specific inhibitor of protein kinase C [119], was subsequently found to be a strong inhibitor of MKP-1 expression [120]. In mesangial cells, pretreatment of the cells with

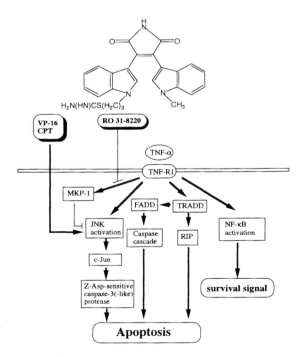

Fig. 5. JNK and apoptosis

Ro31-8220 blocked expression of MKP-1 induced by TNF-α. This treatment also prolonged JNK activation and caused apoptosis [121]. Thus, a sustained JNK activation was associated with the initiation of apoptosis. Furthermore, the expression of MKP-1 was induced by TNF-α, indicating that MKP-1 may be involved in protecting the cells from apoptosis by preventing a prolonged activation of JNK under normal conditions [121] (Fig. 5).

6 Cyclooxygenase-2

Cyclooxygenase (COX)-2 is a key enzyme in the production of prostaglandins and other eicosanoids that affect a number of signal transduction pathways that modulate cellular adhesion, growth, differentiation, and apoptosis [122–124]. Two isoforms of COX have been identified, which will be referred to as COX-1 and COX-2. COX-1 is constitutively expressed in a number of cell types, whereas COX-2 is induced by a variety of factors including cytokines, growth factors, and tumor promoters [125]. COX-2 was initially identified as a member of the immediate early or early growth response family of genes that are rapidly and transiently induced after growth factor or phorbol ester stimulation of quiescent cells. It was demonstrated that adhesion to selected extracellular matrix proteins is enhanced and that apoptosis is inhibited in intestinal epithelial cells that are programmed to overexpress the COX-gene [124]. Overexpression of COX-2 leads to a prolongation of G1, decreased cyclin D1 levels, inhibition of apoptosis, and altered adhesion properties of cells [126]. Furthermore, COX-2 overexpression may contribute to the tumorigenic potential of dysplastic intestinal epithelial cells by enhancing adhesion to the extracellular matrix and by inhibition of apoptosis. Several lines of evidence suggest that non-steroidal anti-inflammatory drugs (NSAIDs) prevent carcinogen-induced colorectal cancer in rodents, and NSAID treatment induces a dramatic regression of adenomas in patients with familial adenomatous polyposis coli [127, 128]. It is also well known that NSAIDs can inhibit COXs, and some studies have shown that NSAIDs can induce apoptosis. The most extensively investigated NSAID for chemoprevention is sulindac [128, 129], and sulindac sulfide (SUS) is an active metabolite of sulindac. SUS resulted in death of both colorectal cancer cell lines, HCT116 and SW480, in a dose- and time-dependent fashion. This death appeared to be apoptotic, as it was accompanied by nuclear condensation and fragmentation [130]. These results raised the possibility that sulindac functions through inhibiting the production of prostaglandins. However, treatment of cells with prostaglandin E2 (PGE2), the major COX product produced by colonic tumors did not inhibit SUS-induced apoptosis. Similarly, PGA2 and PGJ2 failed to protect against NSAID-induced apoptosis and in some instances potentiated apoptosis, indicating that prostaglandins do not affect NSAID-induced death. Another possibility is that induction of apoptosis by SUS might be due to increased levels of the prostaglandin precursor arachidonic acid (AA) resulting from inhibition of COX. In both SW480 and HCT116 cells, apoptosis-inducing doses of SUS increased AA by 5 fold at 48 h. If increased AA is the true mediator of SUS

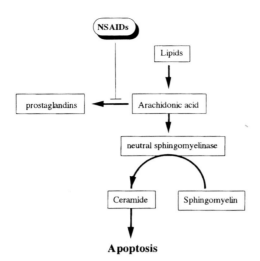

Fig. 6. Inhibition of COX and induction of apoptosis by NSAIDs

activity then AA should mimic its effect. 200 μM AA was found to be a potent inducer of apoptosis. Thus, the effect of SUS is not likely to be related to a reduction in prostaglandins but rather are due to the elevation of AA. AA is known to stimulate neutral sphingomyelinase, which catalyzes the conversion of sphingomyelin to ceramide, a known inducer of apoptosis. SUS dramatically stimulated the production of ceramide. Therefore, SUS appears to affect tumor growth by inhibiting COX activity, causing a build-up of the COX substrate AA, and activating sphingomyelinase activity leading to production of the powerful apoptosis inducer ceramide (Fig. 6). Indomethacin (INDO), an NSAID structurally distinct from sulindac, also induces apoptosis and increases AA and ceramide concentrations. This ability of NSAID is not limited to colon cancer cells; primary fibroblasts and immortalized keratinocytes also increase ceramide and undergo apoptosis in response to NSAIDs [130].

7   Calmodulin-Dependent Kinase III

Calmodulin (CaM) is a major intracellular calcium receptor in eukaryotic cells that regulates diverse intracellular processes including cellular proliferation, cyclic nucleotide and glycogen metabolism, calcium concentrations, and cellular morphology and motility [131–133]. The intracellular content of CaM is often increased in rapidly proliferating cell lines and in a variety of chemically and virally transformed cells [134–136]. Pharmacological studies support the role of CaM in cellular proliferation. For example, CaM antagonists are selective for proliferating cells, block cell cycle transit at the G1-S and G2-M interfaces and augment the cytotoxicity of chemotherapeutic drugs. A link between growth factor stimulation and CaM signaling is beginning to emerge. For example, CaM-mediated pathways are activated in response to

Fig. 7. Structure of rottlerin

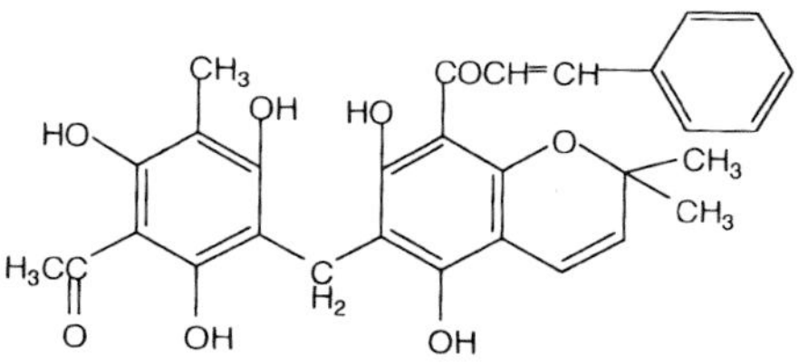

several mitogens through the activation of phospholipase C, resulting in the subsequent generation of inositol[1,4,5]tris phosphate [137]. Inositol[1,4,5]tris phosphate raises the cytosolic concentration of calcium to levels that bind to and activate CaM. In the calcium-saturated state, CaM can interact with target proteins. However, the downstream elements through which CaM mediates the proliferative response to mitogens remain unknown. CaM-dependent kinases are central to many of the functions mediated by CaM [138]. The presence of several CaM-dependent kinases in malignant gliomas has been demonstrated, including CaM kinase I, II, III and IV [139]. Although the activity and subcellular distribution of all of these kinases differed somewhat between glioma cells and normal glial tissues, the apparent selective activation of CaM kinase III in the cytosol of C6 and 9L glioma, as well as in other cell lines were observed. CaM kinase III is found in several mammalian tissues and cell lines and is distinguished from other CaM-dependent kinase by its unique substrate, eEF-2 [140, 141]. Phosphorylation of eEF-2, by CaM kinase III, decreases its affinity for the ribosome and transiently arrests protein synthesis. The activity of CaM kinase III can also be regulated by mitogens. For example, phosphorylation of eEF-2 has been shown to increase rapidly in quiescent fibroblasts exposed to EGF, vasopressin, and bradykinin [142]. Thus, CaM kinase III may provide an important link between transient increases in intracellular calcium seen in cancer cells in response to mitogens, protein synthesis, and cell division. Recently, rottlerin has been shown to inhibit CaM kinase III in pancreatic cytosol [143]. Rottlerin is a 5,7-dihydroxy-2,2-dimethyl-6-(2,4,6-trihydroxy-3-methyl-5-acetylbenzyl)-8-cinnamoyl-1,2-chromene isolated from the pericarps of *Mallotus phillippinensis* (Fig. 7). Rottlerin inhibits the activity of CaM kinase III partially purified from rabbit reticulocytes, with an IC_{50} of approximately 2.5 μM. Rottlerin is cytotoxic to glioma cells at drug concentrations that inhibit CaM kinase III activity. Rottlerin produces changes in cell morphology in C6 cells that are reminiscent of programmed cell death [144]. These include the appearance of a hypodiploid population of cells prior to G0-G1, an accumulation of cytoplasmic vacuoles, packaging of cellular components within membranes, cell shrinkage, and alterations in nuclear structure. Rottlerin did not produce DNA laddering in glioma cells, even when measured for 72 h after exposure to cytotoxic concentrations of the drug. In contrast, rottlerin produced DNA laddering in the human leukemia cell line, HL-60, after 18 h. These data suggest that inhibition of CaM kinase III by rottlerin can induce apoptotic cell death in certain cell types and similar form of demise in glioma cells.

8 PI-3 Kinase-Akt

Extracellular stimuli are transmitted intracellularly by signaling cascades that involve the interaction of macromolecules and/or the generation of second messenger molecules that transduce signaling events over a distance between the origin and target of a signal. One class of such second messenger molecules is generated via phosphorylation of phosphoinositides on the D-3 position by phosphoinositide 3-kinase (PI-3K) [145]. These products of PI-3K can act on multiple downstream effectors that include Src homology-2(SH2) and Pleckstrin homology (PH) domains of serine/threonine and tyrosine kinases and various cytoskeletal proteins. The role of PI-3K in intracellular signaling has been underscored by its implication in a plethora of biological responses. Although it is unlikely that these multiple responses will be explained by the action of a single downstream target, recent research from several laboratories indicates that a signaling pathway from PI-3K to the serine/threonine protein kinase Akt/PKB may mediate some cellular responses of PI-3K, including protection from apoptosis [146–151]. Exciting insights into the function of Akt/PKB have been revealed by studies investigating its function in PI-3K-dependent pathways that are involved in the regulation of cell survival. Previous studies have demonstrated that PI-3K is involved in serum-dependent survival of PC1cells [152]. Differentiation of PC12 cells was induced by incubating the cells with NGF-containing medium for 6–7 days. Apoptosis in these differentiated cells was induced by removal of NGF and incubation of the cells in serum-free medium. Insulin-like growth factor 1(IGF-1) was capable of preventing apoptosis. The PI-3K inhibitors wortmannnin [153] and LY294002 [154] blocked the effect of IGF-1. Wortmannin, a fungal metabolite, demonstrates a substantial degree of specificity for PI-3K compared with a num-

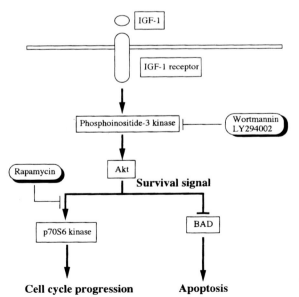

Fig. 8. Model of survival signaling by IGF-1

ber of other lipid kinases. PI-3K comprises an 85-kDa regulatory subunit and a 110-kDa catalytic subunit that phosphorylates phosphatidylinositol at the D-3-hydroxyl of the inositol ring. Wortmannin binds irreversibly to the catalytic subunit (p110), thereby inhibiting the signaling pathway following PI-3K activation. Wortmannin enhanced apoptosis and inhibited the effect of IGF-1 when added to PC12 cells, suggesting that PI-3K is involved in the regulation of apoptosis. Support for this hypothesis comes from the use of the synthetic PI-3K inhibitor LY294002, which gave essentially the same results as wortmannin. Withdrawal of survival factors leads to the rapid and synchronous apoptosis of cerebellar neurons, and the apoptosis could be inhibited by IGF-1. Wortmannin and LY294002 also inhibited IGF-1-dependent survival of cerebellar neurons, indicating that a pathway that includes PI-3K is important for the survival of cerebellar neurons [147]. IGF-1 activation of PI-3K triggered the activation of two protein kinases, the serine-threonine kinase Akt and the p70 ribosomal protein S6 kinase (p70^{S6K}). In cerebellar neurons, IGF-1 activates Akt, and this activation was blocked by wortmannnin and LY294002, which suggested that Akt activation was dependent on PI-3K. When cerebellar neurons were transfected with Akt or with control vector and were deprived of survival factors after 1 day, expression of Akt markedly reduced the amount of apoptosis. The ability of exogeneously expressed Akt to block apoptosis was not reduced in the presence of LY294002, consistent with Akt acting downstream of PI-3K. Because both PI-3K and Akt can promote activation of p70^{S6K}, p70^{S6K} seems to be a potential mediator of the survival effects of IGF-1. Originally identified as a ribosomal protein S6 kinase, p70^{S6K} has since been shown to regulate progression from the G1 to the S phase of the cell cycle. Rapamycin, which blocks activation of p70^{S6K} by inhibiting phosphorylation of p70^{S6K} [155], had no effect on the promotion of survival by IGF-1. Thus, Akt but not p70^{S6K} mediates PI-3K-dependent survival [147, 151]. BAD is a distant member of the Bcl-2 family that promotes cell death. Recently, it has been shown that Akt phosphorylates BAD and blocks the BAD-induced death [156–157]. Thus, the proapoptotic function of BAD is regulated by the PI-3K-Akt pathway (Fig. 8).

9 Cell Cycle Regulator

Recent evidence suggests that intracellular signals governing cell proliferation and cell cycle progression also mediate apoptosis. Therefore, the development of potent inhibitors of G1 progression would represent a relatively novel approach to the inhibition of tumor cell growth. This would provide a pharmacological means to compensate the loss of tumor suppressor genes or the presence of aberrant endogenous cell cycle regulation that contributes to the tumorigenic process.

9.1 Inostamycin

Inostamycin, a novel polyether compound, was isolated from *Streptomyces* sp. MH816-AF15 as an inhibitor of CDP-DG:inositol transferase which catalyzes phosphatidylinositol

(PtdIns) synthesis [158, 159]. It does not inhibit PtdIns-specific phospholipase C, PtdIns kinase, protein kinase C, mitogen-activated protein kinase, casein kinase II, CDK2 and tyrosine kinases and macromolecular synthesis directly [160, 161]. Inostamycin inhibited cell proliferation in normal rat kidney cells by blocking cell cycle progression at the G1 phase, presumably through the inhibition of PtdIns turn-over-mediated signal transduction. Treatment of exponentially proliferating human small cell lung carcinoma Ms-1 cells with low concentrations of inostamycin accumu-lated cells in the G1 phase. Inostamycin decreased cyclin D1, and increased cyclin-dependent kinase (CDK) inhibitors such as p21^{WAF1} and p27^{KIP1} in Ms-1 cells [162]. On the other hand, higher concentrations of inostamycin induce apoptosis in Ms-1 cells without inducing G1 arrest. These results raise the possibility that inostamycin-induced cell cycle arrest may be involved in apoptotic signals, and as a consequence, cells undergo apoptosis before accumulating in G1 phase. Therefore, it is likely that inostamycin-induced inhibition of cyclin D1 expression, and accumulation of CDKIs leading to hypophosphorylation of pRb are responsible for apoptosis [162].

The ratio of Bcl-2 to Bax dictates the susceptibility of cells to apoptotic stimuli. However, inostamycin does not affect the expression levels of Bcl-2, and Bax, and does not change the Bcl-2:Bax ratio. Activation of caspases appears to be a critical step in the control of apoptosis in mammalian cells. Activation of caspase-3 was induced by higher concentrations of inostamycin, as indicated by the formation of an active p17 subunit on Western blots. Furthermore, a caspase-3 inhibitor, Ac-DEVD-CHO, completely inhibited inostamycin-induced apoptosis. These results indicate that inostamycin-induced apoptosis is mediated by caspase-3. During Fas-mediated apoptosis, caspase-3 is cleaved and activated by caspase-1 [163]. However, the in-hibitor of caspase-1, Ac-YVAD-CHO, failed to prevent inostamycin-induced apoptosis, suggesting that caspase-3 is activated by mechanisms other than caspase-1. An inhibitor of caspase-3 did not affect inostamycin-induced suppression of cyclin D1 expression. Therefore, activation of caspase-3 is not required for G1 arrest induced by inostamycin. These results do not exclude the possibility that inostamycin-in-duced apoptosis is independent of cell cycle arrest, however, it is possible that an imbalance in the regulation of cell cycle by inostamycin causes activation of caspase-3 leading to apoptosis [162].

9.2 Butyrolactone I

Butyrolactone I is a highly selective inhibitor of CDKs. It inhibits the kinase activity of CDK2 as well as that of CDC2, while scarcely affecting PKC, protein kinase A, casein kinase, mitogen-activated protein kinase, and EGF-receptor tyrosine kinase [164, 165]. The addition of butyrolactone I in the culture medium resulted in a dose-dependent inhibition of the cell growth and induction of apoptosis [166]. The induc-tion of apoptosis by butyrolactone I was associated with greatly decreased levels of phosphorylated pRB in HL60 cells, which lack p53. The results from flow cytometry indicated that apoptosis induced by butyrolactone I occurred in G1 and S-phase cells. These results suggest that aberrant status of the cell cycle, where cells lacking the normal G1 checkpoint overriding the G1-S boundary, even in the presence of signals for G1 arrest, could lead to a self limitation by apoptotic cell death [166].

9.3 Anguinomycins

The retinoblastoma tumor suppressor protein (pRB) plays a central role in mammalian cell cycle control and is inactivated during the development of a wide variety of human cancers [167]. The human papillomaviruses (HPV) are highly associated with human cervical cancers and carry E6 and E7 oncoproteins, which bind and inactivate the tumor suppressors p53 and pRB, respectively [168]. In the course of screening for antitumor antibiotics, which selectively induce apoptosis against pRB-inactivated cells by using HPV E7 transformed rat glia cells, Hayakawa et al. found new members of the leptomycin/anguinomycin family, anguinomycins C and D [169]. At very low concentrations, anguinomycins C and D, as well as anguinomycins A and B induced growth arrest against normal cells and induced cells death against pRB-inactivated Rat glia cells. The effects of anguinomycins were further investigated using normal and transformed 3Y1 rat fibroblast. Anguinomycins A to D induced growth arrest against normal 3Y1 cells and v-H-ras-transformed cells, and caused cell death against 3Y1 cells transformed with v-src, simian virus 40, adenovirus type 12 and its E1A gene. Flow cytometric analysis revealed that anguinomycins C and D arrested the cell cycle of 3Y1 cells mainly at the G1 phase. Except for src-transformed cells, cell lines highly sensitive to the killing effect of anguinomycins commonly express viral oncoproteins including HPV E7, adenovirus E1A and simian virus 40 large T antigen, which can bind and inactivate pRB. Elevation of p53 is known to cause G1 arrest in normal cells and apoptotic cell death in pRB-inactivated cells. The activities of anguinomycins resemble those of p53, although they induced cell death against cells with p53 inactivated by HPV E6, adenovirus E1B or simian virus 40 large T antigen. It is possible that anguinomycins might activate a signal pathway after p53 [169]. Recently, CRM1 was found to be an export receptor for leucine-rich nuclear export signals [170, 171]. Schizosaccharomyces pombe CRM1 is the target of leptomycin B [172, 173]. In mammalian cells, leptomycin B was shown to bind to CRM1 protein, specifically inhibiting interaction of CRM1 with the nuclear export signal (NES). Since anguinomycins are family members of the leptomycins, anguinomycin-induced apoptosis would be due to the inhibition of nuclear transport of survival factor by anguinomycins.

10 p53

The p53 protein is a transcription factor that enhances the rate of transcription of six or seven known genes that carry out, at least in part, the p53-dependent functions in a cell [174, 175]. The p53 plays a role in triggering apoptosis in certain cell types. Stimuli such as DNA damage, withdrawal of growth factors, and expression of Myc or E1A can also cause p53-dependent apoptosis [176–180]. The pro-apoptotic proteins Bax and IGF-BP3 are transcriptional targets of p53 [181], suggesting that transactivation by p53 is important in inducing apoptosis in some circumstances. In addition, the anti-apoptotic proteins Bcl-2 and the adenovirus 19kDa E1B protein can prevent p53-mediated apoptosis. Normally, in a cell p53 protein is kept at a low concentration by its relatively short half-life. The proteases responsible for this are not known, but some evidence has suggested that ubiquitin-mediated proteolysis plays a role. The p53 is one of the most commonly mutated proteins found in tumor cells. Mutated p53 proteins interfere in the function of wild type p53 and may also serve as a dominant oncogene. The vast majority of p53 mutations result in a protein of altered conformation and prolonged half-life. Most mutated, but not wild type, p53 proteins are able to associate with Hsp70/72 [182–186], perhaps contributing to their increased stability. One report has implicated Hsp70/72 in the regulation of p53 conformation [187]. Geldanamycin, an antibiotic of the benzoquinone ansamycin class has been shown to interact with an Hsp70/72 class of heat shock proteins and can interfere with the function of heat shock proteins [188]. Brief exposure to geldanamycin destabilized the p53 protein of several breast, prostate and leukemic cell lines harboring mutated p53 alleles, resulting in a significant reduction in p53 steady state level and half-life [189]. In contrast to its effects on mutated p53, geldanamycin altered neither the steady state level nor inducibility of the wild type protein. In addition to its effects on p53 stability, geldanamycin also altered the conformation of mutated p53, so that it was no longer detectable with a mutant conformation-specific antibody. Finally, mutated p53 protein isolated from geldanamycin-treated cells regained partial ability to bind a wild type-specific p53 DNA consensus sequence. These data indicate the feasibility of pharmacologic intervention for altering the mutated p53 phenotype [189]. However, transcriptional activation is not always sufficient since, under some conditions, p53 can induce apoptosis in the absence of transcription [179, 190, 191]. Inhibition of transcription by actinomycin D or translation by cycloheximide does not always affect p53-dependent apoptosis. Post-translational modification by phosphorylation is thought to be an important mechanism that regulates p53 function. P53 is phosphorylated at multiple sites in vitro and in vivo. Indeed, inhibitors of protein phosphatases induce p53-dependent apoptosis in the absence of transactivation. Exposure of Balb/c 3T3 cells to okadaic acid, an inhibitor of protein phosphatases 1 and 2A, increased the phosphorylation of p53 without changing p53 levels [192]. Okadaic acid treatment enhanced the binding of p53 to a consensus DNA target sequence and caused a 5–8 fold increase in p53 transcriptional activity. Transient expression of SV40 small tumor antigen, a specific inhibitor of protein phosphatase 2A, caused a 4-fold increase in p53 transcriptional activity. Incubation of Balb/c 3T3 cells with okadaic acid also induced programmed cell death in a dose- and time-dependent manner. Decreases in viability, morphological changes, and the appearance of DNA fragmentation were dependent on p53 since cells lacking functional p53 were resistant to okadaic acid-induced apoptosis. However, incubation of Balb/c 3T3 cells with actinomycin D did not reduce okadaic acid-induced apoptosis, indicating that the p53-dependent apoptosis induced by okadaic acid did not require p53 transcriptional activity. The fact that SV40 small tumor antigen did not induce apoptosis provides additional evidence that p53 transcriptional activity is not sufficient for p53-

mediated apoptosis. These results indicate that signaling pathways involving protein phosphorylation play critical roles in controlling the apoptotic activity of p53. Furthermore, a basal level of protein phosphatase 1 or 2A activity is necessary to prevent p53-dependent apoptosis.

11 Hydroxymethyl Glutaryl CoA Reductase

Lovastatin blocks the rate-limiting step in the cholesterol-synthesis pathway, the formation of mevalonic acid from hydroxymethyl glutaryl (HMG) CoA [193–195]. Lovastatin induces apoptosis of the prostate cancer cell line LNCaP. These effects were prevented by the simultaneous addition of mevalonate to the culture medium. Lovastatin induced a proteolytic activity that was able to cleave the enzyme poly(ADP-ribose) polymerase (PARP) and the substrate Z-DEVD-AFC, which is modeled after the P1–P4 amino acids of the PARP cleavage site. Caspase-7, but not caspase-3, underwent proteolytic activation during lovastatin-induced apoptosis, an effect prevented by mevalonate [196]. Caspase-7 was the only detected caspase with DEVD cleavage activity that exhibited lovastatin-induced mRNA up-regulation. Again, mevalonate blocked this effect. Lovastatin-induced apoptosis also was prevented when the caspase inhibitors Z-DEVD-CH$_2$F or Z-VAD-CH$_2$F were added to the me-

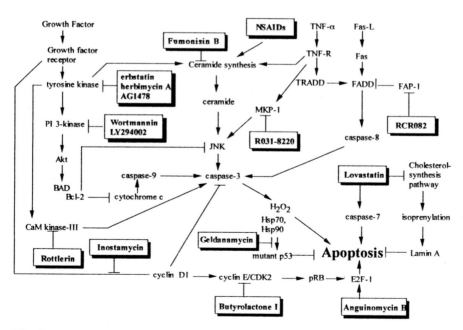

Fig. 9. Apoptosis mechanisms and bioprobes

dium. The mechanism whereby lovastatin triggers apoptosis is unclear. Because lovastatin inhibits HMG-CoA reductase, metabolites of the cholesterol pathway, such as farnesyl and geranyl Ppi that are needed for isoprenylation of critical proteins, are reduced. Lamin A, an important constituent of the nuclear membrane, is one protein that may be impacted by lovastatin because it requires farnesylation to reach its final mature form. Lamin A is recognized as a substrate that is degraded during apoptosis, and its cleavage may be a prerequisite for apoptotic nuclear disassembly. Ras is another protein that may trigger the apoptotic response in cells treated with lovastatin, because it is prevented from attaining its mature conformation after treatment with this HMG-CoA reductase inhibitor [196]. Major bioprobes for apoptosis regulation are summarized in Fig. 9.

References

1. White E (1993) Regulation of apoptosis by the transforming genes of the DNA tumor virus adenovirus. Proc Sco Exp Biol Med 204:30–39
2. White E, Gooding LR (1994) Regulation of apoptosis by human adenoviruses. In: D. Tomei and F. Cope (eds), Apoptosis: The molecular Basis for Cell Death II, pp111–141. Cold Spring Harbor, NY: Cold Spring Harbor Laboratory
3. Evan GI, Wylie AH, Gilbert CS, Littlewood T, Land H, Brooks M, Waters CM, Penn LZ, Hancock DC (1992) Induction of apoptosis in fibroblasts by c-myc protein. Cell 69:119–128
4. Lin Y, Benchimol S. (1995) Cytokines inhibit p53-mediated apoptosis but not p53-mediated G1 arrest. Mol Cell Biol 15:6045–6054
5. Rao L, Debbas M, Sabbatini P, Hockenberry D, Korsmeyer S, White E (1992) The adenovirus E1A proteins induce apoptosis which is inhibited by the E1B 19K abd Bcl-2 proteins. Proc Natl Acad Sci USA 89:7742–7746
6. Sakamuro D, Eviner V, Elliott KJ, Showe L, White E, Prendergast GC (1995) c-Myc induces apoptosis in epithelial cells by both p53-dependent and p53-independent mechanism. Oncogene 11:2411–2418
7. Wagner AJ, Kokontis JM, Hay N (1994) Myc-mediated apoptosis requires wild-type in a manner independent of cell cycle arrest and the ability of p53 to induce p21$^{WAF1/Cip1}$. Genes Dev 8:2817–2830
8. Yonish-Rouach E, Resnitzky D, Lotem J, Sachs L, Kimchi A, Oren M (1991) Wild-type p53 induces apoptosis of myeloid leukaemic cells that is inhibited by interleukin-6. Nature 352:345–347
9. Clarke AR, Purdie CA, Harrison DJ, Morris RG, Bird CC, Hooper ML, Wyllie AH (1993) Thymocyte apoptosis induced by p53-dependent and independent pathways. Nature 362:849–852
10. Han Z, Chatterjee D, He DM, Early J, Pantazis P, Wyche JH, Hendrickson EA (1995) Evidence for a G2 checkpoint in p53-independent apoptosis induction by X-irradiation. Mol Cell Biol 15:5849–5857
11. Lowe SW, Schmitt EM, Smith SW, Osborn BA, Jacks T (1993) p53 is required for radiation-induced apoptosis in mouse thymocytes. Nature 362:847–849
12. Lowe SW, Ruley HE, Jacks T Houseman DE (1993) p53-dependent apoptosis modulates the cytotoxicity of anticancer agents. Cell 74:957–967

13. Horvits HR, Shahan S, Hengartner MO (1994) The genetics of programmed cell death in the nematode Caenorthabditis elegans. Cold Spring Harb Symp Quant Biol 59:377–385

14. Yuan JY, Horvitz RH (1990) Genetics mosaic analysis of ced-3 ced-4, two genes that control programmed cell death in the nematode C. elegans. Dev Biol 138:33–41

15. Hengartner MO, Ellis RE, Horvitz RH (1992) Caenorhabditis elegans gene ced-9pro-tects cells from programmed cell death. Nature 356:494–499

16. Alnemri ES, Fernandes TF, Haldar S, Croce CM, Litwack G (1992) Involvement of BCL-2 in glucocorticoid-induced apoptosis of human pre-B-leukemias. Cancer Res 52:491–495

17. Hengartner MO, Horvitz RH (1994) C. elegans cell survival gene ced-9 encodes a functional homolog of the mammalian protooncogene bcl-2. Cell 76:665–676

18. White E (1996) Life, death, and the pursuit of apoptosis. Genes Dev 10:1–15

19. Raff MC (1992) Social controls on cell survival and cell death. Nature 356:397–400

20. Hockenberry D, Nunez G, Milliman CL, Schreiber RD Korsmeyer SJ (1990) Bcl-2 is an inner mitochondrial membrane protein that blocks programmed cell death. Nature 377:695–701

21. Vaux DL, Cory S, Adams TM (1988) Bcl-2 promotes the survival of haemopoietic cells and cooperates with c-myc to immortalize pre-B cells. Nature 335:440–442

22. Boise LH, Gonzalez-Garcia M, Postema CE, Ding L, Lindsten T, Turka LA, Mao X, Nunez G, Thompson CB (1993) Bcl-X, a Bcl-2-related gene that functions as a dominant regulator of apoptotic death. Cell 74:597–608

23. Gibson L, Holmgreen SP, Huang DC, Bernard O, Copeland NG, Jenkins NA, Sutherland GR, Baker E, Adams JM, Cory S (1996) Bcl-W, a novel member of the Bcl-2 family, promotes cell survival. Oncogene 13:665–675

24. Kozopas KM, Yang T, Buchan HL, Zhou P, Craig RW (1993) MCL1, a gene expressed in programmed myeloid cell differentiation, has sequence similarity to BCL-2. Proc Natl Acad Sci USA 90:3516–3520

25. Oltvai ZN, Milliman CL, Korsmeyer SJ (1993) Bcl-2 heterodimerizes in vivo with a conserved homolog, Bax, that accelerates programmed cell death. Cell 27:609–619

26. Chittenden T, Harrington EA, O'Connor R, Flemington C, Lutz RJ, Evan GI, Guild BC (1995) Induction of apoptosis by the Bcl-2 homologue Bak. Nature 374:733–736

27. Kiefer MC, Brauer MJ, Powers VC, Wu JJ, Umansky SR, Tomei LD, Barr PJ (1995) Modulation of apoptosis by the widely distributed Bcl-2 homologue Bak. Nature 374:736–739

28. Yang E, Zha J, Jackel J, Boise LH, Thompson CB, Korsemeyer SJ (1995) Bad, a heterodimeric partner for Bcl-XL and Bcl-2, displaces Bax and promotes cell death. Cell 80:285–291

29. Boyd JM, Gallo GJ, Elangovan B, Houghton AB, Malstrom S, Avery BJ, Ebb RG, Subramanian T, Chittenden T, Lutz RJ (1995) Bik, a novel death-inducing protein shares a distinct sequence motif with Bcl-2 family proteins and interacts with viral and cellular survival proteins. Oncogene 11:1921–1928

30. Han J, Sabbatini P, White E (1996) Induction of apoptosis by human Nbk/Bik, a BH3-containing protein that interacts with E1B 19K. Mol Cell Biol 16:5857–5864

31. Wang K, Yin X-M, Chao DT, Milliman CL, Korsmeyer SJ (1996) BID: a novel BH3 domain-only death agonist. Genes Dev 10:2859–2869

32. Minn AJ, Boise LH, Thompson CB (1996) Bcl-Xs antagonizes the protective effects of Bcl-XL. J Biol Chem 271:6306–6312

33. Chresta CM, Masters JRW, Hickman JA (1996) Hypersensitivity of human testicular tumors to etoposide-induced apoptosis is associated with functional p53 and a high Bax:Bcl-2 ratio. Cancer Res 56:1834–1841
34. Vaux DL (1997) Ced-4-The third horseman of apoptosis. Cell 90:389–390
35. Reed JC (1997) Double identity for proteins of the Bcl-2 family. Nature 387:773–776
36. Kruck RM, Bossy-Wetzel E, Grren DR (1997) The release of cytochrome c from mitochondria: a primary site for bcl-2 regulation of apoptosis. Science 275:1132–1136
37. Yang J, Liu X, Bhalla K, Kim CN, Ibrado AM, Cai J, Peng, I-I, Jones DP, Wang, X (1997) Prevention of apoptosis by bcl-2: release of cytochrome c from mitochondria blocked. Science 275:1129–1132
38. Yuan J, Shaham S, Ledoux S, Ellis HM, Horvitz HR (1993) The C. elegans cell death gene ced-3 encodes a protein similar to mammalian interleukin 1β-converting enzyme. Cell 75:641–652
39. Alnemri ES, Livingston DJ, Nicholson DW, Salvesen G, Thornberry NA, Wong WW, Yuan J (1996) Human ICE/CED-3 protease nomenclature. Cell 87:171
40. Nicholson DW (1996) ICE/CED3-like proteases as therapeutic targets for the control of inappropriate apoptosis. Nature Biotech 14:297–301
41. Salvesen GS and Dixit VM (1997) Caspases: Intracellular signaling by proteolysis. Cell 91:443–446
42. Thornberry NA, Rano TA, Peterson EP, Rasper DM, Timkey T, Garcia-Calvo M, Houtzager VM, Nordstrom PA, Roy S, Vaillancourt JP, Chapman KT, Nicholson DW (1997) A combinatorial approach defines specificities of member of the caspase family and granzyme B. J Biol Chem 272:17907–17911
43. Talanian RV, Quinlan C, Trautz S, Hackett MC, Mankovich JA, Banach D, Ghayur T, Brady KD, Wong WW (1997) Substrate specificities of caspase family proteases. J Biol Chem 272:9677–9682
44. Kaufmann SH, Desnoyers S, Ottaviano Y, Davidson NE, Poirier GG (1993) Specific proteolytic cleavage of poly(ADP-ribose) polymerase: an early marker of chemotherapy-induced apoptosis. Cancer Res 53:3976–3985
45. Kothakota S, Azuma T, Reinhard C, Klippel A, Tang J, Chu K, McGarry TJ, Kirschner MW, Koths K, Kwaitkowski DJ, Williams LT (1997) Caspase-3-generated fragment of gelsolin: effector of morphological change in apoptosis. Science 278:294–298
46. Song Q, Lees-Miller SP, Kumar S, Zhang Z, Chan DW, Smith GC, Jackson SP, Alnemri ES, Litwack G, Khanna KK, Lavin MF (1996) DNA-dependent protein kinase catalytic subunit: a target for an ICE-like protease in apoptosis. EMBO J 15:3238–3246
47. Na S, Chuang T-H, Cunningham A, Turi, TG, Hanke JH, Bokoch GM, Danley DE (1996) D4-GDI, a substrate of CPP32, is proteolyzed during Fas-induced apoptosis. J Biol Chem 271:11209–11213
48. Goldberg YP, Nicholson DW, Rasper DM, Kalchman MA, Koide NA, Vaillancourt JP, Havden MR (1996) Cleavage of huntingtin by apopain, a proapoptotic cysteine protease, is modulated by the polyglutamine tract. Nat Genet 13:442–449
49. Ghayur T, Hugunin M, Talanian RV, Ratnofsky S, Quinlan C, Emoto Y, Pandey P, Datta R, Huang Y, Kharbanda S, Allen H, Kamen R, Wong W, Kufe D (1996) Proteolytic activation of protein kinase C δ by an ICE/CED 3-like protease induces characteristics of apoptosis. J Exp Med 184:2399–2404
50. Datta R, Kojima H, Yoshida K, Kufe D (1997) Caspase-3-mediated cleavage of protein kinase C θ in induction of apoptosis. J Biol Chem 272:20317–20320

51. Chen W, Otterson GA, Lipkowitz S, Kheif SN, Coxon AB, Kaye FJ (1997) Apoptosis is associated with cleavage of a 5 kDa fragment from RB which mimics dephosphorylation and modulates E2F binding. Oncogene 14:1243–1248

52. Rudel T, Bokoch GM (1997) Membrane and morphological changes in apoptotic cells regulated by caspase-mediated activation of PAK2. Science 276:1571–1574

53. Liu X, Zou H, Slaughter C, Wang X (1997) DFF, a heterodimeric protein that functions downstream of caspase-3 to trigger DNA fragmentation during apoptosis. Cell 89:175–184

54. Enari M, Sakahira H, Yokoyama H, Okawa K, Iwamatsu A, Nagata S (1998) A caspase-activated DNase that degrades DNA during apoptosis, and its inhibitor ICAD. Nature 391:43–50

55. Sakahira H, Enari M, Nagata S (1998) Cleavage of CAD inhibitor in CAD activation and DNA degradation during apoptosis. Nature 391:96–99

56. Shaham S, Horvitz HR (1996a) Developing Caenorhabditis elegans neurins may contain both cell-death protective and killer activities. Genes Dev 10:578–591

57. Shaham S, Horvitz HR (1996b) An alternative spliced C. elegans ced-4 RNA encodes a novel cell death inhibitor. Cell 86:201–208.

58. Zou H, Henzel WJ, Liu X, Wang X (1997) Apaf-1, a human protein homologous to C. elegans CED-4, participates in cytochrome c-dependent activation of caspase-3. Cell 90:405–413

59. Liu X, Kim CN, Yang J, Jemmerson R, Wang X (1996) Induction of apoptotic program in cell free extract: requirement for dATP and cytochrome c. Cell 86:147–157

60. Li P, Nijhawan D, Budihardjo I, Srinivasula SM, Ahmad M, Alnemri ES, Wang X (1997) Cytochrome c and dATP-dependent formation of Apaf-1/caspase-9 complex initiates an apoptotic protease cascade. Cell 91:479–489

61. Nagata S, Golstein P (1995) The Fas death factor. Science 267:1449–1456

62. Smith CA, Farrah T, Goodwin RG (1994) The TNF receptor superfamily of cellular and viral proteins: activation, costimulation and death. Cell 76:959–962

63. Segal RA, Greenberg ME (1996) Intracellular signaling pathways activated by neurotrophic factors. Annu Rev Neurosci. 19:463–489

64. Stewart CE, Rotwein P (1996) Growth, differentiation, and survival: multiple physiological functions for insulin-like growth factors. Physiol Rev 76:1005–1026

65. Ataliotis P, Mercola M (1997) Distribution and functions of platelet-derived growth factors and their receptors during embryogenesis. Int Rev Cytol 172:95–127

66. Evans CA, Oven-Lynch PJ, Whetton AD Dive C (1993) Activation of the Abelson tyrosine kinase activity is associated with suppression of apoptosis in hemapoietic cells. Cancer Res 53:1735–1738

67. McGahon AJ, Nishioka WK, Martin SJ, Mahboubi A, Cotter TG, Green DR (1995) Regulation of the Fas apoptotic cell death pathway by Abl. J Biol Chem 270:22625–22631

68. Meydan N, Grunberger T, Dadi H, Shahar M, Arpaia E, Lapidot Z, Leeder JS, Freedman M, Cohen A, Gazit A, Levitzki A, Roifman CM (1996) Inhibition of acute lymphoblastic leukaemia by a Jak-2 inhibitor. Nature 379:645–648

69. Umezawa H, Imoto M, Sawa T, Isshiki K, Matsuda N, Uchida T, Iinuma H, Hamada M, Takeuchi T (1986) Studies on a new epidermal growth factor-receptor kinase inhibitor, erbstatin, produced by MH435-hF3. J Antibiot. 39:170–173

70. Imoto M, Umezawa K, Sawa T, Takeuchi T, Umezawa H (1987) In situ inhibition of tyrosine protein kinase by erbstatin. Biochem. Int., 15:989–995

71. Imoto M, Nakamura T, Tanaka S, Umezawa K (1994) Inhibition of T cell receptor-mediated signal transduction by erbstatin. Biosci Biotech Biochem 58:1700–1701

72. Simizu S, Imoto M, Masuda N, Takada N, Umezawa K (1996) Involvement of hydrogen peroxide production in erbstatin-induced apoptosis in human small lung carcinoma cells. Cancer Res 56:4978–4982

73. Simizu S, Umezawa K, Takada M, Imoto M (1998) Induction of hydrogen peroxide production and Bax expression by CPP32(-like) proteases in tyrosine kinase inhibitor-induced apoptosis in human small cell lung carcinoma cells. Exp Cell Res 238:197–203

74. Uehara Y, Hori M, Takeuchi T, Umezawa H (1985) Screening of agents which convert "transformed morphology" of Rous sarcoma virus-infected rat kidney cells to "normal morphology": identification of an active agent as herbimycin and its inhibition of intracellular src kinase. Jpn J Cancer Res 76:672–675

75. Fine HA (1994) The basis for current treatment recommendations for malignant gliomas. J Neurooncol 20:111–120

76. Feun LG, Savaraj N, Landy HJ (1994) Drug resistance in brain tumors. J Neurooncol 20:165–176

77. Nagane M, Huang H-JS, Cavenee WK (1997) Advances in the molecular genetics of gliomas. Curr Opin Oncol 9:215–222

78. Furnari FB, Lin H, Huang H-JS, Cavenee WK (1997) Growth suppression of glioma cells by PTEN requires a functional phosphatase catalytic domain. Proc Natl Acad Sci USA 94:12479–12484

79. Ekstrand AJ, Sugawa N, James CD, Collins VP (1992) Amplified and rearranged epidermal growth factor receptor genes in human glioblastomas reveal deletions of sequences encoding portions of the N- and/or C-terminal tails. Proc Natl Acad Sci USA 89:4309–4313

80. Nishikawa R, Ji XD, Harmon RC, Lazar CS, Gill GN, Cavenee WK, Huang HJ (1994) A mutant epidermal growth factor receptor common in human glioma confers enhanced tumorigenicity. Proc Natl Acad Sci USA 91:7727–7731

81. Huang H-JS, Nagane M, Klingbeil CK, Lin H, Nishikawa R, Ji XD, Huang CM, Gill GN, Wiley HS, Cavenee WK (1997) The enhanced tumorigenic activity of a mutant epidermal growth factor receptor common in human cancers is mediated by threshold levels of constitutive tyrosine phosphorylation and unattenuated signaling. J Biol Chem 272:2927–2935

82. Nagane M, Coufal F, Lin H, Bogler O, Cavenee WK, Huang HJ (1996) A common mutant epidermal growth factor receptor confers enhanced tumorigenicity on human glioblastoma cells by increasing proliferation and reducing apoptosis. Cancer Res 56:5079–5086

83. Nagane M, Levitzki A, Gazit A, Cavenee WK, Huang HJ (1998) Drug resistance of human glioblastoma cells conferred by a tumor-specific mutant epidermal growth factor receptor through modulation of Bcl-XL and caspase-3-like proteases. Proc Natl Acad Sci USA 95:5724–5729

84. Tsai CM, Levitzki A, Wu LH, Chang KT, Cheng CC, Gazit A, Perng RP (1996) Enhancement of chemosensitivity by tyrphostin AG825 in high-p185(neu) expressing non-small cell lung cancer cells. Cancer Res 56:1068–1074

85. Hannun YA (1994) The sphingomyelin cycle and the second messenger function of ceramide. J Biol Chem 269:3125–3128

86. Hannun YA, Linardic CM (1994) Sphingolipid breakdown products: anti-proliferative and tumor-suppressor lipids. Biochim Biophys Acta 1154:223–236

87. Kolesnick R, Golde DW (1994) The sphingomyelin pathway in tumor necrosis factor and interleukin-1 signaling. Cell 77:325–328

88. Liu B, Obeid LM, Hannun YA (1997) Sphingomyelinases in cell regulation. Cell Dev Biol 8:311–322
89. Hannun YA (1996) Functions of ceramide in coordinating cellular responses to stress. Science 274:1855–1859
90. Martin SJ, O'Brien GA, Nishioka WK, McGahon AJ, Mahboubi A, Saido TC, Green DR (1995a) Proteolysis of fodrin (non-erythroid spectrin) during apoptosis. J Biol Chem 270:6425–6428
91. Zhang J, Alter N, Reed JC, Borner C, Obeid LM, Hannun YA (1996) Bcl-2 interrupts the ceramide-mediated pathway of cell death. Proc Natl Acad Sci USA 93:5325–5328
92. Martin SJ, Reutelingsperger CP, McGahon AJ, Rader JA, van Schie RC, LaFace DM, Green DR (1995b) Early redistribution of plasma membrane phosphatidylserine is a general feature of apoptosis regardless of the initiating stimulus: inhibition by overexpression of Bcl-2 and Abl. J Exp Med. 182:1545–1556
93. Dobrowsky RT, Hannun YA (1992) Ceramide stimulates a cytosolic protein phosphatase. J Biol Chem 267:5048–5051
94. Wolff RA, Dobrowsky RT, Bielawska A, Obeid LM, Hannun YA (1994) Role of ceramide-activated protein phosphatase in ceramide-mediated signal transduction. J Biol Chem 269:19605–19609
95. Joseph CK, Byun H-S, Bittman R, Kolesnick RN (1993) Substrate recognition by ceramide-activated protein kinase. J Biol Chem 268:20002–20006
96. Mathias S, Dressler KA, Kolensnick RN (1991) Characterization of a ceramide-activated protein kinase: Stimulation by tumor necrosis factor α. Proc Natl Acad Sci USA 88:10009–10013
97. Faucher M, Girones N, Hannun YA, Bell RM, Davis RJ (1988) Regulation of the epidermal growth factor receptor phosphorylation state by sphingosine in A431 human epidermoid carcinoma cells. J Biol Chem 263:5319–5327
98. Goldkorn T, Dressler KA, Muindi J, Radin NS, Mendelsohn J, Menaldino D, Liotta D, Kolesnick RN (1991) Ceramide stimulates epidermal growth factor receptor phosphorylation in A431 human epidermoid carcinoma cells. J Biol Chem 266:16092–16097
99. Liu J, Mathias S, Yang Z, Kolesnick RN (1994) Regulation and tumor necrosis factor-α stimulation of a 97-kDa ceramide-activated protein kinase. J Biol Chem 269:3047–3052
100. Cifone MG, Roncaioli P, De Maria R, Camarda G, Santoni A, Ruberti G, Testi R (1995) Multiple pathways originate at the Fas/APO-1 (CD95) receptor: sequential involvement of phosphatidylcholine-specific phospholipase C and acidic sphingomyelinase in the propagation of the apoptotic signal. EMBO J 14:5859–5868
101. Wang E, Norred WP, Bacon CW, Riley RT, Merrill AH Jr (1991) Inhibition of sphingolipid biosynthesis by fumosines. J Biol Chem 266:14486–14490
102. Suzuki A, Iwasaki M, Kato M, Wagai N (1997) Sequential operation of ceramide synthesis and ICE cascade in CPT-11-initiated apoptotic death signaling. Exp Cell Res 233:41–47
103. Itoh N, Yonehara S, Ishii A, Yonehara M, Mizushima S, Sameshima M, Hase A, Seto Y, Nagata S (1991) The polypeptide encoded by the cDNA for human cell surface antigen Fas can mediate apoptosis. Cell 66:233–243
104. Sato T, Irie S, Kitada S, Reed JC (1995) FAP-1: a protein tyrosine phosphatase that associates with Fas. Science 268:411–415
105. Chinnaiyan AM, O'Rourke K, Tewari M, Dixit VM (1995) FADD. a novel death domain-containing protein interacts with the death domain of Fas and initiates apoptosis. Cell 81:513–523

106. Boldin MP, Varfolomeev EE, Pancer Z, Meett IL, Camonis JH, Wallach D (1995) A novel protein that interacts with the death domain of Fas/APO1 contains a sequence motif related to the death domain. J Biol Chem 270:7795–7798

107. Kischkel FC, Hellbardt S, Behrmann I, Germer M, Pawlita M, Krammer PH, Peter ME (1995) Cytotoxicity-dependent APO-1 (Fas/CD95)-associated proteins form a death-inducing signaling complex (DISC) with the receptor. EMBO J 14:5579–5588

108. Stanger BZ, Leder P, Lee TH, Kim E, Seed B (1995) RIP, a novel protein containing a death domain that interacts with Fas/APO-1(CD90) in yeast and causes cell death. Cell 81:513–523

109. Itoh N, and Nagata S (1993) A novel protein domain required for apoptosis. J Biol Chem 268:10932–10937

110. Kyriakis JM, Banerjee P, Nikolakaki E, Dai T, Rubie EA, Ahmad MF, Avruch J, Woodgett JR (1994) The stress-activated protein kinase subfamily of c-Jun kinases. Nature 369:156–160

111. Derijard B, Hibi M, Wu L-H, Barrett T, Su B, Deng T, Karin M, Davis RJ (1994) JNK1: a protein kinase stimulated by UV light and Ha-Ras that binds and phosphorylates the c-Jun activation domain. Cell 76:1025–1037

112. Sanchez I, Hughes RT, Mayer BJ, Yee K, Woodgett JR, Avruch J, Kyriakis JM, Zon LI (1994) Role of SAPK/ERK kinase-1 in the stress-activated pathway regulating transcription factor c-Jun. Nature 372:794–798

113. Yan M, Dai T, Deak JC, Kyriakis JM, Zon LI, Woodgett JR, Templeton DJ (1994) Activation of stress-activated protein kinase by MEKK1 phosphorylation of its activator SEK1. Nature 372:798–800

114. Liu Y, Gorospe M, Yang C, Holbrook NJ (1995) Role of mitogen-activated protein kinase phosphatase during the cellular response to genotoxic stress. J Biol Chem 270:8377–8380

115. Hibi M, Lin A, Smeal T, Minden A, Karin M (1993) Identification of an oncoprotein- and UV-responsive protein kinase that binds and potentiates the c-Jun activation domain. Genes Dev 7:2135–2148

116. Seimiya H, Mashima T, Toho M, Tsuruo T (1997) c-Jun NH_2-terminal kinase-mediated activation of interleukin-1β converting enzyme/CED-3-like protease during anticancer drug-induced apoptosis. J Biol Chem 272:4631–4636

117. Beyaert R, Fiers W (1994) Molecular mechanisms of tumor necrosis factor-induced cytotoxicity. What we do understand and what we do not. FEBS Lett 340:9–16

118. Guo YL, Baysal K, Kang B, Yang LJ, Williamson JR (1998a) Correlation between sustained c-Jun N-terminal protein kinase activation and apoptosis induced by tumor necrosis factor-a in rat mesangial cells. J Biol Chem 273:4027–4034

119. Wilkinson SE, Parker PJ, Nixon JS (1993) Isoenzyme specificity of bisindolylmaleimides, selective inhibitors of protein kinase C. Biochem J 294:335–337

120. Beltman J, McCormick F, Cook SJ (1996) The selective protein kinase C inhibitor, Ro-31-8229, inhibits mitogen-activated protein kinase phosphatase-1(MKP-1) expression, induces c-Jun expression, and activates Jun N-terminal kinase. J Biol Chem 271:27018–27024

121. Guo YL, Kang B, Williamson JR (1998) Inhibition of the expression of mitogen-activated protein phosphatase-1 potentiates apoptosis induced by tumor necrosis factor-a in rat mesangial cells. J Biol Chem 273:10362–10366

122. Bronestein J, Bull A (1993) The correlation between 13-hydroxyoctadecadienoate dehydrogenase (13-HODE dehydrogenase) and intestinal cell differentiation. Prostaglandins 46:387–395

123. Eberhart CE, DuBois RN (1995) Eicosanoids and the gastrointestinal tract. Gastroenterology 109:285–301

124. Tsuji M, DuBois RN (1995) Alterations in cellular adhesion and apoptosis in epithelial cells overexpressing prostaglandin endoperoxide synthase 2. Cell 83:493–501

125. Williams CS, DuBois RN (1996) Prostaglandin endoperoxide synthase: why two isoform? Am J Physiol 270:G393–G400.

126. DuBois RN, Shao J, Tsuji M Sheng H, Beauchamp RD (1996) G1 delay in cells overexpressing prostaglandin endoperoxide synthase-2. Cancer Res 56:733–737

127. Thun MH, Namboodiri MM, Heath CW (1991) Aspirin use and reduced risk of fatal colon cancer. N Engl J Med 325:1593–1596

128. Giardiello FM, Hamilton SR, Krush AJ (1993) Treatment of colonic and rectal adenomas with sulindac in familial adenomatous polyposis. N Engl J Med 328:113–1316

129. Waddell WR, Ganser GF, Cerise EJ, Loughry RW (1989) Sulindac for polyposis of the colon. Am J Sug 157:175–179

130. Chan TA, Morin PJ, Vogelstein B, Kinzler KW (1998) Mechanism underlying nonsteroidal antiinflammatory drug-mediated apoptosis. Proc Natl Acad Sci USA 95:681–686

131. Means AR (1984) Calcium, calmodulin and cell cycle regulation. FEBS Lett 347:1–4.

132. Lu KP, Means AR (1993) Regulation of the cell cycle by calcium and calmodulin. Endocr Rev 14:40–58

133. Means AR (1988) Molecular mechanisms of action of calmodulin. Recent Prog Hormone Res 44:223–262

134. Chafouleas JG, Pardue RL, Brinkley BR, Dedman JR, Means AR (1981) Regulation of intracellular levels of calmodulin and tublin in normal and transformed cells. Proc Natl Acad Sci USA 78:996–1000

135. MacManus JP, Braceland BM, Rixon RH (1981) An increase in calmodulin during growth of normal and cancerous liver in vivo. FEBS Lett 133:99–102

136. Veigl ML, Vanaman TC, Brance ME, Sedwick WD (1984) Differences in calmodulin levels of normal and transformed cells as determined by culture conditions. Cancer Res 44:3184–3189

137. Rasmussen CD, Means AR (1987) Calmodulin is involved in regulation of cell proliferation. EMBO J 6:3961–3968

138. Sheng M, Thompson MA, Greenberg ME (1991) CREB: a Ca^{2+} regulated transcription factor phosphorylated by calmodulin-dependent kinases. Science (Washington DC) 252:1427–1430

139. Cheng EHC, Gorelick FS, Czemik AJ, Halt WN (1995) Calcium/calmodulin-dependent binding proteins and kinase activity in murine and human glioblastoma. Cell growth & Differ 6:1–7

140. Narin AC, Palfrey HC (1987) Identification of the major Mr 100,000 substrate for calmodulin-dependent protein kinase III in mammalian cells as elongation factor 2. J Biol Chem 262:17299–17303

141. Ryazanov AG, Shestakova EA, Natapov PG (1988) Phosphorylation of elongation factor 2 by EF-2 kinase affects rate of translation. Nature (Lond.) 334:170–173

142. Palfrey HC, Nairn AC, Muldoon L, Villereal ML (1987) Rapid activation of calmodulin-dependent protein kinase III in mitogen-stimulated human fibroblasts. J Biol Chem 262:9785–9792

143. Gschwendt M, Kittstein W, Marks F (1994) Elongation factor-2 kinase: effective inhibition by the novel protein kinase inhibitor rottlerin and relative insensitivity towards staurosporine. FEBS Lett 338:85–88

144. Parmer TG, Ward MD, Hait WN (1997) Effects of Rottlerin, an inhibitor of calmodulin-dependent kinase III, on cellular proliferation, viability, and cell cycle distribution in malignant glioma cells. Cell Growth & Differ 8:327–334
145. Carpenter CL, Cantley LC (1996) Phosphoinositide kinases. Curr. Opin. Cell Biol 5:153–158
146. Ahmed NN, Grimes HL, Bellacosa A, Chan TO, Tsichlis PN (1997) Transduction of interleukin-2 antiapoptic and proliferative signals via Akt protein kinase. Proc Natl Acad Sci USA 94:3627–3632
147. Dudek H, Datta SR, Franke TF, Birnbaum MJ, Yao R, Cooper GM, Segal RA, Kaplan DR, Greenberg ME (1997) Regulation of neuronal survival by the serine-threonine protein kinase Akt. Science 275:661–665
148. Kauffmann-Zeh A, Rodriguez-Viciana P, Ulrich E, Gilbert C, Coffer P, Downward J, Evan G (1997) Suppression of c-Myc-induced apoptosis by Ras signaling through PI(3)K and PKB. Nature 385:544–548
149. Kennedy SG, Wagner AJ, Conzen SD, Jordan J, Bellacosa A, Tsichlis PN, Hay N (1997) The PI 3-kinase/Akt signaling pathway delivers an anti-apoptotic signal. Genes Dev. 11:701–713
150. Khwaja A, Rodriguez-Viciana P, Wennstrom S, Warne PH, Downward J (1997) Matrix adhesion and Ras transformation both activate a phosphoinositide 3-OH kinase and protein kinase B/Akt cellular survival pathway. EMBO J 16:2783–2793
151. Kulik G, Kippeal A, Weber MJ (1997) Antiapoptotic signaling by the insulin-like growth factor I receptor, phosphatidylinositol 3-kinase and Akt. Mol Cell Biol 17:1595–1606
152. Parrizas M, Saltie AR, LeRoith D (1997) Insulin-like growth factor 1 inhibits apoptosis using the phosphatidylinositol 3'-kinase and mitogen-activated protein kinase pathways. J Biol Chem 272:154–161
153. Ui M, Okada T, Hazeki K, Hazeki O (1995) Wortmannin as unique probe for an intracellular signaling protein, phosphoinositide 3-kinase. Trends Biochem Sci 20:1456–1462
154. Vlahos CJ, Matter WF, Hui KY, Brown RF (1994) A specific inhibitor of phosphatidylinositol 3-kinase, 2-(4-morpholinyl)-8-phenyl-4H-1-benzopyran-4-one (LY294002). J Biol Chem 269:5241–5248
155. Chou MM, Blenis J (1995) The 70kDa S6 kinase: regulation of a kinase with multiple roles in mitogenic signalling. Curr Opin Cell Biol 7:806–814
156. Datta SR, Dudek H, Tao X, Masters S, Fu H, Gotoh Y, Greenberg ME (1997) Akt phosphorylation of BAD couples survival signals to the cell-intrinsic death machinery. Cell 91:231–241
157. del Peso L, Gonzalez-Garcia M, Page C, Herrera R, Nunez G (1997) Interleukin-3-induced phosphorylation of BAD through the protein kinase Akt. Science 278:687–689
158. Imoto M, Umezawa K, Takahashi Y, Naganawa H, Iitaka Y, Nakamura H, Koizumi Y, Sasaki Y, Hamada M, Sawa T, Takeuchi T (1990) Isolation and structure determination of inostamycin, a novel inhibitor of phosphatidylinositol turnover. J Nat Prod, 53:825–829
159. Imoto M, Taniguchi Y, Umezawa K (1992) Inhibition of CDP-DG:inositol transferase by inostamycin. J Biochem 112:299–302
160. Imoto M, Morii T, Deguchi A, Umezawa K (1994) Involvement of phosphatidylinositol synthesis in the regulation of S phase induction. Exp Cell Res 215:228–233
161. Deguchi A, Imoto M, Umezawa K (1996) Inhibition of G1 cyclin expression in normal rat kidney cells by inostamycin, a phosphatidylinositol synthesis inhibitor. J Biochem 120:1118–1122

162. Imoto M, Tanabe K, Simizu S, Tashiro E, Takada M, Umezawa K (1998) Inhibition of cyclin D1 expression and induction of apoptosis by inostamycin in small cell lung carcinoma cells. Jpn J Cancer Res 89:315–322

163. Enari M, Talanian RV, Wong WW, Nagata S (1996) Sequential activation of ICE-like and CPP32-like proteases during Fas-mediated apoptosis. Nature 380:723–726

164. Kitagawa M, Okabe T, Ogino H, Matsumoto H, Suzuki-Takahashi I, Kokubo T, Higashi H, Saitoh S, Taya Y, Yasuda H, Okuyama A (1993) Butyrolactone I: a selective inhibitor of cdk2 and cdc2 kinase. Oncogene 8:2425–2432

165. Kitagawa M, Higashi H, Takahashi IS, Okabe T, Ogino H, Taya Y, Hishimura S, Okuyama A (1994) A cyclin-dependent kinase inhibitor, butyrolactone I, inhibits phosphorylation of RB protein and cell cycle progression. Oncogene 9:2549–2557

166. Shibata Y, Nishimura S, Okuyama A, Nakamura T (1996) p53-independent induction of apoptosis by cyclin-dependent kinase inhibition. Cell Growth Differ 7:887–891

167. Cobrinik D, Dowdy SF, Hinds PW, Mittnacht S, Weinberg RA (1992) The retinoblastoma protein and the regulation of vell vyvling. Trends Biochem Sci 17:312–315

168. Vousden K (1993) Interaction of human papillomavirus transforming proteins with the products of tumor suppressor genes. FASEB J 7:872–879

169. Hayakawa Y, Sohda K, Shin-ya K Hidaka T, Seto H (1995) Anguinomycin C and D, new antitumor antibiotics with selective cytotoxicity against transformed cells. J Antibiot 48:954–961

170. Fornerod M, Ohno M, Yoshida M, Mattaj IW (1997) CRM1 is an export receptor for leucine-rich nuclear export signals. Cell 90:1051–1060

171. Stade K, Ford CS, Guthrie C, Weis K (1997) Exportin 1(Crm1p) is an essential nuclear export factor. Cell 90:1041–1050

172. Hamamoto T, Gunji S, Tsuji H, Beppu T (1983) Leptomycins A and B, new antifungal antibiotics I. Taxonomy of the producing strain and their fermentation, purification, and characterization. J Antibiot 36:639–645

173. Nishi K, Yoshida M, Fujiwara D, Nishikawa M, Horinouchi S, Beppu T (1994) Leptomycin B targets a regulatory cascade of CRM1, a fission yeast nuclear protein, involved in control or higher order chromosome structure and gene expression. J Biol Chem 269:6320–6324

174. Levine AJ (1997) p53, the cellular gatekeeper for growth and division. Cell 88:323–331

175. Agarwal ML, Taylor WR, Chernov MV, Chernova OB, Stark GR (1998) The p53 network. J Biol Chem 273:1–4

176. Debbas M, White E (1993) Wild-type p53 mediates apoptosis by E1A which is inhibited by E1B. Genes Dev 7:546–554

177. Lowe SW, Ruley HE, Jacks T, Housman DE (1993) p53-Dependent apoptosis modulates the cytotoxicity of anticancer agents. Cell 74:957–968

178. Lowe SW, Schmitt EM, Smith SW, Osborne BA, Jacks T (1993) p53 is required for radiation induced apoptosis in mouse thymocytes. Nature 362:847–849

179. Wagner AJ, Kokontis JM, Hay N (1994) MYC-mediated apoptosis requires wild-type p53 in a manner independent of cell cycle arrest and the ability of p53 to induce p21$^{waf1/cip1}$. Genes Dev 8:2817–2830

180. Yonish-Rouach E, Resnitzky D, Lotem J, Sachs L, Kimchi A, Oren M (1991) Wild-type p53 induces apoptosis of myeloid leukaemic cells that is inhibited by interleukin-6. Nature 352:345–347

181. Ko LJ, Prives C (1996) p53: puzzle and paradigm. Genes Dev 10:1054–1072

182. Pinhasi-Kimhi O, Michalovitz D, Ben-Zeev A, Oren M (1986) Specific interaction between the p53 cellular tumour antigen and major heat shock proteins. Nature 320:182–185

183. Hinds PW, Finlay C, Frey A, Levine AJ (1987) Immunological evidence for the association of p53 with a heat shock protein, hsc70, in p53-plus-ras-transformed cell lines. Mol Cell Biol 7:2863–2869

184. Finlay CA, Hinds PW, Tan TH, Eliyahu D, Oren M, Levine AJ (1988) Activating mutations for transformation by p53 produce a gene product that forms an hsc70-p53 complex with an altered half-life. Mol Cell Biol 8:531–539

185. Sturzbecher HW, Addison C, Jenkins JR (1988) Characterization of mutant p53-Hsp72/73 protein-protein complexes by transient expression in monkey COS cells. Mol Cell Biol 8:3740–3747

186. Sturzbecher HW, Chumakov P, Welch WJ, Jenkins JR (1987) Mutant p53 proteins bind Hsp 72/73 cellular heat shock-related proteins in SV40-transformed monkey cells. Oncogene 1:201–211

187. Hainaut P, Milner J (1992) Interaction of heat-shock protein 70 with p53 translated in vitro: evidence for interaction with dimeric p53 and for a role in the regulation of p53 conformation. EMBO J 11:3513–3520

188. Whitesell L, Mimnaugh EG, De Costa B, Myers CE, Neckers LM (1994) Inhibition of heat shock protein Hsp90-pp60v-src heteroprotein complex formation by benzoquinone ansamycins: essential role for stress proteins in oncogenic transformation. Proc Natl Acad Sci USA 91:8324–8328

189. Blagosklonny MV, Toretsky J, Neckers L (1995) Geldanamycin selectively destabilizes and conformationally alters mutated p53. Oncogene 11:933–939

190. Caelles C, Helmberg A, Karin M (1994) p53-dependent apoptosis in the absence of transcriptional activation of p53-target genes. Nature 370:220–223

191. Haupt Y, Rowan S, Shaulian E, Vousden KH, Oren M (1995) Induction of apoptosis in HeLa cells by trans-activation-deficient p53. Genes Dev 9:2170–2183

192. Yan Y, Shay JW, Wright WE, Mumby MC (1997) Inhibition of protein phosphatase activity induces p53-dependent apoptosis in the absence of p53 transactivation. J Biol Chem. 272:15220–15226

193. Goldstein JL, Helgeson JA, Brown MS (1979) Inhibition of cholesterol synthesis with compactin renders growth of cultured cells dependent on the low density lipoprotein receptor. J Biol Chem 254:5403–5409

194. Alberts AW, Chen J, Kuron G, Hunt V, Huff J, Hoffman C, Rothrock J, Lopez M, Joshua H, Harris E, Patchett A, Monaghan R, Currie S, Stapley E, Albers-Schonberg G, Hensens O, Hirshfield J, Hoogsteen K, Liesch J, Springer J (1980) Mevinolin: a highly potent competitive inhibitor of hydroxymethylglutaryl-coenzyme A reductase and a cholesterol-lowering agent. Proc Natl Acad Sci USA 77:3957–3961

195. Keyomarsi K, Sandoval L, Band V, Pardee AB (1991) Synchronization of tumor and normal cells from G1 to multiple cell cycles by lovastatin. Cancer Res 51:3602–3609

196. Marcelli M, Cunningham GR, Haidacher SJ, Padayatty SJ, Sturgis L, Kagan C, Denner L (1998) Caspase-7 is activated during lovastatin-induced apoptosis of the prostate cancer cell line LNCap. Cancer Res 58:76–83

5
Immune Cell Functions

KAZUO NAGAI and TAKAO KATAOKA

1 Introduction: Immune Network

The immune response, as expressed in its most sophisticated form in higher verte-
brates including humans, is a self defense mechanism that works to protect the host
by distinguishing and neutralizing or destroying the non-self substances, such as
abnormal cells or invading infectious agents. All cells participating in immune reactions are derived from pluripotent hematopoietic stem cells produced in the bone
marrow.The stem cells differentiate into myeloid progenitors and common lymphoid
progenitors.The myeloid progenitors are the precursor of the granulocytes and macrophages. There are three types of granulocytes, neutrophils, eosinophils and baso-
phils. The common lymphoid progenitors give rise to the lymphocytes which are
divided into two major groups, B lymphocytes and T lymphocytes. These lymphocytes have antigen receptors, the surface immunoglobulin of B cells and the antigen
receptor of T cells (TCR). B cells can detect antigen outside the cell, while on the
other hand, T cells recognize antigens that are generated inside the cell and expressed on the cell surface as molecules combined with a major histocompatibility
complex (MHC).Upon activation, B lymphocytes differentiate into plasma cells, which
secrete antibodies. There are two main classes of T lymphocytes, helper T cells (TH),
which activate immune cells such as B cells and macrophages, and cytotoxic T lymphocytes (CTLs), which kill virus-infected cells and the cells of transplanted tissues
or organs bearing altered MHC-peptides.

The immune system is composed of two major responses, innate immunity and
adaptive immunity. Innate immunity is typically observed during bacterial invasion,
and is mediated by the granulocytes, which include macrophages and neutrophils.
Those cells have surface receptors that can recognize and bind to common components of the bacterial cell surface.They then engulf the bacteria and become induced
to secrete cytokines. The released cytokines then activate other immune cells.

Adaptive immune responses are dependent on the functions of lymphocytes.
Naive antigen-specific T cells are activated by antigen-presenting cells which express MHC-peptides as well as co-stimulatory molecules such as B7 on their cell
surface. There are three types of antigen presenting cells: macrophages, dendritic
cells and B cells. Macrophages engulf infectious agents through phagocytosis and

express MHC-class II molecules bearing peptides derived from the digested agents and co-stimulatory molecules. Dendritic cells constitutively express MHC class II-peptide molecules and co-stimulatory activity. B cells bind to soluble protein antigens through their surface immunoglobulin, internalize the antigen, and then express a MHC class II-peptide which, when presented with co-stimulatory molecules, can activate corresponding T cells. Resting naive T cells, when they encounter a specific antigen in the presence of co-stimulatory signal, enter into the G1 phase of the cell cycle and are induced to synthesize interleukin-2 (IL-2) and the high affinity receptor for IL-2. Binding of IL-2 to the high affinity receptor further stimulates T cells to proliferate, which enables clonal expansion. The differentiated T cells then perform effector functions against the target cells expressing specific antigen without the need for co-stimulatory signals. The effector functions of TH are the activation of macrophages by TH1 cells to express cytotoxicity and the activation of B cells by TH1 and TH2 to produce varieties of antibodies. The activation signals are mediated by direct cellular interactions as well as by soluble factors such as cytokines. Another effector function is the killing of infected cells by CD8 CTL. Thus, effector T cells play important roles to produce both cell-mediated immunity and humoral immune responses. These interactions among the cells which comprise a network of immune response are briefly summarized in Fig. 1.

In the process of activation of the immune network, immune cells specifically receive signals through receptors on the cell surface, transduce signals into the cytoplasm, and modify and activate transcription factors allowing transcription of the genes required for cell proliferation, cytokine production, cytokine receptor formation. They become capable of carrying out effector functions. Any compound that

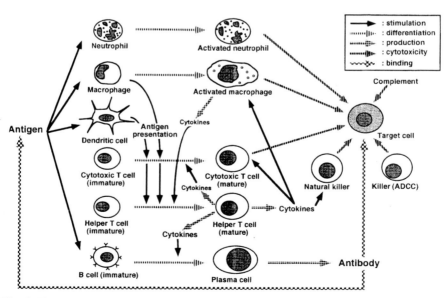

Fig. 1. The immune network

interacts and modifies one of these steps, therefore, may positively or negatively affect the immune reactions. The mode of action as well as the biological effects of such low molecular weight compounds will be discussed in this chapter.

2 Bioprobes for Lymphocyte Proliferation

There are several immunosuppressants whose mode of action and molecular targets have been well documented [1–3]. Thus, these compounds can be used as effective bioprobes for studying the biochemical processes involved in the expression of certain biological functions of immune cells. Moreover, some of these agents are clinically useful or are under clinical trials for suppression of unfavorable immune responses such as rejection of transplanted tissues or organs, autoimmune diseases, allergy, and inflammatory reactions, etc. [4].

Targets include biochemical pathways which are unique to lymphocytes and are thus more important for the proliferation, maturation and expression of specific functions of the lymphocytes than for other proliferating cells. These are steps in signal binding, signal transduction cascades, transcription factors, gene expression including transcription, translation and transport of effector molecules and, finally, cellular interactions between immune cells. Some of the steps or molecular targets and the compounds affecting these processes are summarized in Fig. 2.

2.1 Therapeutic Potential of Inhibitors of Enzymes for Purine and Pyrimidine Metabolism

The finding that 6-mercaptopurine (6-MP) inhibits antibody production, causing strong immunosuppressive activity, suggested that purines are involved in lymphocyte functions [5]. The importance of purine biosynthesis in immune function is supported by genetic information. Patients deficient of adenosine deaminase (ADA) [6] and purine nucleoside phosphorylase (PNP) [7], two essential enzymes acting sequentially in the purine salvage pathway, suffer from serious immunodeficiency [8].

Mizolibine (MZ) selectively inhibits inosine monophosphate dehydrogenase (IMPDH), resulting in suppression of immune cell function without producing myelotoxicity [9,10]. This suggests that the de novo pathway for purine biosynthesis is more important for the proliferative response of lymphocytes than for other rapidly dividing cells. De novo pyrimidine biosynthesis also participates in lymphocyte proliferation after stimulation. Inhibitors of the pathway have been developed. The specificity of the pyrimidine biosynthetic pathway in lymphocyte proliferation, however, remains to be elucidated.

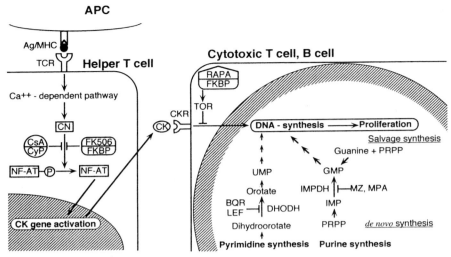

Fig. 2. Schematic description of the sites of action of immunosuppressants. A foreign antigen (*Ag*) plus the major histocompatibility antigen (*MHC*) class II on an antigen presenting cell (*APC*) is presented to the T cell receptor (*TCR*) on helper T cells. The calcium-dependent intracellular signal pathway for the release of cytokines including IL-2 in the helper T cell is then mediated by calcineurin (*CN*), which is inhibited by cyclosporin A (*CsA*)-cyclophilin (*CyP*) and the FK506-FK-binding protein (*FKBP*) complexes. In the cytotoxic T cell, rapamycin (*RAPA*), as the RAPA-FKBP complex, inhibits the intracellular signal pathway from the IL-2 receptor to DNA synthesis by interacting with the target binding protein of RAPA (*TOR*). Mizoribine (*MZ*) and the active metabolite of mycophenolate mofetil, mycophenolic acid (*MPA*), inhibit inosine monophosphate dehydrogenase (*IMPDH*). This inhibits de novo purine synthesis which is actively functioning in the activated lymphocytes. Brequinar (*BQR*) and leflunomide (*LEF*) inhibit pyrimidine synthesis by inhibiting dihydroorotate dehydrogenase (*DHODH*). As a consequence, these drugs inhibit DNA synthesis and thereby cell proliferation

2.2 De Novo Pathway is Major Way to Synthesize Purine in Lymphocytes

Most cells have two major pathways for purine biosynthesis, the de novo pathway and the salvage pathway. IMPDH catalyzes the conversion of purine inosine mono-phosphate (IMP) to xanthosine monophosphate (XMP) and is the rate-limiting en-zyme in the de novo synthesis of guanosine monophosphate (GMP). GMP is the precursor for guanosine triphosphate (GTP) and deoxyguanosine triphosphate (dGTP) which are incorporated into RNA and DNA respectively. Guanosine and deoxyguanosine, the degradation products of RNA and DNA or GTP-containing intermediates that participate in biochemical pathways including protein glycosylation, can be recycled to form GTP and dGTP via the salvage pathway. The fact, however, that children lacking hypoxanthine-guanine phosphoribosyltransferase (HGPRT), which participates in salvage synthesis, have essentially normal numbers and functions of lymphocytes [11] suggests that the salvage pathway for GTP and dGTP biosynthesis is limited and that most of the GTP and dGTP is supplied by the de novo pathway in these lymphocytes. This suggests that IMPDH is a crucial enzyme for lymphocyte proliferation.

It has also been shown that antigenic or mitogenic stimulation of lymphocytes causes de novo purine synthesis to increase significantly. Therefore, inhibitors of IMPDH are expected to be powerful immunosuppressants by blocking specific lymphocyte proliferation induced by antigen stimulation [8].

2.3 Mizoribine (bredinin; 4-carbonyl-1-β-D-ribofuranosyl imidazolium-5-olate) (MZ)

MZ, an imidazole nucleoside antibiotic, was the first drug found to selectively inhibit de novo purine synthesis and to suppress immune cell function. MZ is phosphorylated to MZ 5'-monophosphate by adenosine kinase [9] and competitively inhibits IMPDH and GMP synthetase. Thus, MZ blocks the conversion of IMP to GMP in the de novo pathway which is responsible for purine biosynthesis [10]. The compound elicits immunosuppressive activity on antibody production, delayed-type hypersensitivity, and cellular immunity in experimental animals [12].

Although MZ has no suppressive effect on IL-2 production, it profoundly inhibits thymocyte proliferation induced by concanavalin A and IL-2 [10]. Stimulation of purified human peripheral blood T lymphocytes by phorbol ester and ionomycin leads to a five-fold increase in guanine ribonucleotide levels. The addition of MZ to these cultures at concentrations that are achieved in vivo leads to a dose-dependent inhibition of proliferation and a concomitant decrease in guanine ribonucleotide levels. Similar effects are seen with direct stimulation of the CD3/TCR complex. Inhibition of proliferation occurs at the transition of the G1 to S phase in the cell cycle and is reversible in the presence of guanosine, which repletes the pool of GTP [13].

IgM production by purified human B cells stimulated with *Staphylococcus aureus* Cowan I (SA) plus IL-2 is suppressed by MZ at pharmacologically attainable concentrations [14]. Cell cycle analysis disclosed that MZ decreased the proportion of B cells in the S+G2+M phase. MZ did not decrease GTP levels in SA-stimulated B cells, in contrast to the case in activated T cells. The suppressive effects of MZ on IgM production of SA-stimulated B cells was not reversed by the addition of GMP at the concentrations that overcome the inhibitory effects of MZ on the proliferation of anti-CD3-stimulated T cells. MZ markedly suppressed the expression of cyclin A in SA-activated B cells without affecting the expression of CD25 and cdc2c kinase. Thus MZ interferes with the cell cycle progression of activated B cells by suppressing the expression of cyclin A by a mechanism distinct from guanine ribonucleotide depletion [14].

2.4 Mycophenolic Acid (MPA) and Mycophenolate Mofetil (MMF)

MMF is a prodrug that, when adsorbed from the gastrointestinal tract, is hydrolyzed by liver esterases to produce the biologically active component MPA [15]. The mofetil group was added to increase the bioavailability of MPA [16].

MPA is a noncompetitive, reversible inhibitor of eucaryotic IMPDH [17] which blocks the de novo synthesis of the purine guanosine. Two distinct cDNAs encoding IMPDH have been cloned from a human spleen cDNA library and expressed in *E. coli* [18]. The clones encode the closely related proteins IMPDH type I and type II. Further studies [19] indicated a predominant expression of the type I gene in resting human lymphocytes and a strong expression of the type II gene in T cells after stimulation with phytohemagglutinin (PHA) and in B cells transformed by Epstein-Barr virus. Both isoforms of the enzyme expressed in *E. coli* have been purified. The type II isoform is four to five times more sensitive to MPA than the type I isoform [20]. MPA has more potent cytostatic effects on human peripheral lymphocytes responding to stimulation by T-cell mitogens and a B cell mitogen than on other cell types, including human fibroblasts and endothelial cells proliferating in response to growth factors [21]. This is due partly to the dependence of lymphocytes on de novo dGTP synthesis and partly to the expression in these cells of the type II isoform IMPDH. It would, therefore, be expected that MPA would more efficiently deplete GTP and dGTP pools in lymphocytes than in other cell types. In fact, clinically attainable concentrations of MPA (1–10μM) significantly deplete GTP in human lymphocytes and monocytes but not in neutrophils [22].

An additional effect of MPA on T lymphocytes may be to reduce glycosylation of adhesion molecules such as the vascular cell adhesion molecule (VCAM)-1 which is required for the T cells to interact with other inflammatory cells and allograft target cells during acute rejection [21,22].

2.5 Inhibitors of De Novo Pyrimidine Synthesis: Brequinar (BQR) and Leflunomide (LEF)

As the rate-limiting enzyme in the de novo synthesis of pyrimidine in lymphocytes, dihydroorotate dehydrogenase (DHODH) is considered to be the molecular target of the compounds that affect immune cell functions. Two inhibitors of DHODH, BQR and LEF, have been selected and used as immunosuppressive drugs capable of suppressing autoimmune diseases and controlling the allograft and xenograft rejection [23].

By using radioactive de novo pyrimidine synthesis precursors, the primary target of these agents in proliferating human T-lymphocytes was shown to be pyrimidine biosynthesis at the level of DHODH [24]. LEF at 25 and 50μM and BQR at 0.5 and 1 μM were cytostatic, not cytotoxic, with proliferation being halted in the G1 phase. At these concentrations, both drugs restricted the normal 4 to 8-fold mitogen-stimulated increase of pyrimidine pools to concentrations found in nonstimulated T cells. Addition of uridine (50µM) restored this increase in the pools [24] and partly antagonized the inhibition of T cell proliferation [25]. From the IC_{50} values against recombinant rat and human DHODH, it was deduced that BQR is a more potent inhibitor of the human enzyme activity than of the rat enzyme [26]. The rat enzyme was influenced by LEF to a greater extent than the human enzyme. These results may explain the difference in susceptibility to these drugs of cultured cell lines: in comparison to cells derived from human tissues, rat and other rodent cells were more susceptible to LEF and less susceptible to BQR. The interaction of LEF and BQR with their target was shown by spectral analyses of the functional flavin with the native human DHODH to interfere with the transfer of electrons from the flavin to the quinone [27].

At 100 µM LEF or 1 µM BQR, ATP and GTP pools in the mitogen-stimulated T cells markedly decreased, which would have serious consequences for the ATP-dependent enzymes essential for the immune response. Uridine was unable to reverse the effect of these drugs at high concentration. Under this condition, BQR inhibited protein tyrosine phosphorylation in anti-CD3-stimulated murine T lymphocytes. In an in vitro system, BQR inhibited kinase activity of $p56^{lck}$ and $p59^{fyn}$ [28]. Inhibition of protein tyrosine phosphorylation by LEF has also been well documented [29]. Addition of uridine was shown to restore proliferation and IgM secretion in LEF-treated lipopolysaccharide (LPS)-stimulated B cells but it did not completely restore secretion of IgG antibody. Treatment of B cells with LPS plus IL-4 induces IgG1 secretion. In this model of isotype switch, the signal transduction pathway involves tyrosine phosphorylation of the IL-4 receptor, JAK1, JAK3 and STAT6 proteins induced by IL-4 binding to the IL-4R. LEF diminished tyrosine phosphorylation of JAK3 and STAT6 as well as IgG production in the absence or presence of uridine. Thus, it has been suggested that LEF acts as a tyrosine kinase inhibitor to block IgG1 production [30].

It has recently been reported that treatment of a human T cell line (Jurkat) with LEF blocks TNF-mediated NF-κB activation in a dose- and time-dependent manner, with maximum inhibition at 5–10µM. The inhibition of NF-κB activation by LEF was also observed when the cells were induced to activate by treatment with other inflammatory agents, including phorbol ester, LPS, H_2O_2, okadaic acid, and ceramide. LEF blocked the degradation of IκBα and the subsequent nuclear translocation of the p65 subunit, steps essential for NF-κB activation. This correlated with inhibition of dual specificity-mitogen-activated protein kinase kinase as well as an Src protein tyrosine kinase, $p56^{lck}$, by LEF. LEF completely suppressed the TNF-induced gene expression dependent on NF-κB activation. Thus LEF was suggested to be a potent inhibitor of NF-κB activation induced by a wide variety of inflammatory stimuli, and this provides the molecular basis for its anti-inflammatory and immunosuppressive effects [31].

3   Signal Transduction and Gene Expression in the Process of T Cell Activation

Ligation of the surface TCR of T cells in the presence of additional signals stimulates them to proliferate and function. When the antigen-MHC complex binds to the TCR, tyrosine kinase is activated and phospholipase C-γ1 on the cytoplasmic side of the cell membrane is activated. This hydrolyses membrane lipid phosphatidyl inositol 4,5-bisphosphate, releasing inositol 1,4,5-triphosphate (IP_3) and diacylglycerol (DG) from the membrane. DG leads to the activation of protein kinase C (PKC) [32]. Activated PKC in turn induces or activates transcription factors AP1 and NF-κB. IP_3 binds to intracellular IP_3 receptors (IP_3R), changes their conformation and opens the Ca^{2+} channel causing an increase of intracellular Ca^{2+} [33]. Receptor-mediated Ca^{2+} is released from internal stores, often followed by Ca^{2+} influx across the plasma membrane [34].

In parallel, Ca^{2+} influx activates calcineurin (CN), a Ca^{2+}/calmodulin-dependent Ser/Thr phosphatase. CN is a dimer made up of a 59kDa enzymatic A subunit with a binding site for calmodulin and a 19kDa regulatory B subunit (CN-B) [35]. Ca^{2+}/calmodulin- activated CN dephosphorylates a subunit, NFATp, of the nuclear factor in activated T cells (NFAT) [36]. Dephosphorylated NFATp then moves from the cytoplasm to the nucleus, where it forms NFAT with the AP1 proteins Fos and Jun and thereby activates the IL-2 promoter. CN also inactivates IκB [37], thus enhancing the nuclear localization of NF-κB and its transcription activity. All these transcription factors are required for maximal expression of the IL-2 gene and other cytokine genes.

Cytokine production is a critical step in T cell activation. Cytokines that engage their receptors drive the T cell from G1 to S and through the cell cycle, causing clonal expansion, with the emergence of sufficient numbers of activated and differentiated T cells to mediate an effective immune reaction.

3.1   Cyclosporin A (CsA), FK506 and Rapamycin (RAPA)

CsA (see p.2), a cyclic undecapeptide produced by *Poypocladium inflatum* was originally reported as an antifungal antibiotic. Since the discovery of its immunosuppressing activities, it has been widely used for the treatment and prevention of graft rejection after solid organ transplantation and graft-versus-host disease in bone marrow recipients. The findings that CsA suppressed mixed lymphocyte reactions (MLR) promoted a research group of Fujisawa Pharmaceutical to screen novel microbial metabolites inhibiting MLR. They succeeded in isolation of a macrocyclic compound FK506 (see p.7) from a culture broth of *Streptomyces tsukubaensis*. A structural analogue of FK506, RAPA (see p.20), was produced by fermentation of *Streptomyces hygroscopicus* and initially found to have antifungal, antibacterial and antitumor activities [38].

These three immunosuppressants bind target proteins, the so-called immunophilins cyclophilins A and B for CsA, FKBP12 and FKBP12.6 for FK506 and

FKBP12 for RAPA [4,38–40]. Since all immunophilins have cis-trans peptidyl-prolyl isomerase (PPIase) activity and binding to CsA, FK506 or RAPA results in loss of their activity, PPIase activity was thought to be critical for expression of T cell functions [41]. This turned out not to be the case, however, because many other PPIase inhibitors exert no immunosuppression.

Further study revealed that immunophilins bound to CsA and FK506, but not to RAPA, associate with the CN/calmodulin complex on the CN-B binding domain [42] and inactivate the phosphatase activity of CN. They subsequently inhibit the activation and nuclear localization of NFATp [36] and NF-κB [37]. Thus the transcription of the genes of cytokines IL-2, -3, -4, -5, interferon-γ, tumor necrosis factor-α, and granulocyte-macrophage colony stimulation factor, as well as the receptors for IL-2 and IL-7, is suppressed by CsA and FK506 [43].

RAPA, like FK506, binds to FKBP12 and inactivates PPIase activity but neither associates with the CN/calmodulin complex nor inhibits phosphatase activity of CN. Therefore, RAPA does not prevent early events in activated T lymphocytes including the expression of cytokine genes. The RAPA-FKBP12 complex can interact with a cellular protein called TOR (target binding protein of RAPA) or FRAP (FKBP-RAPA associated protein)/RAFT [44,45]. TOR displays a kinase activity to phosphorylate the protein kinase p70^{S6K} [46], the functions of which are essential for G1 progression. RAPA-FKBP12-TOR complex leads to dephosphorylation and inactivation of p70^{S6k} [47]. Critical substrates for p70^{S6k} are the S6 protein of the small ribosomal subunit and the transcription activator CREM [48]. Phosphorylation of the S6 is considered to be essential to the signal transduction cascade in IL-2-induced T cell proliferation [49]. CREM is required for transcriptional induction of the proliferating cell nuclear antigen (PCNA) gene [50] which participates in progression of cells into the S phase. As a consequence, RAPA blocks IL-2- induced T cell proliferation .

3.2 15-Deoxyspergualin (DSG)

15-Deoxyspergualin (DSG) is a synthetic analogue of spergualin which was isolated from the culture broth of *Bacillus laterosporus* as an antitumor substance [51]. Further study of the activity of DSG revealed its immunosuppressive properties in animal models of allotransplant rejection. DSG inhibits antibody response against T cell dependent antigens as well as T cell-independent antigens. These results indicate that DSG is effective not only on T cells, but also on other effector cells in the immune response [52,53].

DSG, unlike CsA and FK506, does not inhibit T cell responses such as cytokine secretion and proliferation induced by mitogens such as ConA and PHA. However, DSG inhibits the generation of effector cells such as CTLs in MLR. When DSG is added at the time or 24 hours after the initiation of the MLR, similar inhibition of cell proliferation is demonstrated, whereas CsA inhibition of the MLR decreased dramatically if CsA is withheld for 24 hours. The mechanism of action of DSG, therefore, appears to be different from that of CsA [54].

DSG has been shown to bind specifically to Hsc70, the constitutive or cognate member of the heat shock protein 70 (Hsp70) protein family [54,55]. The members of the Hsp70 family of heat shock proteins are important for many cellular processes, including the binding and stabilization of immunoglobulin heavy chains before the binding of light chains [56] and antigen processing and presentation with MHC class II molecules [57]. The heat shock proteins may, therefore, represent a class of binding proteins, or immunophilins, distinct from PPIase. FK506 binds to a putative heat shock protein termed p59 [58]. This protein is found in a complex with the glucocorticoid receptor, as well as members of the Hsp90 and Hsp70 family of heat shock proteins. Modulation of activity of these proteins by binding of immunosuppressants may result in suppression of immune cell functions.

4   Cell-Mediated Cytotoxicity and Its Bioprobes

4.1 An In Vivo Role for Cell-Mediated Cytotoxicity

T cells have plural lethal weapons to kill target cells such as virus-infected and transformed cells and even perform suicide under certain circumstances (Fig. 3.). Gene targeting experiments clearly showed that CTL and natural killer (NK) cells mainly utilize two distinct killing pathways, one which depends on perforin and the second on the Fas ligand (FasL). Perforin-deficient mice are unable to clear lymphocytic choriomeningitis virus infection and show reduced tumor rejection. In vitro, the killing activity of CTLs and NK cells is greatly impaired, thus indicating that the perforin-dependent pathway plays a crucial role in cell-mediated cytotoxicity. The

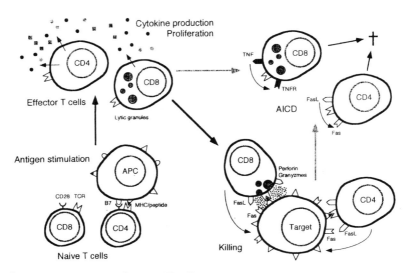

Fig. 3. Immune responses of mature T cells

remaining cytolytic pathway is mediated by Fas, since no killing activity remained in the short-term assay when both pathways were inactivated. However, in the long-term assay, CTLs still have residual cytotoxic activity which was shown to be TNF-dependent [59, 60].

The mouse spontaneous mutants *lpr* (lymphoproliferation) and *gld* (generalized lymphoproliferative disease) carry mutations in Fas and FasL, respectively. These mice develop lymphoadenopathy and splenomegaly, and autoimmune diseases such as nephritis [61]. In secondary lymphoid organs, abnormal T cells having CD4⁻CD8⁻ B220⁺Thy1⁺ accumulate, apparently having escaped from deletion in the periphery [61]. In humans, dominant-negative Fas mutations were found in patients with autoimmune lymphoproliferative syndrome. Recently, the Fas/FasL system has been reported to be involved in two more diseases: hepatitis [62] and thyroiditis [63]. Fulminant hepatitis often occurs upon the infection of hepatitis B or C virus. By employing mouse models for hepatitis, it has been shown that Fas-expressing hepatocytes are targets for CTLs [62]. Hashimoto's thyroiditis is an autoimmune disease which results from dysfunction of thyroid gland and accelerated apoptosis during the pathogenesis phases. Fas and FasL interactions among thyrocytes were shown to contribute to the pathogenesis of Hashimoto's thyroiditis [63].

In immune privileged sites, such as the eyes and testes, immune responses and inflammation are suppressed to avoid tissue destruction. FasL was found to be expressed constitutively in these particular tissues. It appears to kill activated inflammatory cells that invade the privileged sites. In fact, tumor cells adopt this strategy for defeating the host immune system. Tumor cells such as melanoma cells express functional FasL on their surface and gain the function to counterattack CTLs [64–66]. In addition, tumor cells downregulate Fas expression and upregulate antiapoptotic proteins such as FLIP (FLICE-inhibitory protein) to become resistant to Fas-mediated cytotoxicity [67]. Thus, these properties of tumor cells might contribute to tumorigenesis.

Mature T cells in the periphery respond to antigen stimulation and acquire proliferation ability and cytokine production (Fig. 3.). However, previously activated T cells undergo apoptosis upon stimulation, termed activation-induced cell death (AICD). AICD is believed to have two roles: the elimination of autoreactive T cells which escape from thymic selection and the downregulation of activated T cells after exerting their task. Fas and FasL are involved in AICD of T cells, and TNF-α contributes to AICD of most CD8⁺ T cells through TNF receptor 2. The lack of AICD in *lpr* or *gld* mice would explain the accumulation of abnormal T cells and the pathogenesis of autoimmune diseases. Again, the Fas/FasL system plays a crucial role in the regulation of the immune system, and its dysfunction leads to various diseases.

4.2 Regulation of Cell-Mediated Cytotoxicity

Cell-mediated cytotoxicity can be divided into five different stages (Fig. 4.): a) storage of effector molecules, b) recognition of target cells, c) signal transduction, d) expression of effector functions (granule exocytosis and FasL expression), and e) apoptosis of target cells (granzyme B- and Fas-dependent pathways).

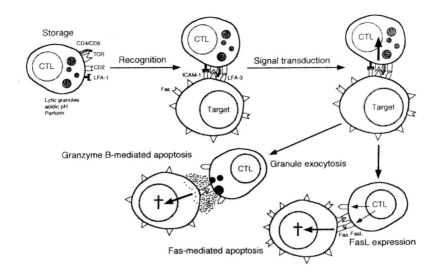

Fig. 4. Regulation of cell-mediated cytotoxicity

a) *Storage of effector molecules.* CTLs and NK cells harbor specific organelles termed lytic granules which contain perforin and a series of serine proteases (granzymes). The lytic granules have a property of lysosomes, in that they contain lysosomal proteins such as cathepsin D [68,69], and its internal pH is estimated to be around 5.5 [68,70,71]. We have shown that acidification is essential to maintain the structure and function of lytic granules [71–75]. Upon neutralization of the internal pH in the lytic granules, perforin is inactivated and proteolytically degraded [72,75].

b) *Recognition.* CTLs recognize target cells through TCR which binds to peptides in the context of MHC class I or II molecules. CD4 and CD8 which bind to MHC class II β2 domain and class I α3 domain, respectively, not only support efficient affinity but transmit signals through protein tyrosine kinase Lck which associates with them. Other surface molecules of CTLs also contribute to exert effector functions. LFA (lymphocyte function-associated antigen)-1 (CD11a/CD18) and CD2 can bind to ICAM (intercellular adhesion molecule)-1 (CD54) and LFA-3 (CD58) on target cells, respectively, and play an essential role in efficient recognition and adhesion.

c) *Signal transduction.* TCR ligation activates the protein tyrosine kinases Lck and Fyn which phosphorylate ITAM (immunoreceptor tyrosine based activation motif) of CD3 subunits such as CD3ζ. Another protein tyrosine kinase, ZAP-70, binds to CD3ζ via its phosphorylated ITAMs, and then activates downstream substrates essential for signal transduction. As a result, phospholipase Cγ1 is activated through phosphorylation, which generates secondary messengers (IP_3 and diacylglycerol) that elevate intracellular Ca^{2+} and activate PKC, respectively. These two events, possibly in concert with other signals (e.g. the MAP kinase pathway), lead to granule exocytosis and FasL expression.

d) *Expression of effector functions.* Upon activation, lytic granules are released into the interface between CTLs and target cells. Perforin undergoes polymerization on the membrane of target cells, and granzyme B, possibly with other cytotoxic molecules, enters the cytoplasm of target cells to activate the apoptotic machinery in a perforin-dependent manner. On the other hand, FasL is newly synthesized and transported to the cell surface.

e) *Granzyme B-mediated apoptosis.* Among cytotoxic molecules which reside in the lytic granules, granzyme B plays a major role in the rapid induction of apoptosis, since granzyme B-deficient CTLs lack the ability to induce DNA fragmentation in short-term assays [76]. Granzyme B directly cleaves and activates several caspases [77] which results in the activation of cytoplasmic apoptotic machinery, leading finally to DNA fragmentation and nuclear condensation. However, it has been recently reported that target cell lysis by granule exocytosis is independent of caspase family [78] and that granzyme B directly cleaves several downstream caspase substrates [79]. Thus, both caspase-dependent and independent mechanisms might be operative in granzyme B-mediated apoptosis.

f) *Fas-mediated apoptosis.* Members of the TNF/NGF receptor family regulate cell survival, differentiation and cell death. To date, this family consists of more than 20 members and two of the best characterized receptors are TNF receptor 1 and Fas [80]. TNF receptor 1 appears to play a major role in inflammatory responses, whereas Fas is an essential receptor transmitting death signals in cell-mediated cytotoxicity. Fas contains a cytoplasmic domain called the death domain (DD) responsible for recruiting adaptor proteins. Fas-associated death domain (FADD) which contains a DD and death effector domain (DED) binds to Fas through DD-DD interactions. FADD is also able to associate with the most upstream caspase, procaspase-8, which is composed of N-terminal DED and C-terminal caspase domains. In response to Fas ligation, the recruited procaspase-8 becomes activated by self-cleavage, and then cleaves other caspases (e.g. caspase-3) and downstream substrates (e.g. Bid). In addition to FADD, Daxx (Fas death domain-associated protein) [81] which does not have any homology with DD, can bind specifically to Fas DD and activate the Jun N-terminal kinase (JNK) pathway. Thus, FADD and Daxx are assumed to mediate two distinct pathways downstream of Fas.

Mitochondria have been shown to play a central role in apoptosis. Apoptotic stimuli cause mitochondrial permeability transition (PT) which allows the disruption of inner transmembrane potential ($\Delta\Psi m$) and the release of proapoptotic factors, that is, cytochrome C and AIF(apoptosis-inducing factor) [82]. The electrical and volume homeostasis in mitochondria which is regulated by Bcl-XL might have a role in transmitting apoptotic signals. Apaf(apoptotic protease activating factor)-1 which is composed of CARD (caspase recruitment domain), CED (cell death abnormal)-4-like domain and

WD-40 repeats has been shown to activate caspase-9 in a cytochrome C/dATP-dependent manner, followed by caspase-3 activation [83]. Activated caspase-3 is able to cleave many substrates including PARP (poly(ADP-ribose) polymerase) and DNA-PK (DNA-dependent protein kinase), and PKCδ. Most importantly, a caspase-activated DNAase (CAD) that degrades DNA during apoptosis and its partner (ICAD) have been found [84]. CAD is produced as a complex with ICAD, and caspase-3 cleaves and inactivates ICAD, which allows the translocation of CAD to the nucleus.

Distinct anti-apoptotic proteins have been identified: the Bcl-2 family and FLIP [67]. So far, more than dozen members of the Bcl-2 family have been reported, and they can be divided into two major groups: anti-apoptotic (e.g. Bcl-2 and Bcl-XL) and pro-apoptotic (e.g. Bax, Bik, and Bad). Bcl-2 and Bcl-XL are able to support survival upon apoptotic stimuli including growth factor withdrawal, irradiation, and anti-cancer drug treatment. However, Bcl-2 is barely able to antagonize Fas-mediated apoptosis, with the exception of some cells which may depend mainly on mitochondrial functions for apoptotic signaling [85]. FLIP, which contains the DED domain and an inactive caspase-like domain, interacts with FADD and caspase-8, thereby inhibiting the apoptotic pathways mediated by death receptors [67]. FLIP expression is correlated with the susceptibility of T cells to FasL after activation [67], and IL-2 seems to downregulate FLIP [86]. Thus, Fas-mediated apoptosis appears to be regulated not only by the expression of Fas and FasL but also by the balance of anti-apoptotic and pro-apoptotic proteins in the cell.

4.3 Assay System for Perforin/Fas-Dependent CTL-Mediated Cytotoxicity

We have screened inhibitors for perforin- and Fas-dependent cytotoxicity from natural products [87,88]. To dissect these two pathways, we employed different pairs of

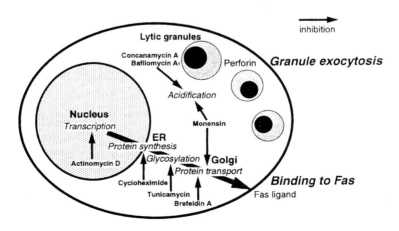

Fig. 5. Regulation of granule exocytosis and FasL expression upon activation

CTL clones and target cells. The perforin-dependent pathway was measured by cytolysis of Fas-low expressing P815 through an alloantigen-specific CD8+ CTL clone, OE4, which exerts perforin- and Fas-dependent killing. The Fas-dependent pathway was measured by cytolysis of Fas-positive B cells A20.HL through a CD4+ CTL clone BK-1 which lacks perforin expression. [51Cr] sodium chromate- or [3H] thymidine-labeled target cells were cocultured with CTLs for 4 to 6 h, and then the release of radioactivity into the supernate was measured. Representative data are summarized in Table 1 [87]. Perforin-dependent cytotoxicity was insensitive to actinomycin D, cycloheximide, brefeldin A, and tunicamycin, but was strongly and selectively inhibited by inhibitors of vacuolar type H^+-ATPase. In contrast, Fas-dependent cytotoxicity was markedly blocked by actinomycin D, cycloheximide, brefeldin A, tunicamycin, and monensin, but totally insensitive to inhibitors of vacuolar type H^+-ATPase. Taken together, the following conclusions were drawn (Fig. 5.): 1) A sufficient amount of perforin is stored in lytic granules and de novo synthesis is unnecessary for its immediate use in target cell lysis, 2) Acidification is essential to maintain the structure and function of lytic granules and the integrity of perforin, and 3) Fas-dependent cytotoxicity depends on the de novo RNA and protein synthesis and intracellular transport through the Golgi apparatus.

Table 1. Summary of bioprobes for perforin- and Fas-dependent CTL-mediated cytotoxicity

Groups	Compounds	IC_{50} (μM) Perforin	Fas	Primary targets
Macromolecular synthesis	Mitomycin C	>100	>100	DNA synthesis
	Actinomycin D	>10	0.53	RNA synthesis
	Cycloheximide	>10	8.5	Protein synthesis
Secretion	Brefeldin A	>10	1.6	Intracellular transport
	Tunicamycin	>10	2.9	Glycosylation
	Castanospermine	>100	>100	Glycosidase
	Swainsonine	>100	>100	Mannosidase
Acidification	Concanamycin A	0.002	>1	Vacuolar type H^+-ATPase
	Bafilomycin A1	0.041	>1	Vacuolar type H^+-ATPase
	Monensin	1.4	0.013	Na+/H+ polyether ionophore
Cytoskeleton	Cytochalasin D	5.7	0.8	Actin
	Colchicine	>1000	>1000	Tubulin
	Taxol	>10	>10	Tubulin
Protein kinases	Herbimycin A	3.9	>10	Tyrosine kinase
	Genistein	>320	170	Tyrosine kinase
	H-89	>10	>10	PKA
	Calphostin C	0.21	0.15	PKC
	ML-9	>10	>10	MLCK
	Wortmannin	1.8	partial	PI-3 kinase
	K-252a	0.22	0.47	Broad specificity
Others	FK506	partial	partial	Calcineurin
	Cyclosporin A	partial	partial	Calcineurin
	Rapamycin	>10	>10	FRAP/RAFT
	Camptothecin	>10	>10	Topoisomerase I
	Etoposide	>10	>10	Topoisomerase II

OE4 was incubated with compounds for 2h, and then incubated with [51Cr]-labeled P815 for 4h (perforin-dependent cytotoxicity). On the other hand, BK-1 was incubated with compounds for 2h, and then incubated with [51Cr]-labeled KLH-pulsed A20.HL for 6 h (Fas-dependent cytotoxicity). Radioactivity released into supernatants was measured.

4.4 Cell-Free Assay Systems for Perforin and Granzyme B

We tried to obtain specific inhibitors to perforin and granzyme B in cell-free conditions, since they are major executors in target cell lysis mediated by granule exocytosis. Perforin activity was measured by the lysis of sheep red blood cells (SRBC). We prepared a membrane fraction from a large granular lymphocyte clone SPB2.4 which contains high levels of perforin. The fraction readily lysed SRBC without any further purification [75]. Since perforin polymerization depends on calcium, pretreatment with calcium and EGTA are known to inhibit its hemolytic activity. Granzyme A and B are serine proteases which have distinct specificities. Granzyme A has trypsin-like specificity and granzyme B cleaves after aspartic acid, a proteolytic specificity shared with the caspases that have direct roles in apoptosis. We prepared the cell lysate from the CD8+ CTL clone OE4 and measured the activity of granzyme A and B by using two different peptide substrates, CBZ-Gly-Arg-thiobenzylester and Boc-Ala-Ala-Asp-thiobenzylester, respectively, together with 5,5'-dithio-bis-(2-nitrobenzoic acid) for color reaction [75]. Although we have applied these systems to known reagents, microbial metabolites and plant extracts, we have not yet found any potentially active compound.

4.5 Assay System for Fas-Mediated Apoptosis

Since FasL expression in CTLs depends on de novo synthesis and intracellular transport upon activation [73,87,88], it seems difficult to obtain specific inhibitors for Fas-dependent CTL-mediated cytotoxicity by random screening. Therefore, we focused on Fas-dependent apoptosis of target cells, and set up a random screening system. [^{3}H] thymidine-labeled Fas-positive A20.2J was pretreated with samples for 2 h, and then incubated with anti-Fas antibody Jo2 for 4 h. The radioactivity released into the supernatant after Triton X-100 treatment was measured. To exclude DNA intercalators which seemingly prevent DNA fragmentation (e.g. actinomycin D and doxorubicin), we also assessed cell viability by MTT assay. Several compounds with different structures have been identified that block Fas-mediated apoptosis (i.e. 12-O-tetradecanoyl phorbol 13-acetate (TPA), cytochalasin D, FD-891, patulin, and penicillic acid; unpublished).

4.6 Bioprobes for Cell-Mediated Cytotoxicity

4.6.1 Concanamycin A (CMA)

CMA, which belongs to the 18-membered macrolides, specifically inhibits vacuolar H+-ATPase and perturbs the functions of acidic organelles. CMA completely blocks the perforin-based killing pathway by a wide range of CTLs at 10–100 nM [72,73]. However, the Fas-dependent killing pathway was hardly affected even at the higher concentrations [73]. Thus, CMA was proven to be a potent and specific inhibitor to the perforin-dependent killing pathway. CMA did not affect conjugate formation between CTLs and target cells, and only marginally decreased granule exocytosis in response to CD3 activation [72]. CTLs and NK cells harbor lytic granules which contain effector molecules such as perforin and granzymes. These granules are acidic organelles comparable to lysosomes, and have an internal pH estimated to be about 5.5 [68,70,71]. CMA raised the pH of the lytic granules to around neutral [71], and induced drastic morphologic changes in these granules [71–73]. Indeed, perforin was inactivated and proteolytically degraded within lytic granules, although CMA did not decrease the granzyme activity [72,75]. Bafilomycin A1, which belongs to the 16-membered macrolides, has a similar activity on CTLs as CMA, as far as has been tested [74]. CMA induced a marked DNA fragmentation and nuclear condensation characteristic of apoptosis in a CD8$^+$ CTL clone, but not in a CD4$^+$CTL clone, when incubated for longer periods [89]. Anti-CD3 or PMA accelerated the cell death of the CD8$^+$ CTL clone, but not the CD4$^+$ CTL clone, without accompanying DNA fragmentation and nuclear condensation [89]. These data suggest that vacuolar type H$^+$-ATPase activity is essential for survival of CD8$^+$ CTLs especially when activated. It has been reported that CMA-induced apoptosis in murine B cell clone WEHI-231 is antagonized by CD40 signaling or by constitutive expression of Bcl-2 [90].

4.6.2 Prodigiosin 25-C (PRG)

Although PRG has long been known as a red pigment produced by *Streptomyces* and *Serratia*, we initially identified it as an active compound that preferentially inhibits the proliferation of concanavalin A (ConA)-stimulated murine spleen cells more than that of LPS-stimulated one [91]. In vivo administration of PRG to mice immunized with allogeneic tumors markedly inhibited the induction of CD8$^+$ T cells with only marginal effects on the populations and functions of CD4$^+$ T cells and B cells [92–95]. However, in in vitro cell culture, PRG did not exhibit any selectivity to murine T cell clones and lymphoma cell lines so far tested, and the toxic effect was shown to be augmented synergistically by Con A [96,97]. In human lymphocytes, PRG was reported to block the phosphorylation of retinoblastoma proteins, and to prevent the induction of some genes such as *cdk-2* and *cdk-4* [98]. However, the immunosuppressive effects of PRG have not been fully understood. As one of the molecular target(s) of PRG, we have recently shown that PRG blocks the translocation of protons by vacuolar type H$^+$-ATPase without inhibiting its ATPase activity [99]. PRG thereby inhibits vacuolar acidification and affects glycoprotein processing in intact cultured cells [99]. In a CD8$^+$ CTL clone, PRG inhibited vacuolar acidification and inhibited the perforin-dependent killing [74,95]. However, in the PRG-treated CTL clone, perforin was not proteolytically degraded, but only inactivated [74].

4.6.3 FK506 and CsA

FK506 and CsA are potent immunosuppressants sharing similar biological activities, although they possess quite different structures. FK506 binds to FKBPs, whereas CsA binds to cyclophilins. The drug-protein complexes, i.e. FK506-FKBP12 and CsA-cyclophilin A, inhibit a single target, a calcium-dependent serine/threonine phosphatase calcineurin. FK506 and CsA inhibit the translocation of NF-AT from the cytosol to the nucleus through calcineurin inhibition, and thereby prevent expression of lymphokine genes such as IL-2. Several groups have employed FK506 and/or CsA to clarify the involvement of calcineurin in granule exocytosis and FasL expression [100–105]. CsA partially inhibits FasL expression in response to stimulation, suggesting that calcineurin-dependent and independent pathways are operative in CTLs [103,105]. CsA and FK506 strongly inhibit granule exocytosis with no inhibition or marginal inhibition of actual killing activity [101,102,104]. We observed that FK506 inhibits granule exocytosis in a CD8$^+$ CTL clone at the IC_{50} value of 3 nM (unpublished). A significant amount of granule exocytosis was still operative even in the presence of excess FK506, although IFN-γ production was completely inhibited at the IC_{50} value of 0.01 nM (unpublished). Taken together, these data suggest that a calcineurin-independent TCR-mediated granule exocytosis pathway exists in CD8$^+$ CTLs and plays a role in the target cell lysis.

4.6.4 Cytochalasin D

The microtubule poisons, taxol and colchicine, failed to inhibit perforin-dependent cytotoxic activity [87], although lytic granules have been shown to move on microtubules in a kinesin-dependent manner [106]. In contrast, a microfilament poison, cytochalasin D, was reported to inhibit degranulation and cytolysis [107]. Some groups who employed cytochalasin B or D for analysis of T cell activation [108–110] concluded that sustained signaling leading to T cell activation requires a functional cytoskeleton, and that a tyrosine-phosphorylated T cell receptor ζ chain associates with the actin cytoskeleton upon activation. Thus, actin cytoskeleton plays a critical role in signal transduction leading to the effector functions of T cells.

4.6.5 Gliotoxin

Gliotoxin is one possible etiologic agent that is synthesized by *Aspergillus fumigatus* and other pathogenic fungi, and exhibits a variety of immunosuppressive activities in vivo and in vitro. Thus, gliotoxin has been reported to inhibit the generation of alloreactive CTLs [111], prevent the activation of transcription factor NF-κB in T and B cells [112], and induce apoptosis in a variety of cells [113,114]. During the course of our screening, we identified gliotoxin as an active compound which inhibits CTL-mediated cytotoxicity [115]. Gliotoxin markedly inhibited both perforin- and Fas-dependent killing pathways at an IC_{50} value of 150–250 nM [115]. The effector/target conjugate formation was inhibited in a dose-dependent manner only when the effec-

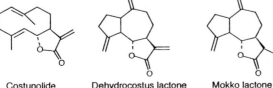

Costunolide   Dehydrocostus lactone   Mokko lactone

Fig. 6. Structures of costunolide, dehydrocostus lactone, and mokko lactone

tor CTL clone was treated with gliotoxin, although the expression of surface molecules of the CTL clone such as CD3 was unaffected under the same conditions (unpublished). Gliotoxin also blocked granule exocytosis and the production of inositol phosphates in response to anti-CD3 stimulation, most likely because the gliotoxin-treated CTL clone was incapable of binding to immobilized anti-CD3 (unpublished). Thus, gliotoxin seems to prevent the ability of CTLs to interact with target cells.

4.6.6 α-Methylene-γ-Butyrolactones

Costunolide and dehydrocostus lactone (Fig. 6.) were isolated from an extract of mokko (*Saussurea lappa* Clarke) as inhibitors of CTL-mediated killing [116]. From the same extract, mokko lactone (Fig. 6.) was also isolated as an inactive compound [116]. Costunolide and dehydrocostus lactone, but not mokko lactone, have α-methylene-γ-butyrolactone which is known to be an alkylating agent and undergo a Michael reaction with biological nucleophiles such as L-cysteine and SH-containing enzymes. Since glutathione completely neutralized the inhibitory effect of costunolide [116], it seems that these compounds inhibit the killing pathway by binding to SH-containing proteins. Costunolide markedly prevented granule exocytosis and the production of inositol phosphates in response to anti-CD3 stimulation [116]. Tyrosine phosphorylation was significantly inhibited by costunolide [116]. Thus, our data suggest that costunolide primarily inhibits an increase in tyrosine phosphorylation in response to T cell activation, thereby inhibiting cytotoxicity.

4.6.7 Protein Tyrosine Kinase Inhibitors

Protein tyrosine kinases such as Lck, Fyn, and ZAP-70 are involved in early signal transduction through T cell receptors. Protein tyrosine kinase inhibitors such as genistein and herbimycin A have frequently been used for analysis of early signal events through T cell activation. Genistein and herbimycin A were reported to inhibit T cell-mediated cytotoxicity, granule exocytosis, and FasL expression [103,117,118].

4.6.8 Phorbol Esters

Phorbol esters (e.g. TPA and 12,13-phorbol dibutyrate) inhibit Fas-mediated apoptosis [119–122]. Inhibition of Fas-mediated apoptosis by phorbol esters is independent of protein synthesis but dependent on an increase in intracellular superoxide anion [119]. Since tumor cells typically exhibit a prooxidant state, this might be one natural mechanism that induces resistance to Fas-mediated apoptosis. We identified several compounds that inhibit Fas-mediated apoptosis in our screening system: i.e. TPA, cytochalasin D, penicillic acid, patulin, FD-891 (unpublished), although they have quite different structures. However, all the compounds increased the oxidation states and their inhibitory effects on Fas-mediated apoptosis were antagonized by a radical scavenger, N-acetyl cysteine [119; unpublished]. More recently, phorbol esters have been reported to block caspase-3 activation, but not caspase-8 activation, in response to Fas stimulation [120,121]. We also observed the blockage of caspase-3 activation by TPA and FD-891 (unpublished). Thus, molecular event(s) downstream of caspase-8 and upstream of caspase-3 might be a target for TPA-dependent inhibition. Sphingosine 1-phosphate [121] which is generated via PKC-mediated activation of sphingosine kinase, and the upregulation of MAP-kinase signaling [122] are implicated in the suppression of Fas-mediated apoptosis.

5   Conclusions

The immune system comprises two groups of cells, one major group including granulocytes such as macrophages and other phagocytic cells, and the other B and T lymphocytes and NK cells. These cells can function autonomously, however most of the immune responses are carried out by cooperative work after proliferation and activation of the immune cells. Signals are received through direct interaction between cells as well as through soluble factors including cytokines. Low molecular weight substances that affect the process involved in signal transduction for cell proliferation, transcription and translation of specific genes required for expression of immune cell functions, therefore, will positively and/or negatively modify immune responses. The enzymes for purine and pyrimidine metabolism have been shown to be good targets since they support the efficient proliferation of the lymphocytes receiving the signal. IMPDH, involved in de novo biosynthesis of GTP and dGTP, is a target of MZ and MPA, whereas DHODH, involved in pyrimidine synthesis is a target for BQR and LEF. The pathways for activation or modification of transcription factors are affected by complexes between the cellular proteins, immunophilins, and CsA and FK506. Similar complexes containing RAPA affect different process required for lymphocyte proliferation. The target of DSG, an immunosuppressant, is shown to be Hsc70 and thus the heat shock proteins may play a significant role in immunoregulatory pathways. CTL and NK cells, after activation by receiving signals, kill target cells by two distinct pathways, one dependent on perforin and another on Fas L. The former pathway was specifically suppressed by the treatment of the cells with inhibitors of vacuolar type H^+-ATPase including CMA and bafilomycin A1 because inhibition of acidification of the lytic granules results in inactivation of perforin. Inhibitors of gene transcription, translation, and protein transport which did not affect perforin-mediated cytotoxicity significantly suppressed the Fas-mediated cytotoxic pathway of CTL. Thus it is possible to separately assay those two killing pathways by using different bioprobes. Further studies on the mechanism of action of immunomodulating substances will give us new targets for development of pharmaceutically useful agents.

Acknowledgments

We thank Drs. Moselio Schaechter and Michael Hahne for critical reading of this manuscript and Dr. Kunio Ando for helpful suggestions.

References

1. Morris RE (1993) New small molecule immunosuppressants for transplantation: review of essential concepts. J Heart Lung Transplant 12:S275–286
2. Morris RE (1995) Mechanism of action of new immunosuppressive drugs. The Drug Monit 17:564–569
3. Halloran PF (1996) Molecular mechanism of new immunosuppressive agents. Clin Transplantation 10:118–123
4. Gerber DA, Bonham CA, Thompson AW (1998) Immunosuppressive agents: Recent developments in molecular action and clinical application. Transplant Proc 30:1573–1579
5. Schwartz RS, Stack J, Dameshek W (1958) Effect of 6-mercaptopurine on antibody production. Proc Soc Exp Biol Med 88:164–167
6. Gilblett EL, Anderson JE, Cohen F, Plara B, Neuwissen HJ (1972) Adenosine deaminase deficiency in two patients with severely impaired cellular immunity. Lancet ii:1067–1069
7. Gilblett EL, Ammann AJ, Wara DW, Sadma R, Diamond LK (1975) Nucleoside phosphorylase deficiency in child with severely defective T-cell immunity and normal B-cell immunity. Lancet i:1010–1011
8. Georgiev VST (1993) Enzymes of the purine metabolism: Inhibition and therapeutic potential. Ann N Y Acad Sci 685:207–216
9. Koyama H, Tsuji M (1983) Genetic and biochemical studies on the activation of and cytotoxic mechanism of bredinin, a potent inhibitor of purine biosynthesis in mammalian cells. Biochem Pharmacol 32:3547–3553
10. Ichikawa Y, Ihara H, Takahara S, Takada K, Shrestha GR, Ishibashi M, Arima M, Sagawa S, Sonoda T (1984) The immunosuppressive mode of action of mizoribine. Transplantation 38:262–267
11. Allison AC, Hovi T, Watts RWE, Webster ADB (1975) Immunological observations on patients with Lesch-Nyhan syndrome, and on the role of de novo purine synthesis in lymphocyte transformation. Lancet ii:1179–1183
12. Kayama K, Okubo M, Ishigamori E, Masaki Y, Uchida H, Watanabe K, Kashiwagi N (1983) Immunosuppressive effect of bredinin on cell-mediated and humoral immune reactions in experimental animals. Transplantation 35:144–1449
13. Mitchell BS, Dayton JS, Turka LA, Thompson CB (1993) IMP dehydrogenase inhibitors as immunomodulators. Ann N Y Acad Sci 685:217–224
14. Hirohata S, Yanagita T (1995) Inhibition of expression of cyclin A in human B cells by an immunosuppressant mizoribine. J Immunol 155:5175–5183
15. Sweeney MJ, Hoffman DH, Esterman MA (1972) Metabolism and biochemistry of mycophenolic acid. Cancer Res 32:1803–1809
16. Lee WA, Gu L, Miksztal AR, Nancy C, Kwan L, Nelson PH (1990) Bioavailability improvement of mycophenolic acid through amino ester derivatization. Pharm Res 7:161–166
17. Franklin TJ, Cook JM (1969) The inhibition of nucleic acid synthesis by mycophenolic acid. Biochem J 113:515–524
18. Natsumeda Y, Ohno S, Kawasaki H, Konno Y, Weber G, Suzuki K (1990) Two distinct cDNAs for human IMP dehydrogenase. J Biol Chem 265:5292–5295
19. Nagai M, Natsumeda Y, Weber G (1992) Proliferation-linked regulation of the type II IMP dehydrogenase gene in human normal lymphocytes and HL-60 leukemic cells. Cancer Res 52:258–261

20. Carr SF, Rapp E, Wu JC, Natsumeda Y (1993) Characterization of human type I and type II IMP dehydrogenases. J Biol Chem 268:27286–27290
21. Allison AC, Eugui EM (1994) Preferential suppression of lymphocyte proliferation by mycophenolic acid and predicted long-term effects of mycophenolate mofetil in transplantation. Transplant Proc 26:3205–3210
22. Allison AC, Eugui EM (1993) The design and development of an immunosuppressive drug, mycophenolate mofetil. Springer Semin Immunopathol 14:353–380
23. Makowka L, Sher L, Cramer D (1993) The development of brenquinar as an immunosuppressive drug for transplantation. Immunol Rev 136:51–70
24. Ruckemann K, Fairbanks LD, Carrey EA, Hawrylowicz CM, Richards DF, Kirschbaum B, Simmonds HN (1998) Leflunomide inhibits pyrimidine de novo synthesis in mitogen-stimulated T-lymphocytes from healthy humans. J Biol Chem 273:21682–21691
25. Chong AS, Rezai K, Gebel HM, Finnegan A,Foster P, Xu X, Williams JW (1996) Effects of leflunomide and other immunosuppressive agents on T cell proliferation in vitro. Transplantation 15:140–145
26. Bader B, Knecht W, Fries M, Loffler M (1998) Expression, purification, and characterization of histidine-tagged rat and human flavoenzyme dehydroorotate dehydrogenase. Protein Expr Purif 13:414–422
27. Knecht W, Loffler M (1998) Species-related inhibition of human and rat dihydroorotate dehydrogenase by immunosuppressive isoxazol and cinchoninic acid derivatives. Biochem Pharmacol 56:1259–1264
28. Xu X, Williams JW, Shen J, Gong H, Yin DP, Blinder L, Elder R, Sankary H, Finegan A, Chong ASF (1998) In vitro and in vivo mechanisms of action of the antiproliferative and immunosuppressive agents, brequinar sodium. J Immunol 160:846–853
29. Xu X, Williams JW, Bremer EG, Gong H, Finnegan A, Chong ASF (1996) Two activities of the immunosuppressive metabolite of leflunomide, A77 1726. Inhibition of pyrimidine nucleotide synthesis and protein tyrosine phosphorylation. Biochem Pharmcol 52:527–534
30. Siemasko K, Chong ASF, Jack HM, Gong H, Williams JW, Finnegan A (1998) Inhibition of JAK3 and STAT6 tyrosine phosphorylation by the immunosuppressive drug leflunomide leads to a block in IgG1 production. J Immunol 160:1581–1588
31. Manna SK, Aggarwal BB (1999) Immunosuppressive leflunomide metabolite (A77 1726) blocks TNF-dependent nuclear factor-κB activation and gene expression. J Immunol 162:2095–2102
32. Nishizuka Y (1992) Intracellular signaling by hydrolysis of phospholipids and activation of protein kinase C. Science 258:607–614
33. Ferris CD, Snyder H (1992) 1,4,5-Triphosphate activated calcium channels. Ann Rev Physiol 54:469–488
34. Berridge MJ (1995) Capacitative calcium entry. Biochem J 312:1–11
35. Stemmer PM, Klee CB (1994) Dual calcium regulation of calcineurin by calmodulin and calcineurin B. Biochemistry 33:6859–6866
36. Shaw KTY, Ho AM, Raghavan A, Kim J, Jain J, Park J, Surendra S, Rao A, Hogan PG (1995) Immunosuppressive drugs prevent a rapid dephosphorylation of transcription factor NFAT1 in stimulated immune cells. Proc Natl Acad Sci USA 92:11205–11209
37. Frantz B, Nordby EC, Bren G, Steffan N, Paya CV, Kincaid RL, Tocci MJ, O'Keefe SJ O'Neill EA (1994) Calcineurin acts in synergy with PMA to inactivate IκB/MAD3, an inhibitor of NF-κB. EMBO J 13:861–870
38. Quesniauz VF (1993) Immunosuppressants: Tools to investigate the physiological role of cytokines. BioEssays 15:731–739

39. Cai W, Hu L, Foulkes JG (1996) Transcription-modulating drugs: mechanism and selectivity. Curr Opin Biotechnol 7:608–615

40. Bram RJ, Hung DT, Martin PK, Schreiber SL, Crabtree GR (1993) Identification of the immunophilins capable of mediating inhibition of signal transduction by cyclosporin and FK506: role of calcineurin binding and cellular location. Mol Cell Biol 13:4760–4769

41. Galat A (1993) Peptidylproline cis-trans-isomerase: immunophilins. Eur J Biochem 216:689–707

42. Kissinger CR, Parge HE, Knighton DR, Lewis CT, Pelletier LA, Tempczyk A, Ksalish V, Tucker KD, Showalter RE, Moomaw EW et al. (1995) Crystal structure of human calcineurin and the human FKBP12-calcineurin complex. Nature 378:641–644

43. Schreiber SL, Crabtree GR (1992) The mechanism of action of cyclosporin A and FK506. Immunology Today 13:136–142

44. Sabatini DM, Erdjument-Bromage H, Lui M, Tempst P, Snyder SH (1994) RAFT1: a mammalian protein that binds to FKBP12 in a rapamycin-dependent fashion and is homologous to yeast TORs. Cell 78:35–43

45. Brown EJ, Albers MW, Shin TB, Ichikawa K, Keith CT, Lane WS, Schreiber SL (1994) A mammalian protein targeted by G1-arresting rapamycin receptor complex. Nature 369:756–758

46. Brown EJ, Beal PA, Keith CT, Chen J, Shin TB, Schreiber SL (1995) Control of p70 S6 kinase by FRAP in vivo. Nature 377:411–416

47. Pearson RB, Dennis PB, Han JW, Williamson NA, Kozma SC, Wettenhall REH, Thomas G (1995) The principle target of rapamycin-induced p70s6k inactivation is a novel phosphorylation site within a conserved hydrophobic domain. EMBO J 21:5279–5287

48. De Groot RP, Ballou LM, Sassonew-Corsi P (1994) Positive regulation of the cAMP-responsive activator CREM by the p70 S6 kinase: and alternative route to mitogen-induced gene expression. Cell 79:81–91

49. Jefferies HBJ, Reinhard C, Kozma SC, Thomas G (1994) Rapamycin selectively re-presses transition of the polypyrimidine tract mRNA family. Proc Natl Acad Sci USA 91:4441–4445

50. Feuerstein N, Huang D, Prystowsky MB (1995) Rapamycin selectively blocks interleukin-2-induced proliferating cell nuclear antigen gene expression in T lymphocyte. J Biol Chem 270:9454–5458

51. Nishizawa R, Takei Y, Yoshida M, Tomiyoshi T, Saizo K, Nishikawa K, Nemoto K, Takahashi K, Fujii A, Nakamura T, Takita T, Takeuchi T (1988) Synthesis and biological activity of spergualin analogues. I. J Antibiot 41:1629–1643

52. Tepper MA, Betty B, Bursuker I, Pasternak RD, Cleaveland J, Spitalny GL, Schacter B (1991) Inhibition of antibody production by the immunosuppressive agent, 15-deoxyspergualin. Transplant Proc 23:328–331

53. Dickneite G, Schorlemmer HU, Sedlacek HH (1987) Decrease of mononuclear phagocyte cell functions and prolongation of graft survival in experimental transplantation by (+/–)-15-deoxyspergualin. Int J Immunopharmacol 9:559–565

54. Tepper MA, Nadler S, Mazzucco C, Singh C, Kelley S (1993) 15-deoxyspergualin, a novel immunosuppressive drug: Studies of the mechanism of action. Ann N Y Acad Sci 685:136–147

55. Nadler SG, Tepper MA, Schacter B, Mazzucco CE (1992) Interaction of the immuno-suppressant deoxyspergualin with a member of the Hsp70 family of heat shock proteins. Science 258:484–486

56. Suzuki CK, Bonifacino JS, Lin AY, Davis MM, Klausner RD (1991) Regulating the retention of T-cell receptor α variants within the endoplasmic reticulum: calcium-dependent association with BiP. J Cell Biol 114:189–205

57. VanBuskirk AM, deNagel DC, Guagliardi LE, Brodsky FM, Pierce SK (1991) Cellular and subcellular distribution of PBP72/74, a peptide-binding protein that plays a role in antigen processing. J Immunol 146:500–506

58. Tai PK, Alberts MW, Chang H, Faber LE, Schreiber SL (1992) Association of a 59-kilodalton immunophilin with the glucocorticoid receptor complex. Science 256:1315–1318

59. Braun MY, Lowin B, French L, Acha-Orbea H, Tschopp J (1996) Cytotoxic T cells deficient in both functional Fas ligand and perforin show residual cytolytic activity yet lose their capacity to induce lethal acute graft-versus-host disease. J Exp Med 183: 657–661

60. Lee RK, Spielman J, Zhao DY, Olsen KJ, Podack ER (1996) Perforin, Fas ligand, and tumor necrosis factor are the major cytotoxic molecules used by lymphokine-activated killer cells. J Immunol 157: 1919–1925

61. Nagata S, Golstein P (1995) The Fas death factor. Science 267: 1449–56

62. Kondo T, Suda T, Fukuyama H, Adachi M, Nagata S (1997) Essential roles of the Fas ligand in the development of hepatitis. Nat Med 3: 409–413

63. Giordano C, Stassi G, De Maria R, Todaro M, Richiusa P, Papoff G, Ruberti G, Bagnasco M, Testi R, Galluzzo A (1997) Potential involvement of Fas and its ligand in the pathogenesis of Hashimoto's thyroiditis. Science 275: 960–963

64. Hahne M, Rimoldi D, Schröter M, Romero P, Schreier M, French LE, Schneider P, Bornand T, Fontana A, Lienard D, Cerottini JC, Tschopp J (1996) Melanoma cell expression of Fas (Apo-1/CD95) ligand: Implications for tumor immune escape. Science 274: 1363–1366

65. Strand S, Hofmann WJ, Hug H, Müller M, Otto G, Strand D, Mariani SM, Stremmel W, Krammer PH, Galle PR (1996) Lymphocyte apoptosis induced by CD95 (APO-1/Fas) ligand-expressing tumor cells-A mechanism of immune evasion? Nat Med 2: 1361–1366

66. O'Connell J, O'Sullivan GC, Collins JK, Shanahan F (1996) The Fas counterattack: Fas-mediated T cell killing by colon cancer cells expressing Fas ligand. J Exp Med 184: 1075–1082

67. Irmler M, Thome M, Hahne M, Schneider P, Hofmann K, Steiner V, Bodmer JL, Schröter M, Burns K, Mattmann C, Rimoldi D, French LE, Tschopp J (1997) Inhibition of death receptor signals by cellular FLIP. Nature 388: 190–195

68. Burkhardt JK, Hester S, Lapham CK, Argon Y (1990) The lytic granules of natural killer cells are dual-function organelles combining secretory and pre-lysosomal compartments. J Cell Biol 111: 2327–2340

69. Peters PJ, Borst J, Oorschot V, Fukuda M, Krähenbühl O, Tschopp J, Slot JW, Geuze HJ (1991) Cytotoxic T lymphocyte granules are secretory lysosomes, containing both perforin and granzymes. J Exp Med 173: 1099–1109

70. Masson D, Peters PJ, Geuze HJ, Borst J, Tschopp J (1990) Interaction of chondroitin sulfate with perforin and granzymes of cytolytic T-cells is dependent on pH. Biochemistry 29: 11229–11235

71. Kataoka T, Sato M, Kondo S, Nagai K (1996) Estimation of pH and the number of lytic granules in a CD8+ CTL clone treated with an inhibitor of vacuolar type H+-ATPase, concanamycin A. Biosci Biotech Biochem 60: 1729–1731

72. Kataoka T, Takaku K, Magae J, Shinohara N, Takayama H, Kondo S, Nagai, K (1994) Acidification is essential for maintaining the structure and function of lytic granules of CTL. J Immunol 153: 3938–3947

73. Kataoka T, Shinohara N, Takayama H, Takaku K, Kondo S, Yonehara S, Nagai K (1996) Concanamycin A, a powerful tool for characterization and estimation of contribution of perforin- and Fas-based lytic pathways in cell-mediated cytotoxicity. J Immunol 156: 3678–3686

74. Togashi K, Kataoka T, Nagai K (1997) Characterization of a series of vacuolar type H+-ATPase inhibitors on CTL-mediated cytotoxicity. Immunol Lett 55: 139–144
75. Kataoka T, Togashi K, Takayama H, Takaku K, Nagai K (1997) Inactivation and proteolytic degradation of perforin within lytic granules upon neutralization of acidic pH. Immunology 91: 493–500
76. Heusel JW, Wesselschmidt RL, Shresta S, Russell JH, Ley TJ (1994) Cytotoxic lymphocytes require granzyme B for the rapid induction of DNA fragmentation and apoptosis in allogeneic target cells. Cell 76: 977–987
77. Talanian RV, Yang X, Turbov J, Seth P, Ghayur T, Casiano CA, Orth K, Froelich C J (1997) Granule-mediated killing: Pathways for granzyme B-initiated apoptosis. J Exp Med 186: 1323–1331
78. Sarin A, Williams MS, Alexander-Miller MA, Berzofsky JA, Zacharchuk CM, Henkart PA (1997) Target cell lysis by CTL granule exocytosis is independent of ICE/Ced-3 family proteases. Immunity 6: 209–215
79. Andrade F, Roy S, Nicholson D, Thornberry N, Rosen A, Casciola-Rosen L (1998) Granzyme B directly and efficiently cleaves several downstream caspase substrates: Implications for CTL-induced apoptosis. Immunity 8: 451–460
80. Nagata S (1997) Apoptosis by death factor. Cell 88: 355–365
81. Yang X, Khosravi-Far R, Chang HY, Baltimore D (1997) Daxx, a novel Fas-binding protein that activates JNK and apoptosis. Cell 89: 1067–1076
82. Susin SA, Lorenzo HK, Zamzami N, Marzo I, Snow BE, Brothers GM, Mangion J, Jacotot E, Costantini P, Loeffler M, Larochette N, Goodlett DR, Aebersold R, Siderovski DP, Penninger JM, Kroemer G(1999)Molecular characterization of mitochondrial apoptosis-inducing factor. Nature 397:441–446
83. Li P, Nijhawan D, Budihardjo I, Srinivasula SM, Ahmad M, Alnemri ES, Wang X (1997) Cytochrome c and dATP-dependent formation of Apaf-1/caspase-9 complex initiates an apoptotic protease cascade. Cell 91: 479–489
84. Enari M, Sakahira H, Yokoyama H, Okawa K, Iwamatsu A, Nagata S (1998) A caspase-activated DNase that degrades DNA during apoptosis, and its inhibitor ICAD. Nature 391: 43–50
85. Scaffidi C, Fulda S, Srinivasan A, Friesen C, Li F, Tomaselli KJ, Debatin KM, Krammer PH, Peter ME (1998) Two CD95 (APO-1/Fas) signaling pathways. EMBO J 17: 1675–1687
86. Refaeli Y, Van Parijs L, London CA, Tschopp J, Abbas AK (1998) Biochemical mechanisms of IL-2-regulated Fas-mediated T cell apoptosis. Immunity 8: 615–623
87. Kataoka T, Taniguchi M, Yamada A, Suzuki H, Hamada S, Magae J, Nagai K (1996) Identification of low molecular weight probes on perforin- and Fas-based killing mediated by cytotoxic T lymphocytes. Biosci Biotech Biochem 60: 1726–1728
88. Kataoka T, Nagai K (1997) Characterization of perforin-based and Fas-based CTL-mediated cytotoxicity by using low molecular probes. Funatsu K et al (eds.). Animal Cell Technology: Basic & Applied Aspects, 8: 547–552 Kluwer Academic Publisher
89. Togashi K, Kataoka T, Nagai K (1997) Concanamycin A, a vacuolar type H+-ATPase inhibitor, induces cell death in activated CD8+ CTL. Cytotechnology 25: 127–135
90. Akifusa S, Ohguchi M, Koseki T, Nara K, Semba I, Yamato K, Okahashi N, Merino R, Núñez G, Hanada N, Takehara T, Nishihara T (1998) Increase in Bcl-2 level promoted by CD40 ligation correlates with inhibition of B cell apoptosis induced by vacuolar type H+-ATPase inhibitor. Exp Cell Res 238: 82–89
91. Nakamura A, Nagai K, Ando K, Tamura G (1986) Selective suppression by prodigiosin of the mitogenic response of murine splenocytes. J Antibiot 39: 1155–1159

92. Nakamura A, Magae J, Tsuji RF, Yamasaki M, Nagai K (1989) Suppression of cytotoxic T cell induction in vivo by prodigiosin 25-C. Transplant 47: 1013–1016
93. Tsuji RF, Yamamoto M, Nakamura A, Kataoka T, Magae J, Nagai K, Yamasaki M (1990) Selective immunosuppression of prodigiosin 25-C and FK506 in the murine immune system. J Antibiot 43: 1293–1301
94. Tsuji RF, Magae J, Yamashita M, Nagai K, Yamasaki M (1992) Immunomodulating properties of prodigiosin 25-C, an antibiotic which preferentially suppresses induction of cytotoxic T cells. J Antibiot 45: 1295–1302
95. Lee MH, Kataoka T, Magae J, Nagai K (1995) Prodigiosin 25-C suppression of cytotoxic T cells in vitro and in vivo similar to that of concanamycin B, a specific inhibitor of vacuolar type H⁺-ATPase. Biosci Biotech Biochem 59: 1417–1421
96. Kataoka T, Magae J, Nariuchi H, Yamasaki M, Nagai K (1992) Enhancement by concanavalin A of the suppressive effect of prodigiosin 25-C on proliferation of murine splenocytes. J Antibiot 45: 1303–1312
97. Kataoka T, Magae J, Kasamo K, Yamanishi H, Endo A, Yamasaki M, Nagai K (1992) Effects of prodigiosin 25-C on cultured cell lines: its similarity to monovalent polyether ionophores and vacuolar type H⁺-ATPase inhibitors. J. Antibiot 45: 1618–1625
98. Songia S, Mortellaro A, Taverna S, Fornasiero C, Scheiber EA, Erba E, Colotta F, Mantovani A, Isetta AM, Golay J (1997) Characterization of the new immunosuppressive drug undecylprodigiosin in human lymphocytes: retinoblastoma protein, cyclin-dependent kinase-2, and cyclin-dependent kinase-4 as molecular targets. J Immunol 158: 3987–3995
99. Kataoka T, Muroi M, Ohkuma S, Waritani T, Magae J, Takatsuki A, Kondo S, Yamasaki M, Nagai K (1995) Prodigiosin 25-C uncouples vacuolar type H⁺-ATPase, inhibits vacuolar acidification and affects glycoprotein processing. FEBS Lett 359: 53–59
100. Lancki DW, Kaper BP, Fitch FW (1989) The requirements for triggering of lysis by cytolytic T lymphocyte clones II. Cyclosporin A inhibits TCR-mediated exocytosis but only selectively inhibits TCR-mediated lytic activity by cloned CTL. J Immunol 142: 416–424
101. Trenn G, Taffs R, Hohman R, Kincaid R, Shevach EM, Sitkovsky M (1989) Biochemical characterization of the inhibitory effect of CsA on cytolytic T lymphocyte effector functions. J Immunol 142: 3796–3802
102. Dutz JP, Fruman DA, Burakoff SJ, Bierer BE (1993) A role for calcineurin in degranulation of murine cytotoxic T lymphocytes. J Immunol 150: 2591–2598
103. Anel A, Buferne M, Boyer C, Schmitt-Verhulst AM, Golstein P (1994) T cell receptor-induced Fas ligand expression in cytotoxic T lymphocyte clones is blocked by protein tyrosine kinase inhibitors and cyclosporin A. Eur J Immunol 24: 2469–2476
104. Mehta BA, Schmidt-Wolf IG, Weissman, IL, Negrin RS (1995) Two pathways of exocytosis of cytoplasmic granule contents and target cell killing by cytokine-induced CD3⁺ CD56⁺ killer cells. Blood 86: 3493–3499
105. Rogers AM, Thilenius AR, Russell JH (1997) Cyclosporine-insensitive partial signaling and multiple roles of Ca^{2+} in Fas ligand-induced lysis. J Immunol 159: 3140–3147
106. Burkhardt JK, McIlvain JM Jr, Sheetz MP, Argon Y (1993) Lytic granules from cytotoxic T cells exhibit kinesin-dependent motility on microtubules in vitro. J Cell Sci 104: 151–162
107. O'Rourke AM, Apgar JR, Kane KP, Martz E, Mescher MF (1991) Cytoskeletal function in CD8- and T cell receptor-mediated interaction of cytotoxic T lymphocytes with class I protein. J Exp Med 173: 241–249

108. Valitutti S, Dessing M, Aktories K, Gallati H, Lanzavecchia A (1995) Sustained signaling leading to T cell activation results from prolonged T cell receptor occupancy. Role of T cell actin cytoskeleton. J Exp Med 181: 577–584

109. Caplan S, Zeliger S, Wang L, Baniyash M (1995) Cell-surface-expressed T-cell antigen-receptor ζ chain is associated with the cytoskeleton. Proc Natl Acad Sci USA 92: 4768–4772

110. Rozdzial MM, Malissen B, Finkel TH (1995) Tyrosine-phosphorylated T cell receptor ζ chain associates with the actin cytoskeleton upon activation of mature T lymphocytes. Immunity 3: 623–633

111. Müllbacher A, Eichner RD (1984) Immunosuppression in vitro by a metabolite of a human pathogenic fungus. Proc Natl Acad Sci USA 81: 3835–3837

112. Pahl HL, Krauβ B, Schulze-Osthoff K, Decker T, Traenckner EB, Vogt M, Myers C, Parks T, Waring P, Mühlbacher A, Czernilofsky AP, Baeuerle PA (1996) The immuno-suppressive fungal metabolite gliotoxin specifically inhibits transcription factor NF-κB. J Exp Med 183: 1829–1840

113. Sutton P, Newcombe NR, Waring P, Müllbacher A (1994) In vivo immunosuppressive activity of gliotoxin, a metabolite produced by human pathogenic fungi. Infec Immun 62: 1192–1198

114. Waring P, Khan T, Sjaarda A (1997) Apoptosis induced by gliotoxin is preceded by phosphorylation of histone H3 and enhanced sensitivity of chromatin to nuclease diges-tion. J Biol Chem 272: 17929–17936

115. Yamada A, Kataoka T, Nagai K (1997) Inhibition of CTL-mediated cytotoxicity by gliotoxin. Funatsu K et al (eds.). Animal Cell Technology: Basic & Applied Aspects 8: 633–637 Kluwer Academic Publishers

116. Taniguchi M, Kataoka T, Suzuki H, Uramoto M, Ando M, Arao K, Magae J, Nishimura T, Otake N, Nagai K (1995) Costunolide and dehydrocostus lactone as inhibitors of killing function of cytotoxic T lymphocytes. Biosci Biotech Biochem 59: 2064–2067

117. Anel A, Richieri GV, Kleinfeld AM (1994) A tyrosine phosphorylation requirement for cytotoxic T lymphocyte degranulation. J Biol Chem 269: 9506–9513

118. Rosato A, Zambon A, Mandruzzato S, Bronte V, Macino B, Calderazzo F, Collavo D, Zanovello P (1994) Inhibition of protein tyrosine phosphorylation prevents T-cell-mediated cytotoxicity. Cell Immunol 159: 294–305

119. Clément MV, Stamenkovic I (1996) Superoxide anion is a natural inhibitor of Fas-medi-ated cell death. EMBO J 15: 216–225

120. Ruiz-Ruiz MC, Izquierdo M, de Murcia G, Lòpez-Rivas A (1997) Activation of protein kinase C attenuates early signals in Fas-mediated apoptosis. Eur J Immunol 27: 1442–1450

121. Cuvillier O, Rosenthal DS, Smulson ME, Spiegel S (1998) Sphingosine 1-phosphate inhibits activation of caspases that cleave poly(ADP-ribose) polymerase and lamins during Fas- and ceramide-mediated apoptosis in Jurkat T lymphocytes. J Biol Chem 273: 2910–2916

122. Holmström TH, Chow SC, Elo I, Coffey ET, Orrenius S, Sistonen L, Eriksson JE (1998) Suppression of Fas/APO-1-mediated apoptosis by mitogen-activated kinase signaling. J Immunol 160: 2626–2636

6
Bioprobes at a Glance

TAKEO USUI and HIROYUKI OSADA

Using the Bioprobes at a Glance

1. The Bioprobes at a Glance section is ordered alphabetically by entry name.
2. Molecular weight is rounded to the nearest whole number.
3. Solubility is represented by symbols: +++, easily soluble; ++, soluble; +, sparingly soluble; ±, barely soluble; -, insoluble.

Actinomycin D
(=Actinomycin C₁)

Key Words: [Antitumor][Apoptosis][RNA synthesis inhibitor]

Structure:

Molecular Formula: $C_{62}H_{86}N_{12}O_{16}$
Molecular Weight: 1255
Solubility: DMSO, ++ ; H_2O, + ; MeOH, +

Discovery/Isolation:

Actinomycins were isolated as bacteriostatic and bactericidal substances produced by a soil *Actinomyces* [1]. Brockmann and Pfennig purified each actinomycin compound from a mixture by several chromatographies [2,3]. Actinomycin D is also known as actinomycin C_1, which is the main product of some *Actinomyces*. Hackmann discovered the actinomycins as potent antitumor compounds [4,5]. The structure of actinomycins was reviewed by Brockmann [6].

Biological and Chemical Studies:

Actinomycins show antitumor [4,5] and antibacterial activity [1,7]. Reich et al.. investigated the structure-activity correlations of actinomycins and their derivatives [8]. Actinomycins strongly inhibit DNA dependent RNA synthesis [9–13]. The inhibition mechanism of actinomycin is thought to be that actinomycin forms complexes with DNA at the deoxyguanosine moiety and intercalate [8,13]. Actinomycins also inhibit DNA polymerase but the inhibition is weak [10]. Recently, it was reported that actinomycin D induced the rapid apoptotic death of most antibody-secreting B-cell hybridomas [14].

Biological Activity:

0.25–0.75 µg/ml: growth inhibition of *Staphylococcus aureus* [10]

2.76×10^{-8} M: K_i for DNA-dependent RNA polymerase, competitive with DNA [12]

2.7×10^{-7} M: K_i for DNA polymerase, competitive with DNA [12]

2.0 µg/ml: IC_{50} for DNA-dependent RNA synthesis in HeLa nuclear extract [13]

5 µg/ml: apoptosis induction in D5 B-cell hybridomas [14]

References:

1. Waksman SA, & Woodruff HB (1940) Bacteriostatic and bactericidal substances produced by a soil *Actinomyces*. Proc Soc Exp Biol Med 45:609–614
2. Brockmann H, & Pfennig N (1952) Auftrennung von Actinomycin C durch Gegenstromverteilung. Naturwiss 39:429–430
3. Brockmann H, & Pfennig N (1953) Die Gewinnung reiner Actinomycine durch Gegenstromverteilung. Hoppe-Seylers Z Physiol Chem 292:77–88
4. Hackmann C (1954) Untersuchungen über den Einfluβ des Sanamycins (Actinomycin C) auf tierische Organe: Milz, Thymus, Lymphknoten, Nebennieren und Keimdrüsen. Z Krebsforsch Bd 60:250–255
5. Hackmann C (1952) Experimentelle Untersuchungen über die Wirkung von Actinomycin C (HBF386) bei bösartigen Geschwülsten. Z Krebsforsch Bd 58:607–613
6. Brockmann H (1960) Die Actinomycine. Angew Chem 72:939–947
7. Haywood AM, & Sinsheimer RL (1963) Inhibition of protein synthesis in *E. coli* protoplats by actinomycin D. J Mol Biol 6:247–249
8. Reich E, Goldberg IH, & Rabinowitz M (1962) Structure-activity correlations of actinomycins and their derivatives. Nature 196:743–748
9. Reich E, Franklin RM, Shatkin AJ, & Tatum EL (1961) Effect of actinomycin D on cellular nucleic acid synthesis and virus production. Science 134:556–557
10. Kirk JM (1960) The mode of action of actinomycin D. Biochim Biophys Acta 42:167–169
11. Harbers E, & Müller W (1962) On the inhibition of RNA synthesis by actinomycin. Biochem Biophys Res Comm 7:107–110
12. Hurwitz J, Furth JJ, Malamy M, & Alexander M (1962) The role of deoxyribonucleic acid in ribonucleic acid synthesis of ribonucleic acid and deoxyribonucleic acid by actinomycin D and proflavin. Proc Natl Acad Sci USA 48:1222–1230
13. Goldberg IH, & Rabinowitz M (1962) Actinomycin D inhibition of deoxyribonucleic acid-dependent synthesis of ribonucleic acid. Science 136:315–316
14. Perreault J, & Lemieux R (1993) Rapid apoptotic cell death of B-cell hybridomas in absence of gene expression. J Cell Physiol 156:286–293

Anthracycline (Daunomycin/Adriamycin/Respinomycins A1 and A2)

Key Words: [Antitumor][Apoptosis][DNA scission][Differentiation]

Structure:

	Daunomycin	Adriamycin
Molecular Formula:	$C_{27}H_{29}NO_{10}$	$C_{27}H_{29}NO_{11}$
Molecular Weight:	528	544
Solubility:	DMSO, ++ ; H_2O, ++ ; MeOH, ++	

Respinomycin A1

Molecular Formula:	$C_{51}H_{72}N_2O_{20}$
Molecular Weight:	1,032
Solubility:	DMSO, ++ ; H_2O, ++ ; MeOH, ++

Discovery/Isolation:

Brockmann and Bauer isolated an antibiotic compound named rhodomycin from the culture broth of *Streptomyces purpurascens*. Rhodomycin was an antibiotic mixture and all components were glycoside derivatives based on the structure of 9-ethyl-7,8,9,10-tetrahydro-5,12-tetracenequinone. At present, anthracyclines are classified into such glycoside derivatives of 9-ethyl-, 9-acetyl-, 9-hydroxyacetyl-, 9-α-hydoxyethyl, 9-methyl-, or 9-acetonyl-7,8,9,10-tetrahydro-5,12-tetracenequinone.

Daunomycin (daunorubicin) was found as an antibiotic of the rhodomycin group in 1964 from the culture broth of *S. peucetius* [1]. Adriamycin (doxorubicin), a 14-hydroxy derivative of daunomycin, was isolated from the culture broth of *S. peucetius* var. *caesius*, obtained by mutagenic treatment from *S. peucetius*, the daunomycin producing strain [2].

Respinomycins A1 and A2 were isolated as anthracycline antibiotics that induce the terminal differentiation of a human leukemia cell, K-562 [3,4]. The structure of aglycone was unambiguously determined by Long-range Selective Proton decoupling (LSPD) experiments and Nuclear Overhauser and Exchange Spectroscopy (NOESY), and were revealed to have a common skeleton which is distinguished from that of the nogalamycin group [5].

Biological and Chemical Studies:

Marco et al. and other groups found that daunomycin strongly inhibited DNA and RNA synthesis, and that it had a strong tendency to form a complex with DNA both in vitro and in vivo [1, 6–8]. Adriamycin also showed the same effects. Someya and Tanaka discovered that the binding of adriamycin with DNA and the consequent cleavage of DNA. Adriamycin was observed to induce a single strand scission of DNA in the presence of a reducing agent. The DNA cleavage was enhanced by Cu^{2+} and Fe^{2+}, but not significantly by Ni^{2+}, Zn^{2+}, Mg^{2+} and Ca^{2+}. Furthermore, the anthracycline-induced DNA cleavage was stimulated by H_2O_2. These results suggest that both the free radical of anthracycline quinones, produced by reduction and auto-oxidation of the quinone moiety, and the hydroxyl radical produced by H_2O_2 production directly react with DNA strands to participate in the DNA-cutting effect [9].

Someya et al. investigated the interaction with actin and heavy meromyosin (HMM) for elucidating the biochemical mechanism of anthracycline cardiomyopathy. They found that HMM and acto-HMM Mg^{2+}-ATPase reactions were inhibited by daunomycin and adriamycin and that induced G-actin polymerization [10]. Recently, Gamen et al. reported that anthracyclin drugs induced the activation of CPP32 and apoptosis in Jurkat cells by the Fas-independent pathway [11].

Respinomycins A1 and A2 showed high activity on the terminal differentiation of human leukemia K-562 cells, and strong antiphage activity against actinophage B of *Streptomyces griseus* [3].

Biological Activity:

Daunomycin

15–20 mg/kg: LD_{50} for intravenously administered in the mouse and rat [1]
0.01–0.1 mg/ml: complete in vitro inhibition of mitotic activity of normal and neoplastic cells [1]

Adriamycin

0.4 mM with 0.2 mM $CuCl_2$: DNA fragmentation in vitro [9]
$1.4 – 7.2 \times 10^4$ M^{-1}: association constants to actin monomer [10]
1 μM: apoptosis induction in human T-cell leukemia [11]

Respinomycin A1

ca. 1 μg/ml: ED_{50} for induction of differentiation of K-562 cells [4]
approx. 37.5 mg/kg: LD_{50} by intraperitoneal administered in mice [4]

References:

1. Marco AD, Silvestrini R, Gaetani M, et al., & Valentini L (1964) Daunomycin, a new antibiotic of the rhodomycin group. Nature 201:706–707
2. Arcamone F, Cassinelli G, Fantini G, et al., & Spalla C (1969) Adriamycin, 14-hydroxydaunomycin, a new antitumor antibiotic from *S. peucetius* var. *caesius*. Biotechnol Bioeng 11:1101–1110
3. Ubukata M, Osada H, Kudo T & Isono K (1993) Respinomycins A1, A2 B, C and D, a novel group of anthracycline antibiotics. I. Taxonomy, fermentation, isolation and biological activities. J Antibiot 46:936–941
4. Ubukata M, Tanaka C, Osada H & Isono K (1991) Respinomycin A1, a new anthracycline antibiotic. J Antibiot 44:1274–1276
5. Ubukata M, Uzawa J, Osada H & Isono K (1993) Respinomycins A1, A2, B, C and D, a novel group of anthracycline antibiotics. II. Physico-chemical properties and structure elucidation. J Antibiot 46:942–951
6. Hartmann G, Goller H, Koschel K, et al., & Kersten H (1964) Hemmung der DNA-abhangigen RNA-und DNA-Synthesis durch Antibiotica. Biochemische Zeitschrift 341:126–128
7. Calendi E, Marco AD, Reggiani M, et al., & Valentini L (1965) On physico-chemical interactions between daunomycin and nucleic acids. Biochim Biophys Acta 103:25–49
8. Kersten W & Kersten H (1965) Die Bindung von Daunomycin, Cinerubin und Chromomycin A3 an Nucleinsauren. Biochemische Zetschrift 341:174–183
9. Someya A & Tanaka N (1979) DNA strand scission induced by adriamycin and aclacinomycin A. J Antibiot 28:839–845
10. Someya A, Akiyama T, Misumi M & Tanaka N (1978) Interaction of anthracycline antibiotics with actin heavy mermyosin. Biochem Biophys Res Comm 85:1542–1550

11. Gamen S, Anel A, Lasierra P, et al., & Naval J (1997) Doxorubicin-induced apoptosis in human T-cell leukemia is mediated by caspase-3 activation in a Fas-independent way. FEBS Lett 417:360–364

Apoptolidin

Key Words: [Apoptosis]

Structure:

Molecular Formula: $C_{58}H_{96}O_{21}$
Molecular Weight: 1128
Solubility: DMSO, +++; H_2O, ±; MeOH, +++.

Discovery:

Apoptolidin was isolated as an apoptosis inducer in adenovirus E1A-transformed rat glia cells from *Nocardiopsis* sp. by Kim et al. [1]. The structure of apoptolidin was determined by detailed NMR studies including its stereochemistry [2].

Biological Studies:

Several oncogenes including myc, E2F, and the adenovirus E1A have been demonstrated to sensitize cells to apoptosis. Apoptolidin induced apoptotic cell death in cells transformed with the adenovirus type 12 oncogenes including E1A but not in normal cells or 3Y1 rat fibroblasts transformed with other oncogenes [1]. The adenovirus type qw E1B gene encodes two major proteins of 19kDa and 54 kDa, both of which independently can suppress apoptosis induced by E1A. However, expression of these gene products was insufficient to suppress apoptosis induced by apoptolidin. Further studies on the biological activities of apoptolidin are in progress.

Biological Activity:

$>100 \mu g/ml$: IC_{50} for glia cells [1]
10–13 ng/ml: IC_{50} for RG-E1A-7, RG-E1A19K-2, RG-E1A54K-9, or RG-E1-4 cells [1]

References:

1. Kim JW, Adachi H, Shin-ya K, et al., & Seto H (1997) Apoptolidin, a new apoptosis inducer in transformed cells from *Nocardiopsis* sp. J Antibiot 50:628–630
2. Hayakawa Y, Kim JW, Adachi H, et al., & Seto H (1998) Structure of apoptolidin, a specific apoptosis inducer in transformed cells. J Am Chem Soc 120:3524–3525

Arenastatin A

Key Words: [Antitumor][Microtubule inhibitor]

Structure:

Molecular Formula: $C_{34}H_{42}N_2O_8$
Molecular Weight: 606
Solubility: DMSO, + ; H_2O, - ; MeOH, ++

Discovery/Isolation:

Arenastatin A was isolated from the Okinawan murine sponge *Dysidea arenaria*.

Biological and Chemical Studies:

The chemical structure, including parts of the absolute configurations, showed that arenastatin A is a cyclic didepsipeptide [1]. Kobayashi et al. reported an efficient asymmetric synthesis of a cyclic depsipeptide, arenastatin A, and the structure-activity relationship among several stereoisomers [2].

Arenastatin A exhibited extremely potent cytotoxicity [1]. Morita et al. investigated the interaction of arenastatin A with tubulin by the use of [³H]arenastatin A and other microtubule disrupters [3]. Scatchard analysis indicated the presence of one binding site for arenastatin A per tubulin heterodimer. Rhizoxin was a competitive inhibitor of arenastatin A binding, and vinblastine also inhibited arenastatin A binding in a partially competitive manner. Furthermore, arenastatin A had no inhibitory effect on colchicine binding to tubulin [3].

Biological Activity:

5 pg/ml: IC_{50} for cytotoxicity against KB cells [1]
1.8×10^{-6} M: K_d value for binding to tubulin heterodimer [3]

References:

1. Kobayashi M, Aoki S, Ohyabu N, et al., & Kitagawa I (1994) Arenastatin A, a potent cytotoxic depsipeptide from the okinawan marine sponge *Dysidea arenaria.* Tetrahedron Lett 35:7969–7972
2. Kobayashi M, Wang W, Ohyabu N, et al., & Kitagawa I (1995) Improved total synthesis and structure-activity relationship of arenastatin A, a potent cytotoxic spongean depsipeptide. Chem Pharm Bull 43:1598–1600
3. Morita K, Koiso Y, Hashimoto Y, et al., & Iwasaki S (1997) Interaction of arenastatin A with porcine brain tubulin. Biol Pharm Bull 20:171–174

Azatyrosine

Key Words: [Antitumor]

Structure:

Molecular Formula: $C_8H_{10}N_2O_3$
Molecular Weight: 182
Solubility: DMSO, ++ ; H_2O, + ; MeOH, ++

Discovery/Isolation:

Azatyrosine was isolated from *Streptomyces chibanensis* as an antitumor compound [1].

Biological and Chemical Studies:

Azatyrosine inhibited the growth of NIH 3T3 cells transformed by the c-Ha-*ras* gene but did not significantly inhibit the growth of normal NIH 3T3 cells [1]. Surprisingly, azatyrosine induced most of the transformed cells to be apparently normal. These cells grew in the presence of azatyrosine and stopped growing when they reached confluency, and their normal phenotype persisted during prolonged culture in the absence of azatyrosine [2]. Kyprianou and Taylor-Papadimitriou isolated non-transformed revertant clones from the *ras*-transformed MTSV1-7 cell line after treatment with azatyrosine. The azatyrosine-induced revertants were considered non-transformed but all revertants sustain elevated levels of p21*ras* protein. As the expression of the K-rev-1 gene, a known tumour-suppressor gene, was significantly increased in the all revertants compared with the ras-transformed MTSV1-7 cells, they suggested that tumorigenic transformation of human mammary epithelial cells by v-H-*ras* may be influenced by the level of expression of the tumour-suppressor gene, K-rev-1 [3].

Chung et al. demonstrated that azatyrosine inhibited the p21*ras* protein-induced maturation of *Xenopus* oocytes. They concluded that azatyrosine have potent anti-ras effects intracellularly, because azatyrosine did not inhibit oocyte maturation induced by progesterone, which is known to initiate oocyte maturation by ras-independent pathways [4]. However, Campe et al. reported contradictory results that azatyrosine suppressed meiotic maturation in oocytes induced by both progesterone and the combination of [Val12]p21*ras* microinjection and insulin-like growth factor I [5].

Recently, Fujita-Yoshigaki et al. reported that azatyrosine inhibited differentiation-associated growth arrest of PC12 cells induced by oncogenic Ras but not normal

Ras. They suggested that azatyrosine sensitivity was the result of abnormal signal transduction by oncogenic Ras [6].

Biological Activity:

0.5 mg/ml: non-transformed reversion of *ras*-transformed cell line [1–3]
20–250 µM: dose-dependent suppression of meiotic maturation in oocyte [5].
0.5 mg/ml: inhibition of neurite outgrowth of PC12 transfected with [Val^{12}]p21^{ras} or v-*raf* [6]

References:

1. Inouye S, Shomura T, Tsuruoka T, et al., & Watanabe H (1975) L-β-(5-hydroxy-2-pyridyl)-alanine and L-β-(3-hydroxyureido)-alanine from *Streptomyces*. Chem Pharm Bull 23:2669–2677
2. Shindo-Okada N, Makabe O, Nagahara H & Nishimura S (1989) Permanent conversion of mouse and human cells transformed by activated *ras* or *raf* genes to apparently normal cells by treatment with the antibiotic azatyrosine. Mol Carcinog 2:159–167
3. Kyprianou N & Taylor-Papadimitriou J (1992) Isolation of azatyrosine-induced revertants from ras-transformed human mammary epithelial cells. Oncogene 7:57–63
4. Chung DL, Brandt-Rauf P, Murphy RB, et al., & Pincus MR (1991) A peptide from the GAP-binding domain of the ras-p21 protein and azatyrosine block ras- induced maturation of *Xenopus* oocytes. Anticancer Res 11:1373–1378
5. Campa MJ, Glickman JF, Yamamoto K & Chang K-J (1992) The antibiotic azatyrosine suppresses progesterone or [Val^{12}]p21 Ha-*ras*/insulin-like growth factor I-induced germinal vesicle breakdown and tyrosine phosphorylation of *Xenopus* mitogen-activated protein kinase in oocytes. Proc Natl Acad Sci USA 89:7654–7658
6. Fujita-Yoshigaki J, Yokoyama S, Shindo-Okada N & Nishimura S (1992) Azatyrosine inhibits neurite outgrowth of PC12 cells induced by oncogenic Ras. Oncogene 7:2019–2024

Bactobolin
(Bactobolin and Actinobolin)

Key Words: [Antitumor][Immunosuppressant]

Structure:

Molecular Formula: $C_{14}H_{20}N_2O_6Cl_2$
Molecular Weight: 382
Solubility: DMSO, ± ; H_2O, ++ ; MeOH, ++

Discovery:

Bactobolin was isolated as an antitumor antibiotic [1], and is structurally related to actinobolin [2,3].

Biological Studies:

Bactobolin and actinobolin are structurally related, but differ considerably in biological activity including toxicity, bactobolin being more potent than actinobolin [4]. Bactobolin prolonged the survival period of mice bearing leukemia L-1210 in various dose schedules, and showed stronger suppressive action on antibody formation in vitro [5]. Bactobolin also demonstrated a prophylactic and therapeutic effect on autoimmune encephalomyelitis [6]. Suppressive effects of actionbolin on autoimmune myasthenia gravis was also reported [7].

Several bactobolin and actinobolin derivatives were synthesized, and their structure-activity relationships were investigated [8–11].

Biological Activity:

1.3×10^{-7} M: LD_{50} for cell growth of L-1210 [4]
0.02 µg/ml: ID_{50} for suppression of antibody formation [5]
0.5 mg/kg/day: prevention of the inductive phase autoimmune encephalomyelitis [6]

References:

1. Kondo S, Horiuchi Y, Hamada M, et al., & Umezawa H (1979) A new antitumor antibiotic, bactobolin produced by *Pseudomonas*. J Antibiot 32:1069–1071
2. Antosz FJ, Nelson DB, Herald JDL & Munk ME (1970) The structure and chemistry of actinobolin. J Am Chem Soc 92:4933–4942
3. Haskell TH & Bartz QR (1959) Actinobolin, a new broad-spectrum antibiotic. Isolation and characterization. Antibiot Ann 1958/1959:505–509
4. Hori M, Suzukake K, Ishikawa C, et al., & Umezawa H (1981) Biochemical studies on bactobolin in relation to actinobolin. J Antibiot 34:465–468
5. Ishizuka M, Fukasawa S, Masuda T, et al., & Umezawa H (1980) Antitumor effect of bactobolin and its influence on mouse immune system and hematopoietic cells. J Antibiot 33:1054–1062
6. Tabira T, Da-Lin Y, Yamamura T & Aoyagi T (1987) Prophylactic and therapeutic effect of bactobolin on autoimmune encephalomyelitis. Proc Jpn Acad 63, Ser. B.:127–130
7. Ishigaki Y, Sato T, Song DL, et al., & Aoyagi T (1992) Suppression of experimental autoimmune myasthenia gravis with new immunosuppressants: 15-deoxyspergualin and actinobolin. J Neurol Sci 112:209–215
8. Adachi H, Nishimura Y, Kondo S & Takeuchi T (1998) Synthesis and activity of 3-epi-actinobolin. J Antibiot 51:202–209
9. Adachi H, Usui T, Nishimura Y, et al., & Takeuchi T (1998) Synthesis and activities of bactobolin derivatives based on the alteration of the functionality at C-3 position. J Antibiot 51:184–188
10. Munakata T & Okumoto T (1981) Some structure-activity relationships for bactobolin analogs in the treatment of mouse leukemia P388. Chem Pharm Bull 29:891–894
11. Nishimura Y, Kondo S & Takeuchi T (1992) Syntheses and activities of some bactobolin derivatives. J Antibiot 45:735–741

BE16627B

Key Words: [Antitumor] [Metalloproteinase inhibitor]

Structure:

Molecular Formula: $C_{16}H_{29}N_3O_7$
Molecular Weight: 375
Solubility: DMSO, ++ ; H_2O, + ; MeOH, ++

Discovery/Isolation:

BE16627B was isolated as a metalloproteinase (MP) inhibitor from *Streptomyces* sp. Synthesis and structure-activity relationships of gelatinase inhibitors were investigated [1].

Biological and Chemical Studies:

Naito et al. analyzed the effects of BE16627B on the primary cultures of synovial cells from rheumatoid arthritis patients, and showed that BE16627B inhibited MP activity in the primary culture supernatants from synovial cells in a dose-dependent fashion without showing apparent cytotoxicity or affecting the production and secretion of MPs. They also investigated the effects on human tumor cell growth in nude mice and showed that BE16627B inhibited metalloproteinase-dependent human tumor-cell growth as well as lung colonization without showing cytotoxicity in nude mice [2].

Biological Activity:

25 μM: IC_{50} for active collagenolysis before activation by trypsin [2]
>1,000 mg/kg: LD_{50} in mice (*i.p.*) [2]
0.58 μM: IC_{50} for gelatinase A [3]
0.85 μM: IC_{50} for gelatinase B [3]

References:

1. Tamaki K, Tanzawa K, Kurihara S, et al., & Sugimura Y (1995) Synthesis and structure-activity relationships of gelatinase inhibitors derived from matlystatins. Chem Pharm Bull 43:1883–1893
2. Naito K, Nakajima S, Kanbayashi N, et al., & Goto M (1993) Inhibition of metalloproteinase activity of rheumatoid arthritis synovial cells by a new inhibitor [BE16627B; 1-N-(N-hydroxy-2-isobutylsuccinamoyl)-seryl-l-valine]. Agents Actions 39:182–186
3. Okuyama A, Naito K, Morishima H, et al., & Tanaka N (1994) Inhibition of growth of human tumor cells in nude mice by a metalloproteinase inhibitor. Ann NY Acad Sci 732:408–410

Benthocyanins A, B, and C

Key Words: [Free radical scavenger]

Structure: shows benthocyanin A

Molecular Formula: $C_{31}H_{28}N_2O_4$
Molecular Weight: 492
Solubility: DMSO, ++ ; H_2O, ± ; MeOH, ++

Discovery:

Benthocyanin A was isolated as a potent radical scavenger from *Streptomyces pruicolor* [1]. It possesses an unprecedented structure consisting of a highly conjugated furophenazine ring and a geranyl side chain. Benthocyanins B and C also reported as minor congeners of benthocyanin A [2].

Biological Studies:

Benthocyanins A, B, and C inhibited lipid peroxidation in rat liver microsomes at low concentrations, which were 30–70 times as strong as that of vitamin E. They also showed growth inhibition on rat erythrocyte hemolysis [2].

Biological Activity:

0.16 and 0.29 µg/ml: IC_{50} for lipid peroxidation in rat liver microsomes of benthocyanin B and C [2]

References:

1. Shin-ya K, Furihata K, Hayakawa Y, et al., & Clardy J (1991) The structure of benthocyanin A. A new free radical scavenger of microbial origin. Tetrahedron Lett 32:943–946
2. Shin-ya K, Furihata K, Teshima Y, et al., & Seto H (1993) Benthocyanins B and C, new free radical scavengers from *Streptomyces prunicolor.* J Org Chem 58:4170–4172

Bestatin

Key Words: [Aminopeptidase inhibitor][Immunomodulator]

Structure:

Molecular Formula: $C_{16}H_{24}N_2O_4$
Molecular Weight: 308
Solubility: DMSO, ++ ; H_2O, - ; MeOH, +++

Discovery:

Bestatin (Ubenimex®) was isolated as a competitive inhibitor of aminopeptidase B and leucine aminopeptidase [1], and its structure was determined [2]

Biological Studies:

Bestatin is a competitive inhibitor of aminopeptidases, and one bestatin is bound per subunit in leucine aminopeptidase. The mode of binding is slow and competitive [3]. In intact mouse reticulocytes, bestatin accumulated the low molecular weight intermediates in the degradation of protein [4]. Bestatin was also a potent and reversible inhibitor of leukotriene A_4 (LTA$_4$) hydrolase. Bestatin inhibited LTA$_4$ hydrolase selectively; neither 5-lipoxygenase nor 15-lipoxygenase activity in neutrophil lysates was affected. Purified LTA$_4$ hydrolase exhibited an intrinsic aminopeptidase activity, and both LTA$_4$ and bestatin suppressed the intrinsic aminopeptidase activity of LTA$_4$ hydrolase [5]. Bestatin shows antitumor activity in vivo, probably through both the activation of macrophages [6] and the generation of cytotoxic T cells and NK cells in the spleen [7].

The structure of both leucine aminopeptidase and leukotriene A_4 hydrolase complexed with bestatin were reported [8,9].

Biological Activity:

$0.05 \mu g/ml$: IC_{50} for aminopeptidase B [1]

$0.01 \mu g/ml$: IC_{50} for leucine aminopeptidase [1]

$1.1 \times 10^{-7} M$: K_i for formation of the initial and final complexes with leucine aminopeptidase [3]

201 ± 95 mM: K_i value for leukotriene A_4 hydrolase [5]

172 nM: K_i values for the intrinsic aminopeptidase activity of LTA_4 hydrolase [5]

References:

1. Umezawa H, Aoyagi T, Suda H, et al., & Takeuchi T (1976) Bestatin, an inhibitor of aminopeptidase B, produced by actinomycetes. J Antibiot 29:97–99
2. Suda H, Takita T, Aoyagi T & Umezawa H (1976) The structure of bestatin. J Antibiot 29:100–101
3. Taylor A, Peltier CZ, Torre FJ & Hakamian N (1993) Inhibition of bovine lens leucine aminopeptidase by bestatin: number of binding sites and slow binding of this inhibitor. Biochemistry 32:784–790
4. Botbol V & Scornik OA (1979) Degradation of abnormal proteins in intact mouse reticulocytes: accumulation of intermediates in the presence of bestatin. Proc Natl Acad Sci USA 76:710–713
5. Orning L, Krivi G & Fitzpatrick FA (1991) Leukotriene A_4 hydrolase. Inhibition by bestatin and intrinsic aminopeptidase activity establish its functional resemblance to metallohydrolase enzymes. J Biol Chem 266:1375–1378
6. Talmadge JE, Lenz BF, Pennington R, et al., & Tribble H (1986) Immunomodulatory and therapeutic properties of bestatin in mice. Cancer Res 46:4505–4510
7. Ishizuka M, Masuda T, Mizutani S, et al., & Umezawa H (1987) Antitumor cells found in tumor-bearing mice given ubenimex. J Antibiot 40:697–701
8. Burley SK, David PR, Sweet RM, et al., & Lipscomb WN (1992) Structure determination and refinement of bovine lens leucine aminopeptidase and its complex with bestatin. J Mol Biol 224:113–140
9. Tsuge H, Ago H, Aoki M, et al., & Shimizu T (1994) Crystallization and preliminary X-ray crystallographic studies of recombinant human leukotriene A_4 hydrolase complexed with bestatin. J Mol Biol 238:854–856

Bleomycin

Key Words: [Antitumor][Apoptosis][Cell cycle inhibitor]

Structure: Bleomycin A₂ (Main component of clinically used bleomycin (BLM); 55–70%)

Molecular Formula: $C_{55}H_{84}N_{17}O_{21}S_3$
Molecular Weight: 1416
Solubility: DMSO, + ; H_2O, ++ ; MeOH, ++

Discovery/Isolation:

Bleomycin was isolated from the culture broth of *Streptomyces verticillus* as an antitumor phleomycin-like compound [1,2].

Chemical Studies:

BLMs are a mixture of bleomycins [3], which were different in the terminal amine moiety. The activity of synthetic analogues and biosynthetic intermediates were studied [4].

Biological Studies:

BLM showed moderate to marked effects upon Rous sarcoma virus-induced mouse ascites sarcoma [5]. BLM inhibited DNA syntheses in Ehrlich carcinoma and HeLa cells, and induced single strand scission of DNA in the presence of reducing agents in vitro [6,7]. These results suggest that the BLM complexed with Fe(II) and O_2 is activated to active form "BLM-Fe(III)- O_2H"by reduction and induces scission of single strand DNA [8–10]. The distribution of BLM in mouse organs was studied and found at a high concentration in skin [11]. The selective effect of BLM on mouse skin carcinoma might be due to the low ability to inactivate BLM. The activity of BLM-inactivating enzyme, which hydrolyzed the carboxyl amide group in BLMs in animal tissues, was significantly lower in protein extracted from squamous cell carcinoma of mice skin than in that extracted from sarcoma [12]. BLM also induced apoptosis via the CD95 ligand/receptor system [13].

Biological Activity:

0.78 mg/kg/day: 50% inhibition of solid tumor SR-C3H/He sarcoma in C3H/He mice [6]
1.1μM: free base formation from DNA in vitro [9]
3 mg/ml: apoptosis induction in HepG2 cells [13]

References:

1. Umezawa H, Suhara Y, Takita T & Maeda K (1966) Purification of bleomycins. J Antibiot Ser A 19:210–215
2. Umezawa H, Maeda K, Takeuchi T & Okami Y (1966) New antibiotics, bleomycin A and B. J Antibiot Ser A 19:200–209
3. Umezawa H (1974) Chemistry and mechanism of action of bleomycin. Federation Proceedings 33:2296–2302
4. Sugiura Y, Suzuki T, Otsuka M, et al., & Umezawa H (1983) Synthetic analogues and biosynthetic intermediates of bleomycin metal-binding, dioxygen interaction, and implication for the role of functional groups in bleomycin action mechanism. J Biol Chem 258:1328–1336
5. Takeuchi M & Yamamoto T (1968) Effects of bleomycin on mouse transplantable tumors. J Antibiot 21:631–637
6. Suzuki H, Nagai K, Akutsu E, et al., & Umezawa H (1970) On the mechanism of action of bleomycin. Strand scission of DNA caused by bleomycin and its binding to DNA in vitro. J Antibiot 23:473–480
7. Lown JW & Sim S-K (1977) The mechanism of the bleomycin-induced cleavage of DNA. Biochem Biophys Res Comm 77:1150–1157
8. Suzuki H, Nagai K, Yamaki H, et al., & Umezawa H (1969) On the mechanism of action of bleomycin: Scission of DNA strands in vitro and in vivo. J Antibiot 22:446–448
9. Wu JC, Kozarich JW & Stubbe J (1983) The mechanism of free base formation from DNA by bleomycin. J Biol Chem 258:4694–4697
10. Burger RM, Projan SJ, Horwitz SB & Peisach J (1986) The DNA cleavage mechanism of iron-bleomycin. J Biol Chem 261:15955–15959
11. Umezawa H, Ishizuka M, Hori S, et al., & Takeuchi T (1968) The distribution of ^{3}H-bleomycin in mouse tissue. J Antibiot 21:638–642
12. Umezawa H, Takeuchi T, Hori S, et al., & Komai T (1972) Studies on the mechanism of antitumor effect of bleomycin on squamous cell carcinoma. J Antibiot 25:409–420
13. Hug H, Strand S, Grambihler A, et al., & Galle PR (1997) Reactive oxygen intermediates are involved in the induction of CD95 ligand mRNA expression by cytostatic drugs in hepatoma cells. J Biol Chem 272:28191–28193

Brefeldin A

Key Words: [Antifungal][Antiviral][Protein synthesis inhibitor][Transport inhibitor]

Structure:

Molecular Formula: $C_{16}H_{24}O_4$
Molecular Weight: 280
Solubility: DMSO, ± ; H2O, - ; MeOH, ++

Discovery/Isolation:

Brefeldin A (BFA; primarily named decumbin) was isolated from *Penicillium* [1] and its structure were determined by chemical degradation [2]. BFA was reisolated as an antifungal antibiotic by Hayashi et al. [3].

Biological Studies:

Hayashi et al. investigated the antifungal mechanism and found that the activity was due to the inhibition or disturbance of lipid metabolism [3]. BFA is a unique reagent, which impedes protein transport from the ER to the Golgi complex [4]. The non-clathrin coatomer complex binds reversibly to Golgi membranes and ARF (ADP-ribosylation factor) is required for coatomer binding. Recently, several groups reported that BFA inhibited ARF1-GAP. BFA caused the rapid and reversible dissociation of β-COP and ARF [5,6], and inhibited the function of Golgi membranes, which catalysed the exchange of GTP onto ARF [7]. Helms and Rothman reported an enzyme in a Golgi-enriched fraction that catalyses GDP/GTP exchange on ARF-1 protein, and which is inhibited by BFA [8]. Cukierman et al. cloned ARF1-GAP and showed that BFA prevented the localization of ARF1-GAP [9]. The other effects of BFA, such as protein synthesis inhibition and apoptosis induction, were also reported by several groups [10].

Application:

BFA is a good bioprobe for investigating the transport system, especially from the endoplasmic reticulum to the Golgi network. Nuchtern et al. showed that the influenza A matrix protein can be effectively presented to class II-restricted T cells by two pathways, one of which is chloroquine-sensitive and BFA-insensitive, the other being chloroquine-insensitive and BFA-sensitive [11]. Rubartelli et al. suggested a novel secretory pathway for interleukin-1β, a protein lacking a signal sequence [12].

Biological Activity:

1 µg/ml: secretion block in primary culture of rat hepatocytes [4]
0.1–1 mg/ml: protein synthesis inhibition in rat glioma C6 cells by up to 70% [10]
10 µM: 50–60% inhibition of the membrane-dependent nucleotide-exchange reaction [7,8]
25 µM: 50% inhibition of binding of myristoylated ARF to Golgi membrane [8]

References:

1. Singleton VL, Bohonos N & Ullstrup AJ (1958) Decumbin, a new compound from a species of *Penicillium*. Nature 181:1072–1073
2. Sigg HP (1964) 152. Die Konstitution von Brefeldin A. Helv Chim Acta 47:1401–1415
3. Hayashi T, Takatsuki A & Tamura G (1974) The action mechanism of brefeldin A. I. Growth recovery of *Candida albicans* by lipids from the action of brefeldin A. J Antibiot 27:65–72
4. Misumi Y, Misumi Y, Miki K, et al., & Ikehara Y (1986) Novel blockade by brefeldin A of intracellular transport of secretory proteins in cultured rat hepatocytes. J Biol Chem 261:11398–11403
5. Donaldson JG, Lippincott-Schwartz J, Bloom GS, et al., & Klausner RD (1990) Dissociation of a 110-kD peripheral membrane protein from the Golgi apparatus is an early event in brefeldin A action. J Cell Biol 111:2295–2306
6. Serafini T, Stenbeck G, Brecht A, et al., & Wieland FT (1991) A coat subunit of Golgi-derived non-clathrin-coated vesicles with homology to the clathrin-coated vesicle coat protein β-adaptin. Nature 349:215–220
7. Donaldson JG, Finazzi D & Klausner RD (1992) Brefeldin A inhibits Golgi membrane-catalysed exchange of guanine nucleotide onto ARF protein. Nature 360:350–352
8. Helms JB & Rothman JE (1992) Inhibition by brefeldin A of a Golgi membrane enzyme that catalyses exchange of guanine nucleotide bound to ARF. Nature 360:352–354
9. Cukierman E, Huber I, Rotman M & Cassel D (1995) The ARF1 GTPase-activating protein: Zinc finger motif and Golgi complex localization. Science 270:1999–2002
10. Fishman PH & Curran PK (1992) Brefeldin A inhibit protein synthesis in cultured cells. FEBS Lett 314:371–374
11. Nuchtern JG, Biddison WE & Klausner RD (1990) Class II MHC molecules can use the endogenous pathway of antigen presentation. Nature 343:74–76
12. Rubartelli A, Cozzolino F, Talo M & Sitia R (1990) A novel secretory pathway for interleukin-1β, a protein lacking a signal sequence. EMBO J 9:1503–1510

Butyrolactone I

Key Words: [Antitumor][Cell cycle inhibitor]

Structure:

Molecular Formula: $C_{24}H_{24}O_7$
Molecular Weight: 424
Solubility: DMSO, ++ ; H_2O, - ; MeOH, ++

Discovery/Isolation:

A selective inhibitor of cdc2 kinase was isolated from the cultured medium of *Aspergillus* species using purified mouse cyclin B-cdc2 kinase and a specific substrate peptide for cdc2 kinase, and identified as butyrolactone I [1].

Biological and Chemical Studies:

Butyrolactone I inhibited cdc2 and cdk2 kinases but it had little effect on mitogen-activated protein kinase, protein kinase C, cyclic-AMP dependent kinase, casein kinase II, casein kinase I or epidermal growth factor-receptor tyrosine kinase. Its inhibitory effect was found to be due to competition with ATP. Butyrolactone I selectively inhibited the phosphorylation of H1 histone and retinoblastoma product in vitro [1].

Using this agent, several groups investigated the function of the cdc2 kinase family in cell cycle regulation [2–5], apoptosis [6,7], and microtubule dynamics [8,9]. Nishio et al. also reported the antitumor activity of butyrolactone I [10].

Biological Activity:

0.68 and 94 µM: IC_{50} for cdc2 kinase and MAP kinase [1]
0.36 µM: K_i value against ATP, competitive inhibition [1]
10 µM: prevention of the oscillation of DNA replication in *Xenopus* egg extracts [5]
5 µM: prevention of the Fas-induced apoptosis in HL-60 cells [6]

References:

1. Kitagawa M, Okabe T, Ogino H, et al., & Kokubo T (1993) Butyrolactone I, a selective inhibitor of cdk2 and cdc2 kinase. Oncogene 8:2425–2432
2. Poon RY, Chau MS, Yamashita K & Hunter T (1997) The role of Cdc2 feedback loop control in the DNA damage checkpoint in mammalian cells. Cancer Res 57:5168–5178
3. Kitagawa M, Higashi H, Suzuki-Takahashi I, et al., & Okuyama A (1995) Phosphorylation of E2F-1 by cyclin A-cdk2. Oncogene 10:229–236
4. Kitagawa M, Higashi H, Takahashi IS, et al., & Okuyama A (1994) A cyclin-dependent kinase inhibitor, butyrolactone I, inhibits phosphorylation of RB protein and cell cycle progression. Oncogene 9:2549–2557
5. Someya A, Tanaka N & Okuyama A (1994) Inhibition of cell cycle oscillation of DNA replication by a selective inhibitor of the cdc2 kinase family, butyrolactone I, in *Xenopus* egg extracts. Biochem Biophys Res Commun 198:536–545
6. Furukawa Y, Iwase S, Terui Y, et al., & Kitagawa M (1996) Transcriptional activation of the cdc2 gene is associated with Fas-induced apoptosis of human hematopoietic cells. J Biol Chem 271:28469–28477
7. Lu Y, Yamagishi N, Yagi T & Takebe H (1998) Mutated p21[WAF1/CIP1/SDI1] lacking CDK-inhibitory activity fails to prevent apoptosis in human colorectal carcinoma cells. Oncogene 16:705–712
8. Hosoi T, Uchiyama M, Okumura E, et al., & Hisanaga S (1995) Evidence for cdk5 as a major activity phosphorylating tau protein in porcine brain extract. J Biochem 117:741–749
9. Ookata K, Hisanaga S, Sugita M, et al., & Kishimoto T (1997) MAP4 is the in vivo substrate for CDC2 kinase in HeLa cells: identification of an M-phase specific and a cell cycle-independent phosphorylation site in MAP4. Biochemistry 36:15873–15883
10. Nishio K, Ishida T, Arioka H & Kurokawa H (1996) Antitumor effects of butyrolactone I, a selective cdc2 kinase inhibitor, on human lung cancer cell lines. Anticancer Res 16:3387–3395

Calphostin C

Key Words: [Antitumor][Cell cycle inhibitor]

Structure:

Molecular Formula: $C_{44}H_{38}O_{14}$
Molecular Weight: 790
Solubility: DMSO, ++ ;H_2O, -; MeOH, +

Discovery/Isolation/Structure:

A novel complex of calphostin, which specifically inhibits protein kinase C, was isolated from the culture broth of a fungi *Cladosporium cladosporioides*. This complex was divided into five closely related components, A, B, C, D and I [1,2]. The structures of calphostins were determined by spectral and chemical studies and all of them were found to have a 3,10-perylenequinone skeleton, while calphostin D was revealed to be an antipode of isophleichrome [3].

Biological and Chemical Studies:

Calphostins showed cytotoxic activities against various tumor cells, and these cytotoxicities were proportional to their inhibitory activities against protein kinase C [2]. Calphostin C is a polycyclic hydrocarbon with strong absorbance in the visible and ultraviolet ranges. Interestingly, the inhibition of protein kinase C activity in cell-free systems and intact cells also required light. Light-dependent cytotoxicity was seen at concentrations about 5-fold higher than those inhibiting protein kinase C [4]. Many groups have been using calphostin C for the analysis of the function of protein kinase C in vivo. Chmura et al. report that WEHI-231 undergo apoptosis following exposure to calphostin C, and ceramide production increased over baseline levels with a concurrent decrease in sphingomyelin. They suggested an antagonistic relationship between protein kinase C activity and ceramide in the signaling events preceding apoptosis [5].

Biological Activity:

50 nM: IC_{50} for protein kinase C [2]

0.18 and 0.14 µg/ml: IC_{50} for cytotoxicity of HeLa S3 and MCF-7 [6]

60 nM: IC_{50} for export of vesicular stomatitis virus glycoprotein and vesicle budding from ER in vivo and in vitro [7]

250 nM: apoptosis induction in WEHI-231 cells [5]

References:

1. Kobayashi E, Ando K, Nakano H & Tamaoki T (1989) UCN-1028A, a novel and specific inhibitor of protein kinase C, from *Cladosporium*. J Antibiot 42:153–155
2. Kobayashi E, Ando K, Nakano H, et al., & Tamaoki T (1989) Calphostins (UCN-1028), novel and specific inhibitors of protein kinase C. I. Fermentation, isolation, physico-chemical properties and biological activities. J Antibiot 42:1470–1474
3. Iida T, Kobayashi E, Yoshida M & Sano H (1989) Calphostins, novel and specific inhibitors of protein kinase C. II. Chemical structures. J Antibiot 42:1475–1481
4. Bruns RF, Miller FD, Merriman RL, et al., & Nakano H (1991) Inhibition of protein kinase C by calphostin C is light-dependent. Biochem Biophys Res Commun 176:288–293
5. Chmura SJ, Nodzenski E, Weichselbaum RR & Quintans J (1996) Protein kinase C inhibition induces apoptosis and ceramide production through activation of a neutral sphingomyelinase. Cancer Res 56:2711–2714
6. Kobayashi E, Nakano H, Morimoto M & Tamaoki T (1989) Calphostin C (UCN-1028C), a novel microbial compound, is a highly potent and specific inhibitor of protein kinase C. Biochem Biophys Res Commun 159:548–553
7. Fabbri M, Bannykh S & Balch WE (1994) Export of protein from the endoplasmic reticulum is regulated by a diacylglycerol/phorbol ester binding protein. J Biol Chem 269:26848–26857

Cerulenin

Key Words: [Antifungal][Lipid biosynthesis inhibitor]

Structure:

Molecular Formula: $C_{12}H_{17}NO_3$
Molecular Weight: 223
Solubility: DMSO, + ; H_2O, + ; MeOH, instable

Discovery/Isolation:

Cerulenin was isolated as an antifungal antibiotic from the culture broth of *Cephalosporium caerulens* [1]. The structure was determined by several spectrometries [2,3]. Total synthesis of *dl*-cerulenin was reported [4].

Biological and Chemical Studies:

Cerulenin is a potent and non-competitive inhibitor of fatty acid biosynthesis through irreversible inhibition of β-ketoacyl carrier protein synthetase [5,6]. Funabashi et al. determined the binding site of cerulenin in fatty acid synthetase using [³H]cerulenin and suggested inhibition an mechanism in that cerulenin, forming a hydroxylactam ring, reacts at its epoxide carbon with the SH-group of the cysteine residue in the condensing reaction domain of fatty acid synthetase [7]. Recently, it was shown that the binding site of cerulenin is the highly reactive substrate-binding Cys-204 by detailed investigation [8], and the structure of the complex between the cerulenin and β-ketoacyl carrier protein synthase was revealed [9].

 Cerulenin also shows several inhibitory effects on antigen processing in antigen-presenting cells [10], HIV-1 proteinase in vitro [11], protein palmitoylation and insulin internalization [12].

Biological Activity:

30 µM: IC_{50} for *E. coli* fatty acid synthetase [5]
5 µM: IC_{50} for *E. coli* β-ketoacyl-acyl carrier protein synthetase [5]
2.5 mM: IC_{50} for HIV-1 protease in vitro [11]
0.3 mM: 85% inhibition for insulin internalization in rat adipocytes [12]

References:

1. Sato Y, Nomura S, Kamio Y, et al., & Hata T (1967) Studies on cerulenin, III. Isolation and physico-chemical properties of cerulenin. J Antibiot 20:344–348
2. Omura S, Katagiri M, Nakagawa A, et al., & Nomura S (1967) Studies on cerulenin. V. Structure of cerulenin. J Antibiot 20:349–354
3. Arison BH & Omura S (1974) Revised structure of cerulenin. J Antibiot 27:28–30
4. Boeckman RK, Jr. & Thomas EW (1977) A total synthesis of dl-cerulenin: a novel fatty acid antibiotic and lipid synthesis inhibitor. J Am Chem Soc 99:2805–2806
5. DíAgnolo G, Rosenfeld IS, Awaya J, et al., & Vagelos PR (1973) Inhibition of fatty acid synthesis by the antibiotic cerulenin. Specific inactivation of β-ketoacyl-acyl carrier protein synthetase. Biochim Biophys Acta 326:155–156
6. Vance D, Goldberg I, Mitsuhashi O & Bloch K (1972) Inhibition of fatty acid synthetases by the antibiotic cerulenin. Biochem Biophys Res Commun 48:649–656
7. Funabashi H, Kawaguchi A, Tomoda H, et al., & Iwasaki S (1989) Binding site of cerulenin in fatty acid synthetase. J Biochem 105:751–755
8. Child CJ & Shoolingin-Jordan PM (1998) Inactivation of the polyketide synthase, 6-methylsalicylic acid synthase, by the specific modification of Cys-204 of the beta-ketoacyl synthase by the fungal mycotoxin cerulenin. Biochem J 330:933–937
9. Moche M, Schneider G, Edwards P, et al., & Lindqvist Y (1999) Structure of the complex between the antibiotic cerulenin and its target, β-ketoacyl-acyl carrier protein synthase. J Biol Chem 274:6031–6034
10. Falo LD, Jr., Benacerraf B, Rothstein L & Rock KL (1987) Cerulenin is a potent inhibitor of antigen processing by antigen- presenting cells. J Immunol 139:3918–3923
11. Moelling K, Schulze T, Knoop MT, et al., & Pearl LH (1990) In vitro inhibition of HIV-1 proteinase by cerulenin. FEBS Lett 261:373–377
12. Jochen AL, Hays J & Mick G (1995) Inhibitory effects of cerulenin on protein palmitoylation and insulin internalization in rat adipocytes. Biochim Biophys Acta 1259:65–72

Colchicine

Key Words: [Antitumor][Cell cycle inhibitor][Microtubule inhibitor]

Structure:

Molecular Formula: $C_{22}H_{25}NO_6$
Molecular Weight: 399
Solubility: DMSO, + ; H_2O, + ; MeOH, +++

Discovery/Isolation:

Colchicine is a major alkaloid of *Colchicum autumnale* L, and used for medical treatment of gout. Extraction procedures were written by Chemnitius [1], and the structure were determined by Dewar [2].

Biological and Chemical Studies:

Colchicine has been used for medical treatment of acute gout and its mechanism seems to be that the drug inhibits degranulation of polymorphs which ingested urate crystals [3]. Colchicine is also known to inhibit mitosis in a wide variety of plant and animal cells by interfering with the structure of the mitotic spindle [4]. The binding protein was determined by using [³H]-labeled colchicine as a microtubule protein, tubulin [5–7]. Colchicine forms a complex with tubulin named TC (tubulin-colchicine complex) [8], and this complex inhibits microtubule elongation at both plus and minus ends by binding [9]. The binding site of colchicine on tubulin was determined by the direct photoaffinity labeling method as β-tubulin [10]. Recently, Bai et al. revealed that cysteine 354 of β-tubulin is a part of the binding site for the A ring of colchicine [11].

Biological Activity:

10^{-4} M: endoreduplication induction Chinese hamster cell line [12]
2.5×10^{-7} M: partial disorganization of the spindle in grasshopper neuroblast [13]

References:

1. Chemnitius F (1928) Zur Darstellung des Colchicins. J Prakt Chem 118:29–32
2. Dewar MJS (1945) Structure of colchicine. Nature 155:141–142
3. Rajan KT (1966) Lysosomes and gout. Nature 210:959–960
4. Eigisti OJ & Dustin P, Jr., *Colchicine*. 1955, Ames, Iowa: Iowa State College Press.
5. Wilson L & Friedkin M (1967) The biochemical events of mitosis. II. The in vivo and in vitro binding of colchicine in grasshopper embryos and its possible relation to inhibition of mitosis. Biochemistry 6:3126–3135
6. Borisy GG & Taylor EW (1967) The mechanism of action of colchicine. Binding of colchicine-^{3}H to cellular protein. J Cell Biol 34:525–533
7. Borisy GG & Taylor EW (1967) The mechanism of action of colchicine. Colchicine binding to sea urchin eggs and the mitotic apparatus. J Cell Biol 34:535–548
8. Margolis RL & Wilson L (1977) Addition of colchicine-tubulin complex to microtubule ends: the mechanism of substoichiometric colchicine poisoning. Proc Natl Acad Sci USA 74:3466–3470
9. Bergen LG & Borisy GG (1983) Tubulin-colchicine complex inhibits microtubule elongation at both plus and minus ends. J Biol Chem 258:4190–4194
10. Wolff J, Knipling L, Cahnmann HJ & Palumbo G (1991) Direct photoaffinity labeling of tubulin with colchicine. Proc Natl Acad Sci USA 88:2820–2824
11. Bai R, Pei XF, Boye O, et al., & Hamel E (1996) Identification of cysteine 354 of β-tubulin as part of the binding site for the A ring of colchicine. J Biol Chem 271:12639–12645
12. Rizzoni M & Palitti F (1973) Regulatory mechanism of cell division. I. Colchicine-induced endoreduplication. Exp Cell Res 77:450–458
13. Mueller GA, Gaulden ME & Drane W (1971) The effects of varying concentrations of colchicine on the progression of grasshopper neuroblasts into metaphase. J Cell Biol 48:253–265

Conagenin

Key Words: [Immunomodulator]

Structure:

HO H HO H H
 ... N COOH
   H O   CH$_2$OH

Molecular Formula: C$_{10}$H$_{19}$NO$_6$
Molecular Weight: 249
Solubility: DMSO, ++ ; H$_2$O, ++ ; MeOH, ++.

Discovery:

Conagenin was isolated as an immunomodulator from *Streptomyces roseosporus* [1].

Biological Studies:

Conagenin was isolated as an immunomodulator which stimulates the proliferation of Con A-treated rat splenic T cells [1]. Conagenin exclusively acts on activated T cells and stimulates them to promote DNA synthesis and to produce lymphokines in vivo and in vitro [2]. Conagenin also shows antitumor effects through activation of T cells and enhancement of generation of antitumor effector cells [3]. Kawatsu et al. showed that conagenin prevented reduction in the number of platelets in the peripheral blood of mice given several antitumor drugs through the production of lymphokines, which are responsible for thrombopoiesis [4], and it modulated inflammatory responses induced by 5-FU through production of anti-inflammatory cytokines such as IL-4 and IL-10 [5]. These effects resulted in the inhibition of the side effects of several antitumor drugs without a reduction of the cytotoxicity, and so antitumor activity was improved as seen by the increase in the number of surviving mice [6,7].

Biological Activity:

6.25–25 µg/ml: increased the incorporation of [^{3}H]TdR into Con A-treated cells [1]
5 mg/kg: prevention of number of leukocytes and platelets in mice treated with antitumor drugs [4]
5 mg/kg: improvement of efficacy of antitumor effects in prolonging the survival period [7]

References:

1. Yamashita T, Iijima M, Nakamura H, et al., & Iitaka Y (1991) Conagenin, a low molecular weight immunomodulator produced by *Streptomyces roseosporus*. J Antibiot 44:557–559
2. Kawatsu M, Yamashita T, Osono M, et al., & Takeuchi T (1993) T cell activation by conagenin in mice. J Antibiot 46:1687–1691
3. Kawatsu M, Yamashita T, Osono M, et al., & Takeuchi T (1993) Effect of conagenin in tumor bearing mice. Antitumor activity, generation of effector cells and cytokine production. J Antibiot 46:1692–1698
4. Kawatsu M, Yamashita T, Ishizuka M & Takeuchi T (1994) Effect of conagenin on thrombocytopenia induced by antitumor agents in mice. J Antibiot 47:1123–1129
5. Kawatsu M, Yamashita T, Ishizuka M & Takeuchi T (1997) Modulation by conagenin of inflammatory mediator productions in mice given 5-fluorouracil. Anticancer Res 17:917–922
6. Kawatsu M, Yamashita T, Ishizuka M & Takeuchi T (1996) Improvement of intestinal toxicity of 5-fluorouracil by conagenin, a low molecular immunomodulator. Anticancer Res 16:2937–2941
7. Kawatsu M, Yamashita T, Ishizuka M & Takeuchi T (1995) Improvement of efficacy of antitumor agents by conagenin. J Antibiot 48:222–225

Cryptophycin 1

Key Words: [Antitumor][Apoptosis][Microtubule inhibitor]

Structure:

Molecular Formula: $C_{35}H_{43}ClN_2O_8$
Molecular Weight: 655
Solubility: DMSO, ++ ; H_2O, - ; MeOH, ++

Discovery/Isolation/Structure:

Cryptophycin, the major cytotoxin in the blue-green alga *Nostoc* sp., was isolated as an antifungal and for antitumor activity against solid tumors implanted in mice [1]. The relative and absolute stereochemistry were established using a combination of chemical and spectral techniques [2].

Biological Studies:

Smith et al. found that incubation of cryptophycin resulted in reversible, dose-dependent inhibition of cell proliferation in parallel with the percentage of cells in mitosis in leukemia cells. They found cryptophycin treated cell showed marked depletion of cellular microtubules [3]. Kerksiek et al. investigated the effects of cryptophycin 1 on the microtubules in vitro, and found that the agent is an effective inhibitor of tubulin polymerization, causes tubulin to aggregate, and depolymerizes microtubules. As cryptophycin 1 inhibits vinblastine binding to tubulin, it appears that the cryptophycins may bind to the *Vinca* site in tubulin or to a site that overlaps with the *Vinca* site [4]. Several groups also investigated the effect of cryptophycin and suggested that cryptophycin disrupted the *Vinca* alkaloid site of tubulin; however, the molecular details of this interaction are distinct from those of other antimitotic drugs [5–7]. Panda et al. examined the effects of cryptophycin 1 on the dynamics of individual microtubules assembled to steady state by quantitative video microscopy. They suggested that cryptophycin 1 exerted its antiproliferative and antimitotic activity by binding reversibly and with high affinity to the ends of microtubules, perhaps in the form of a tubulin-cryptophycin 1 complex, resulting in the most potent suppression of microtubule dynamics [8]. Cryptophycin 1 also induces apoptosis [9].

The antitumor activity of synthetic analogs, cryptophycin-8 and -52, were also reported [10–12]

Biological Activity:

<10 pM: EC_{50} for inhibition of proliferation in L1210 leukemia cells [2]
3.9 µM: K_i for binding of [³H]vinblastine to tubulin, noncompetitive [6]
2.1 µM: K_i for binding of [³H]dolastatin 10 to tubulin, competitive [6]
50 pM: apoptosis induction[9]

References:

1. Schwartz RE, Hirsch CF, Sesin DF, et al., & Yudin K (1990) Pharmaceuticals from cultured algae. J Indust Microbiol 5:113–124
2. Trimurtulu G, Ohtani I, Patterson GML, et al., & Demchik L (1994) Total structures of cryptophycins, potent antitumor depsipeptides from the blue-green alga *Nostoc* sp. strain GSV 224. J Am Chem Soc 116:4729–4737
3. Smith CD, Zhang X, Mooberry SL, et al., & Moore RE (1994) Cryptophycin: a new antimicrotubule agent active against drug-resistant cells. Cancer Res 54:3779–3784
4. Kerksiek K, Mejillano MR, Schwartz RE, et al., & Himes RH (1995) Interaction of cryptophycin 1 with tubulin and microtubules. FEBS Lett 377:59–61
5. Smith CD & Zhang X (1996) Mechanism of action cryptophycin. Interaction with the *Vinca* alkaloid domain of tubulin. J Biol Chem 271:6192–6198
6. Bai R, Schwartz RE, Kepler JA, et al., & Hamel E (1996) Characterization of the interaction of cryptophycin 1 with tubulin: binding in the *Vinca* domain, competitive inhibition of dolastatin 10 binding, and an unusual aggregation reaction. Cancer Res 56:4398–4406
7. Mooberry SL, Taoka CR & Busquets L (1996) Cryptophycin 1 binds to tubulin at a site distinct from the colchicine binding site and at a site that may overlap the Vinca binding site. Cancer Lett 107:53–57
8. Panda D, Himes RH, Moore RE, et al., & Jordan MA (1997) Mechanism of action of the unusually potent microtubule inhibitor cryptophycin 1. Biochemistry 36:12948–12953
9. Mooberry SL, Busquets L & Tien G (1997) Induction of apoptosis by cryptophycin 1, a new antimicrotubule agent. Int J Cancer 73:440–448
10. Panda D, DeLuca K, Williams D, et al., & Wilson L (1998) Antiproliferative mechanism of action of cryptophycin-52: kinetic stabilization of microtubule dynamics by high-affinity binding to microtubule ends. Proc Natl Acad Sci USA 95:9313–9318
11. Schultz RM, Shih C, Wood PG, et al., & Ehlhardt WJ (1998) Binding of the epoxide cryptophycin analog, LY355703 to albumin and its effect on in vitro antiproliferative activity. Oncol Rep 5:1089–1094
12. Corbett TH, Valeriote FA, Demchik L, et al., & Moore RE (1996) Preclinical anticancer activity of cryptophycin-8. J Exp Ther Oncol 1:95–108

Cycloheximide

Key Words: [Antitumor][Protein synthesis inhibitor]

Structure:

Molecular Formula: $C_{15}H_{23}NO_4$
Molecular Weight: 281
Solubility: DMSO, ++ ; H_2O, ++ ; MeOH, ++

Discovery/Isolation/Structure:

Cycloheximide (actidione) was isolated from *Streptomyces griseus* as an antifungal antibiotic using *Saccharomyces pastorianus* ATCC 2366 as the test organism [1,2]. Its structure was elucidated by chemical modification [3,4]. Eisenbraun et al. elucidated the absolute configuration of actidione by rotatory dispersion measurements [5]. Johnson et al. reinvestigated the stereochemistry of the cycloheximide both chemically and by means of NMR spectroscopy [6,7].

Biological Studies:

Kerridge reported that cycloheximide inhibited both nucleic acid and protein synthesis in *Saccharomyces carlsbergensis* completely at the minimum growth inhibitory concentration [8]. Siegel and Sisler studied the effects on protein synthesis in vivo and in vitro. They concluded that the primary action of cycloheximide was the inhibition of protein synthesis and that other alterations in cellular metabolism in treated cells is a reflection of disrupted protein synthesis. Using a cell-free system, they suggested that 1) cycloheximide did not interfere with amino acid activation or transfer of activated amino acids to soluble-RNA, 2) the site of action apparently involves the transfer of amino acyl-soluble-RNA to the ribosomes and subsequent protein formation [9,10]. Colombo et al. investigated the effects of cycloheximide on protein synthesis using intact reticulocytes and a lysate cell-free system in detail [11]. Apoptosis inductive activity was also reported [12].

Biological Activity:

0.5–1.0 µg/ml : MIC for *Saccharomyces caqrlsbergensis* [8]
14 µM:IC_{60-70} for protein synthesis inhibition in rabbit reticulocytes [11]
25 µg/ml: apoptosis induction in B-cell hybridoma [12]

References:

1. Leach BE, Ford JH & Whiffen AJ (1947) Actidione, an antibiotic from *Streptomyces griseus*. J Am Chem Soc 69:474
2. Ford JH & Leach BE (1948) Actidione, an antibiotic from *Streptomyces griseus*. J Am Chem Soc 70:1223–1225
3. Kornfeld EC & Jones RG (1948) The structure of actidione, an antibiotic from *Streptomyces griseus*. Science 108:437–438
4. Kornfeld EC, Jones RG & Parke TV (1949) The structure and chemistry actidione, an antibiotic from *Streptomyces griseus*. J Am Chem Soc 71:150–159
5. Eisenbraun EJ, Osiecki J & Djerassi C (1958) On the absolute configuration of the antibiotic actidione. J Am Chem Soc 80:1261–1262
6. Johnson F & Starkovsky NA (1962) Glutarimide antibiotics. part III. The determination of the stereochemistry of the methyl groups of cycloheximide isomers by nuclear magnetic resonance spectroscopy. Tetrahedron Lett 25:1173–1177
7. Johnson F, Starkovsky NA & Carlson AA (1965) Glutarimide antibiotics. IX. The stereochemistry of the dihydrocycloheximides and the configuration of the hydroxyl group of cycloheximide. J Am Chem Soc 87:4612–4620
8. Kerridge D (1958) The effect of actidione and other antifungal agents on nucleic acid and protein synthesis in *Saccharomyces carlsbergensis*. J Gen Microbiol 19:497–506
9. Siegel MR & Sisler HD (1964) Site of action of cycloheximide in cells of *Saccharomyces pastorianus*. II. The nature of inhibition of protein synthesis in a cell-free system. Biochim Biophys Acta 87:83–89
10. Siegel MR & Sisler HD (1965) Site of action of cycloheximide in cells of *Saccharomyces pastorinus*. III. Further studies on the mechanism of action and the mechanism of resistance in *Saccharomyces* species. Biochim Biophys Acta 103:558–567
11. Colombo B, Felicetti L & Baglioni C (1965) Inhibition of protein synthesis by cycloheximide in rabbit reticulocytes. Biochem Biophys Res Commun 18:389–395
12. Perreault J & Lemieux R (1993) Rapid apoptotic cell death of B-cell hybridomas in absence of gene expression. J Cell Physiol 156:286–293

Cyclosporin A

Key Words: [Antifungal][Immunosuppressant][PP2B (calcineurin) inhibitor]

Structure:

Molecular Formula: $C_{62}H_{111}N_{11}O_{12}$
Molecular Weight: 1202
Solubility: DMSO, + ; H_2O, ± ; MeOH, +++

Discovery/Isolation:

Cyclosporins, peptide metabolites from *Trichoderma polysporum* (LINK ex PERS.) *Rifai* were isolated as antifungal antibiotics. The structure of cyclosporin A (CsA), the main cyclosporin, was determined by NMR and Mass spectrometry [1].

Biological and Chemical Studies:

CsA was originally isolated as an antifungal antibiotic; however, its remarkable immunosuppressive activity in several pharmacological models was discovered [1]. CsA inhibits DNA binding activity of NF-AT and causes an absence of interleukin 2 mRNA, which is involved in T cell activation [2,3].

The specific binding protein of CsA, named cyclophilin, was isolated and identified as the peptidyl-prolyl *cis-trans* isomerase (PPI) [4–7]. Furthermore, Lie et al. showed that the complex of CsA/cyclophilin binds to and inhibits calcineurin (PP2B) [8]. The inhibition of calcineurin, not PPI, is sufficient for inhibition of T cell activation [9]. McCaffrey et al. screened CsA-sensitive T cell transcription factor and found that NFATp is a target for calcineurin and CsA [10]. Shibasaki et al. showed that calcineurin and an unknown NF-AT kinase are important for nuclear shuttling of NF-AT4 using CsA [11].

CsA is also a potent inhibitor of the inner membrane permeability transition in mitochondria [12]. CsA binds to cyclophilin D, which is a component of mitochondrial permeability transition pore complex (PTPC), and inhibits Bax-induced apoptosis [13].

Biological Activity:

10 ng/ml: 90% inhibition of NF-AT binding in Jurkat nuclear extract [2]
25–50 pmol/mg protein: IC_{50} for large amplitude swelling of mitochondria [12]

References:

1. von Ruegger A, Kuhn M, Lichti H, et al. & von Wartburg A (1976) Cyclosporin A, ein immunosuppressiv wirksamer Peptidmetabolit aus *Trichoderma polysporum* (LINK ex PERS.) *Rifai.* Helv Chim Acta 59:1075–1092
2. Emmel EA, Verweij CL, Durand DB, et al., & Crabtree GR (1989) Cyclosporin A specifically inhibits function of nuclear proteins involved in T cell activation. Science 246:1617–20
3. Elliott JF, Lin Y, Mizel SB, et al., & Paetkau V (1984) Induction of interleukin 2 messenger RNA inhibited by cyclosporin A. Science 226:1439–41
4. Handschumacher RE, Harding MW, Rice J, et al., & Speicher DW (1984) Cyclophilin: a specific cytosolic binding protein for cyclosporin A. Science 226:544–7
5. Harding MW, Handschumacher RE & Speicher DW (1986) Isolation and amino acid sequence of cyclophilin. J Biol Chem 261:8547–55
6. Fischer G, Wittmann-Liebold B, Lang K, et al., & Schmid FX (1989) Cyclophilin and peptidyl-prolyl cis-trans isomerase are probably identical proteins. Nature 337:476–8
7. Takahashi N, Hayano T & Suzuki M (1989) Peptidyl-prolyl cis-trans isomerase is the cyclosporin A-binding protein cyclophilin. Nature 337:473–5
8. Liu J, Farmer JD, Jr., Lane WS, et al., & Schreiber SL (1991) Calcineurin is a common target of cyclophilin-cyclosporin A and FKBP- FK506 complexes. Cell 66:807–15
9. Clipstone NA & Crabtree GR (1992) Identification of calcineurin as a key signalling enzyme in T-lymphocyte activation. Nature 357:695–7
10. McCaffrey PG, Perrino BA, Soderling TR & Rao A (1993) NF-ATp, a T lymphocyte DNA-binding protein that is a target for calcineurin and immunosuppressive drugs. J Biol Chem 268:3747–52
11. Shibasaki F, Price ER, Milan D & McKeon F (1996) Role of kinases and the phosphatase calcineurin in the nuclear shuttling of transcription factor NF-AT4. Nature 382:370–3
12. Broekemeier KM, Dempsey ME & Pfeiffer DR (1989) Cyclosporin A is a potent inhibitor of the inner membrane permeability transition in liver mitochondria. J Biol Chem 264:7826–30
13. Marzo I, Brenner C, Zamzami N, et al., & Kroemer G (1998) Bax and adenine nucleotide translocator cooperate in the mitochondrial control of apoptosis. Science 281:2027–31

Cytogenin

Key Words: [Angiogenesis inhibitor][Antitumor][Immunomodulator]

Structure:

Molecular Formula: $C_{11}H_{10}O_5$
Molecular Weight: 222
Solubility: DMSO, +++ ; H_2O, - ; MeOH, +

Discovery:

Cytogenin was isolated as an antitumor substance from *Streptomyces eurocidicum* [1], and its biosynthesis was studied by Kumagai et al. [2].

Biological Studies:

Cytogenin was isolated as an antitumor substance, but also exhibits several interesting activities. Cytogenin activates the macrophages and T cells, and shows an antitumor effect via activation of inflammatory effects [3–5]. Cytogenin also has efficacy against both arthritis induced by injection of type II collagen in mice and adjuvant arthritis in rats, either of which is an animal experimental model for human rheumatoid arthritis, one of the angiogenesis-dependent diseases [6,7].

Oikawa et al. investigated the effects of cytogenin on pathological angiogenesis and showed that cytogenin is a novel oral antiangiogenic agent [8]. The mechanism of its antiangiogenic action contributes to its suppressive effects on both tumor growth and rheumatoid arthritis, and it could be developed as a potential therapeutic agent for cancer, rheumatoid arthritis and other angiogenesis-dependent disorders such as diabetic retinopathy.

Biological Activity:

6.3–100 mg/kg/day: retarded growth of solid tumor [1]
30, 100 mg/kg: a potent inhibitory effect on type II collagen-induced arthritis [6]
100 mg/kg oral dose once a day for 5 days: inhibition of angiogenic response 5 days after implantation of a chamber containing S-180 tumor cells, using the mouse dorsal air sac assay system [8]

References:

1. Kumagai H, Masuda T, Ohsono M, et al., & Takeuchi T (1990) Cytogenin, a novel antitumor substance. J Antibiot 43:1505–1507
2. Kumagai H, Amemiya M, Naganawa H, et al., & Takeuchi T (1994) Biosynthesis of antitumor antibiotic, cytogenin. J Antibiot 47:440–446
3. Kumagai H, Masuda T, Ishizuka M & Takeuchi T (1995) Antitumor activity of cytogenin. J Antibiot 48:175–178
4. Kumagai H, Osono M, Iijima M, et al., & Takeuchi T (1995) Action of cytogenin on lymphoid cells and their cytokine production. J Antibiot 48:317–320
5. Kumagai H, Masuda T, Sakashita M, et al., & Takeuchi T (1995) Modulation of macrophage activity in tumor bearing mice by cytogenin. J Antibiot 48:321–325
6. Hirano S, Wakazono K, Agata N, et al., & et al. (1994) Effects of cytogenin, a novel anti-arthritic agent, on type II collagen- induced arthritis in DBA/1J mice and adjuvant arthritis in Lewis rats. Int J Tissue React 16:155–162
7. Abe C, Hirano S, Wakazono K, et al., & et al. (1995) Effects of cytogenin on spontaneous arthritis in MRL/1 mice and on pristane-induced arthritis (PIA) in DBA/1J mice. Int J Tissue React 17:175–180
8. Oikawa T, Sasaki M, Inose M, et al., & Takeuchi T (1997) Effects of cytogenin, a novel microbial product, on embryonic and tumor cell-induced angiogenic responses in vivo. Anticancer Res 17:1881–1886

Cytotrienin A

Key Words: [Apoptosis]

Structure:

Molecular Formula: $C_{37}H_{48}N_2O_8$
Molecular Weight: 648
Solubility: DMSO, +++ ; H_2O, ± ; MeOH, +++

Discovery:

Zhang and Kakeya et al. isolated cytotrienin A and B as apoptosis inducers from *Streptomyces* sp. and the structures were determined by detailed spectral analysis [1,2]. Cytotreinins is the first metabolite with a unique 1-aminocyclopropane-1-carboxylic acid unit. From biosynthetic studies, this unit comes from l-methionine similar to the biosynthetic pathway suggested for higher plants [3].

Biological Studies:

Cytotrienin A was isolated as an apoptosis inducer in HL-60 cells [1,2]. Kakeya et al. investigated the mechanism of apoptosis induction [4]. They found that 36 kDa protein kinase (p36 MBP kinase), which phosphorylated MBP in a gel assay, and was activated in the course of apoptosis induced by cytotrienin. The p36 MBP kinase activation and apoptotic DNA fragmentation were inhibited by antioxidants such as *N*-acetylcysteine and reduced-form glutathione. As a broad specificity inhibitor of caspases (Z-Asp-CH₂-DCB) blocked the activation of p36 MBP kinase induced by cytotrienin A or H_2O_2, but did not inhibit the activation of JNK/SAPK (*see* p.75) and p38 MAPK, p36 MBP kinase activation is the downstream of the activation of Z-Asp-CH₂-DCB-sensitive caspases. The p36 MBP kinase was identified as active proteolytic products of MST1/KRS2 and MST2/KRS1. They also found that p36 MBP kinase was activated in response to anticancer drugs including camptothecin and etoposide as well as cytotrienin A. These results suggest that the p36 MBP kinase is a common component of the diverse signaling pathways leading to apoptosis.

Biological Activity:

10, and >1000 nM: IC_{50} for the proliferation of HL-60 and WI-38 cells, respectively [2]

References:

1. Zhang H-P, Kakeya H & Osada H (1997) Novel triene-ansamycins, cytotrienins A and B, inducing apoptosis on human leukemia HL-60 cells. Tetrahedron Lett 38:1789–1792
2. Kakeya H, Zhang HP, Kobinata K, et al., & Osada H (1997) Cytotrienin A, a novel apoptosis inducer in human leukemia HL-60 cells. J Antibiot 50:370–372
3. Zhang H-P, Kakeya H & Osada H (1998) Biosynthesis of 1-aminocyclopropane-1-carboxylic acid moiety on cytotrienin A in Streptomyces sp. Tetrahedron Lett 39:6947–6948
4. Kakeya H, Onose R & Osada H (1998) Caspase-mediated activation of a 36-kDa myelin basic protein kinase during anticancer drug-induced apoptosis. Cancer Res 58:4888–4894

Delaminomycin A1

Key Words: [Extracellular matrix receptor antagonist][Immunomodulator]

Structure:

Molecular Formula: $C_{29}H_{43}NO_6$
Molecular Weight: 501
Solubility: DMSO, +++ ; H_2O, - ; MeOH, +++

Discovery:

Delaminomycins were isolated as extracellular matrix receptor antagonists and a new class of potent immunomodulator from *Streptomyces albulus* [1], and their structures were determined by analyses of spectral properties and chemical studies [2,3]. The biosynthesis pathway of delaminomycins were also investigated [4].

Biological Studies:

Delaminomycins A, B and C were isolated as cell adhesion inhibitors of components of the extracellular matrix (ECM), fibronectin (FN), laminin (LM) and collagen type IV (CL) [1]. These compounds inhibit the adhesion of tumor cells to components of the ECM. A series of delaminomycins derivatives were prepared and the structure-activity relationships investigated by Ueno et al. [5]. Spiro compounds (A2, B2 and C2) showed stronger inhibitory activity than natural products (A1, B1 and C1) on B16 melanoma cells adhesion assay and ConA-induced proliferation of murine splenic lymphocytes assay. However, natural products showed more potent inhibitory activity than spiro compounds in MLCR and antimicrobial assays.

Biological Activity:

6.5 µg/ml: IC_{50} for adhesion of B16 melanoma cells to laminin [1]
6.0 µg/ml: IC_{50} for adhesion of B16 melanoma cells to fibronectin [1]
3.3 µg/ml: IC_{50} for adhesion of B16 melanoma cells to collagen [1]

References:

1. Ueno M, Amemiya M, Iijima M, et al., & Takeuchi T (1993) Delaminomycins, novel nonpeptide extracellular matrix receptor antagonist and a new class of potent immunomodulator. I. Taxonomy, fermentation, isolation and biological activity. J Antibiot 46:719–727
2. Ueno M, Someno T, Sawa R, et al., & Takeuchi T (1993) Delaminomycins, novel nonpeptide extracellular matrix receptor antagonist and a new class of potent immunomodulator. II. Physico-chemical properties and structure elucidation of delaminomycin A. J Antibiot 46:979–984
3. Ueno M, Someno T, Sawa R, et al., & Takeuchi T (1993) Delaminomycins, novel extra-cellular matrix receptor antagonist. III. Physico-chemical properties and structure eluci-dation of delaminomycins B and C. J Antibiot 46:1020–1023
4. Ueno M, Yoshinaga I, Amemiya M, et al., & Takeuchi T (1993) Delaminomycins, novel extracellular matrix receptor antagonists. V. Biosynthesis. J Antibiot 46:1390–1396
5. Ueno M, Amemiya M, Yamazaki K, et al., & Takeuchi T (1993) Delaminomycins, novel extracellular matrix receptor antagonist. IV. Structure-activity relationships of delaminomycins and derivatives. J Antibiot 46:1156–1162

Dephostatin

Key Words: [Tyrosine phosphatase inhibitor]

Structure:

Molecular Formula: $C_7H_8N_2O_3$
Molecular Weight: 168
Solubility: DMSO, +++ ; H_2O, ± ; MeOH, +++

Discovery:

Dephostatin was isolated from the culture broth of *Streptomyces* sp. as a novel tyrosine phosphatase inhibitor [1], and the structure was elucidated to be 2-(*N*-methyl-*N*-nitroso)hydroquinone by spectral and chemical analyses of dephostatin and its derivatives [2]. Watanabe et al. synthesized dephostatin and its derivatives and developed 3,4-dephostatin (3,4-dihydroxy-*N*-methyl-*N*-nitrosoaniline) [3]. 3,4-Dephosatin is commercially available.

Biological Studies:

Dephostatin inhibited protein tyrosine phosphatase prepared from a human neoplastic T-cell line in a competitive manner against the substrate [1]. Fujiwara et al. reported that 3,4-dephostatin enhanced NGF-induced neurite outgrowth in PC12 cells through sustained NGF-induced tyrosine phosphorylation of several proteins [4]. Several groups investigated the functions of tyrosine phosphorylation in cell regulation [5,6]. Dephostatin and the other phosphatase inhibitors were reviewed by Umezawa [7].

Biological Activity:

7.7 µM: IC_{50} for protein tyrosine phosphatase prepared from a human neoplastic T-cell line [1]
1.8 µg/ml: IC_{50} for Jurkat cells growth [1]
0.1–1 µM: enhancement of NGF-induced neurite outgrowth in PC12h cells [4]

References:

1. Imoto M, Kakeya H, Sawa T, et al., & Umezawa K (1993) Dephostatin, a novel protein tyrosine phosphatase inhibitor produced by *Streptomyces*. I. Taxonomy, isolation, and characterization. J Antibiot 46:1342–1346
2. Kakeya H, Imoto M, Takahashi Y, et al., & Umezawa K (1993) Dephostatin, a novel protein tyrosine phosphatase inhibitor produced by *Streptomyces*. II. Structure determination. J Antibiot 46:1716–1719
3. Watanabe T, Takeuchi T, Otsuka M, et al., & Umezawa K (1995) Synthesis and protein tyrosine phosphatase inhibitory activity of dephostatin analogs. J Antibiot 48:1460–466
4. Fujiwara S, Watanabe T, Nagatsu T, et al., & Umezawa K (1997) Enhancement or induction of neurite formation by a protein tyrosine phosphatase inhibitor, 3,4-dephostatin, in growth factor-treated PC12h cells. Biochem Biophys Res Commun 238:213–217
5. Garver TD, Ren Q, Tuvia S & Bennett V (1997) Tyrosine phosphorylation at a site highly conserved in the L1 family of cell adhesion molecules abolishes ankyrin binding and increases lateral mobility of neurofascin. J Cell Biol 137:703–714
6. Braunton JL, Wong V, Wang W, et al., & Wang YT (1998) Reduction of tyrosine kinase activity and protein tyrosine dephosphorylation by anoxic stimulation in vitro. Neuroscience 82:161–170
7. Umezawa K (1997) Induction of cellular differentiation and apoptosis by signal transduction inhibitors. Adv Enzyme Regul 37:393–401

Depudecin

Key Words: [Angiogenesis inhibitor][Antitumor][Detransforming agent]
[Histone deacetylase inhibitor]

Structure:

Molecular Formula: $C_{11}H_{16}O_4$
Molecular Weight: 212
Solubility: DMSO, ++ ; H_2O, ± ; MeOH, ++

Discovery/Isolation/Structure:

Depudecin was isolated from the culture broth of *Alternaria brassicicola,* inducing
the flat phenotype of *ras-* and *src*-transformed NIH3T3 cells [1]. The structure of
depudecin was determined by its spectroscopic characteristics and X-ray crystallographic analysis.

Biological and Chemical Studies:

Depudecin reversibly induced the flat phenotype of Ki-*ras*-transformed NIH3T3 cells
and actin stress fiber was detected. Almost complete reversion to the flat phenotype
was observed at 6 h after depudecin addition without a decrease of ras-mRNA [2].
Shimada et al. examined the detransformation activity of synthetic derivatives and
revealed the essential role of the epoxide and hydroxyl moieties in depudecin. Kwon
et al. demonstrated that depudecin inhibited histone deacetylase activity effectively
both in vivo and in vitro. Then, they suggested that its ability to induce morphological reversion of transformed cells was the result of its HDAC inhibitory activity [3,4].
      Oikawa et al. showed anti-angiogenic activity in an in vivo assay system involv-
ing the chorioallantoic membrane of growing chick embryo [5].

Biological Activity:

1 µg/ml: inducing the flat phenotype of *ras-* and *src*-transformed NIH3T3 cells [1]
IC_{50}: 4.7 µM (in vitro histone deacetylase assay) [4]
ID_{50}: 320 ng (1.5 nmol) per egg (embryonic angiogenesis) [5]

References:

1. Matsumoto M, Matsutani S, Sugita K, et al., & Yoshida T (1991) Depudecin: A novel compound inducing the flat phenotype of NIH3T3 cells doubly transformed by *ras*- and *src*-oncogene, produced by *Alternaria brassicicola*. J Antibiot 45:879–885
2. Sugita K, Yoshida H, Matsumoto M & Matsutani S (1992) A novel compound, depudecin, induces production of transformation to the flat phenotype of NIH3T3 cells transformed by *ras*-oncogene. Biochem Biophys Res Comm 182:379–387
3. Privalsky ML (1998) Depudecin makes a debut. Proc Natl Acad Sci USA 95:3335–3337
4. Kwon HJ, Owa T, Hassig CA, et al., & Schreiber SL (1998) Depudecin induces morphological reversion of transformed fibroblasts via the inhibition of histone deacetylase. Proc Natl Acad Sci USA 95:3356–3361
5. Oikawa T, Onozawa C, Inose M & Sasaki M (1995) Depudecin, a microbial metabolite containing two epoxide groups, exhibits anti-angiogenic activity in vivo. Biol Pharm Bull 18:1305–1307

Diketocoriolin B

Key Words: [Antitumor][Immunomodulator][(Na$^+$-K$^+$)-ATPase inhibitor]

Structure:

Molecular Formula: $C_{23}H_{32}O_6$
Molecular Weight: 404
Solubility: DMSO, ++ ; H$_2$O, ± ; MeOH, ++

Discovery:

Diketocoriolin B was isolated as an active derivative of coriolin B produced by *Coriolus consors* [1]. Chemical modification and structure-activity relationships were investigated [2–4].

Biological Studies:

Diketocoriolin B showed antitumor activity and prolonged the survival period of mice inoculated intraperitoneally with L-1210 or Ehrlich carcinoma cells [1]. Kunimoto et al. found that the main effect of diketocoriolin B is competitive inhibition of (Na$^+$-K$^+$)-ATPase [5,6].

Different from other antitumor agents, diketocoriolin B increased the number of antibody-forming cells in mouse spleen [7]. Ishizuka et al. studied the mechanism of action of diketocoriolin B to enhance antibody formation, and showed that diketocoriolin B augmented antibody formation in sheep red blood cells in vivo and in vitro. Both treatments of macrophage-rich and lymphocyte-rich cells with diketocoriolin B enhanced antibody formation [8].

Biological Activity:

12.5–100 μg/mouse/day: prolonged the survival period of mice inoculated with L-1210 cells to 162–173% [1]
0.75 μg/ml: LD_{50} for the growth of Yoshida sarcoma cells [6]
0.1 μg/mouse: augmented antibody formation in sheep red blood cells in vivo [8]

References:

1. Takeuchi T, Takahashi S, Iinuma H & Umezawa H (1971) Diketocoriolin B, an active derivative of coriolin B produced by *Coriolus consors*. J Antibiot 24:631–635
2. Takeuchi T, Ishizuka M & Umezawa H (1977) Chemical modification of coriolin B. J Antibiot 30:59–65
3. Nishimura Y, Koyama Y, Umezawa S, et al., & Umezawa H (1980) Syntheses of coriolin, 1-deoxy-1-ketocoriolin and 1,8-dideoxy-1,8-diketocoriolin from coriolin B. J Antibiot 33:404–407
4. Nishmura Y, Koyama Y, Umezawa S, et al., & Umezawa H (1980) Chemical modification of the ester group of diketocoriolin B. J Antibiot 33:393–403
5. Kunimoto T & Umezawa H (1973) Kinetic studies on the inhibition of (Na^++K^+)-ATPase by diketocoriolin B. Biochim Biophys Acta 318:78–90
6. Kunimoto T, Hori M & Umezawa H (1973) Mechanism of action of diketocoriolin B. Biochim Biophys Acta 298:513–25
7. Ishizuka M, Iinuma H, Takeuchi T & Umezawa H (1972) Effect of diketocoriolin B on antibody formation. J Antibiot 25:320–321
8. Ishizuka M, Takeuchi T & Umezawa H (1981) Studies of the mechanism of action of diketocoriolin B to enhance antibody formation. J Antibiot 34:95–102

Dolastatin 10

Key Words: [Antitumor][Microtubule inhibitor]

Structure:

Molecular Formula: $C_{42}H_{68}N_6O_6S$
Molecular Weight: 784
Solubility: DMSO, ++ ; H_2O, ± ; MeOH, +++

Discovery/Isolation/Structure:

Dolastatins 1–9, a series of very potent cell growth inhibitory substances, were isolated from *Dolabella auricularia* [1]. Dolastatins 10, 11, 12, 14, and 15 were also isolated [2–5].

Biological Studies:

Dolastatin 1 showed potent anticancer activity [1]. Dolastatin 10 strongly inhibited microtubule assembly, tubulin-dependent GTP hydrolysis, and the binding of *Vinca* alkaloids to tubulin. A tripeptide segment of dolastatin 10 effectively inhibited tubulin polymerization and GTP hydrolysis. As the tripeptide did not significantly inhibit either vincristine binding or nucleotide exchange, nor did it stabilize colchicine binding, binding sites of dolastatin is located in close to the vinca alkaloid binding site and to the exchangeable GTP site on β-tubulin are suggested[6].

In CHO cells, both dolastatin 15 and dolastatin 10 caused moderate loss of microtubules at the IC_{50} values and complete disappearance of microtubules at concentrations 10-fold higher. Despite its potency and the loss of microtubules in treated cells, the interaction of dolastatin 15 with tubulin in vitro was weak. Nevertheless, its structural similarity to dolastatin 10 indicates that dolastatin 15 may bind weakly in the "*vinca* domain" of tubulin

Biological Activity:

2.7×10^{-7} µg/ml: EC_{50} for dolastatin 3 to P388 cell [1]
240 at 11 µg/kg of dolastatin 1: T/C in the National Cancer Institute's murine B16 melanoma [1]
1.4 µM: K_i for binding of [^{3}H]vincristine to tubulin, noncompetitive [6]
0.4 and 0.5 nM: IC_{50} for dolastatin 10 with the L1210 and CHO cells, respectively[7].
3, 3, and 5 nM: IC_{50} for dolastatin 15 with L1210 murine leukemia cells, human Burkitt lymphoma cells, and CHO cells, respectively[7].
1.2 and 23 µM: IC_{50} for dolastatin 10 and dolastatin 15 to inhibition of glutamate-induced polymerization of tubulin [7]

References:

1. Pettit GR, Kamano Y, Fujii Y, et al., & Michel C (1981) Marine animal biosynthetic constituents for cancer chemotherapy. J Nat Prod 44:482–485
2. Pettit GR, Kamano Y, Kizu H, et al., & Nieman RA (1989) Isolation and structure of the cell growth inhibitory depsipeptides Dolastatins 11 and 12. Heterocycles 28:553–558
3. Pettit GR, Kamano Y, Dufresne C, et al., & Schmidt JM (1989) Isolation and structure of the cytostatic linear depsipeptide Dolastatin 15. J Org Chem 54:6005–6006
4. Pettit GR, Kamano Y, Herald CL, et al., & Kizu H (1990) Antineoplastic agents. 190. Isolation and structure of cyclodepsipeptide Dolstatin 14. J Org Chem 55:2289–2290
5. Pettit GR, Kamano Y, Herald CL, et al., & Bontems RJ (1987) The isolation and structure of a remarkable marine animal antineoplastic constituent: Dolastatin 10. J Am Chem Soc 109:6883–6885
6. Bai RL, Pettit GR & Hamel E (1990) Binding of dolastatin 10 to tubulin at a distinct site for peptide antimitotic agents near the exchangeable nucleotide and vinca alkaloid sites. J Biol Chem 265:17141–17149
7. Bai R, Friedman SJ, Pettit GR & Hamel E (1992) Dolastatin 15, a potent antimitotic depsipeptide derived from *Dolabella auricularia*. Interaction with tubulin and effects of cellular microtubules. Biochem Pharmacol 43:2637–2645

Epolactaene

Key Words: [Apoptosis][Neurite outgrowth inducer]

Structure:

Molecular Formula: $C_{21}H_{27}NO_6$
Molecular Weight: 389
Solubility: DMSO, +++ ; H_2O, ± ; MeOH, +++

Discovery:

Epolactaene was isolated as a neuritogenic compound from *Penicillium* sp., and the structure was determined by detailed spectral analysis [1]. The total synthesis of epolactaene was accomplished [2,3].

Biological Studies:

Epolactaene induced neurite outgrowth in human neuroblastoma SH-SY5Y cells, which lack significant *TRK* family mRNA.

Epolactaene derivatives, a series of 3-substituted 3-pyrrolin-2-ones (named MT compounds) were designed and synthesized, and developed some promising compounds; MT-5, MT-19, and MT-20 for neuritogenic compounds and MT-21 for an apoptosis inducer [4]. MT-5 induced neurite outgrowth without the activation of MAP kinase in SH-SY5Y cells. From the structure-activity relationships, the acetyl group at the 3-position is most important, as is at least one straight long chain alkyl group connected to the γ- lactam ring. In rat pheochromocytoma PC12 cells which express a high level of TRK-A, MT-19 and MT-20 induce neurite outgrowth by activating the downstream target of MAP kinase or by a novel mechanism which is distinct from the NGF-activated pathway [5].

MT-21 induced apoptotic cell death through the production of reactive oxygen species (ROS) and the activation of caspase families in HL-60 cells. Further studies on the biological studies of MT-21 are in progress.

Biological Activity:

10 μg/ml: Neurite outgrowth in SH-SY5Y cells [1]
1 μg/ml: Neurite outgrowth in PC12 cells by MT-19 and MT-20 [5]
40 μM: Apoptosis induction in HL-60 cells by MT-21

References:

1. Kakeya H, Takahashi I, Okada G, et al., & Osada H (1995) Epolactaene, a novel neuritogenic compound in human neuroblastoma cells, produced by a marine fungus. J Antibiot 48:733–735
2. Hayashi Y & Narasaka K (1998) Asymmetric total synthesis of epolactaene. Chem Lett 4:313–314
3. Marumoto S, Kogen H & Naruto S (1998) Absolute configuration and total synthesis of (+)-epolactaene, a neuritogenic agent from *Penicillium* sp. BM 1689-P active in human neuroblastoma cells. J Org Chem 63:2068–2069
4. Osada H (1998) Discovery and mechanism of action of nonpeptide neuritogenic compounds. Drugs of the Future 23:395–399
5. Yao R & Osada H (1997) Induction of neurite outgrowth in PC12 cells by γ-lactam-related compounds via Ras-MAP kinase signaling pathway independent mechanism. Exp Cell Res 234:233–239

Epothilone A and B

Key Words: [Antitumor][Microtubule inhibitor]

Structure:

	Epothilone A (R = H)	Epothilone B (R = CH$_3$)
Molecular Formula:	C$_{26}$H$_{39}$NO$_6$S	C$_{27}$H$_{41}$NO$_6$S
Molecular Weight:	493	507
Solubility:	DMSO, ++ ; H$_2$O, ± ; MeOH, ±	

Discovery/Isolation:

Epothilone A and B was isolated as antifungal and cytotoxic antibiotics from Myxobacteria [1,2]. Bollag et al. reisolated as new microtubule-stabilizing agent by filtration-colorimetric assay [3]. Crystal structure and conformation in solution were studied by Hofle et al. [4].

Biological and Chemical Studies:

Bollag et al. identified epothilones A and B as compounds that possess all the biological effects of paclitaxel both in vitro and in cultured cells. Epothilones are equipotent and exhibit kinetics similar to paclitaxel in inducing tubulin polymerization into microtubules in vitro and in producing enhanced microtubule stability and bundling in cultured cells. As these compounds are competitive inhibitors of [^{3}H]paclitaxel binding, epothilones represent a novel structural class of compounds which not only mimic the biological effects of paclitaxel but also appear to bind to the same microtubule-binding site as paclitaxel [3]. Kowalski et al. examined interactions of the epothilones with purified tubulin and paclitaxel-resistant ovarian carcinoma cells with an altered β-tubulin. They found that the epothilones are competitive inhibitors of the binding of [^{3}H] paclitaxel to tubulin polymers, and that a multidrug-resistant colon carcinoma line and the paclitaxel-resistant ovarian line retained sensitivity to the epothilones.

Biological Activity:

2 ng/ml: IC_{50} for cytotoxicity against mouse fibroblast L929 [4]
0.55 μM: K_D for [^{3}H]taxol binding to preformed microtubule [3]
30 and 39 nM: EC_{50} for mitotic arrest and cytotoxicity in HeLa cells [3]
1.4 and 0.7 μM: K_i for epothilones A and B by Hanes analysis [5]
0.6 and 0.4 μM: K_i for epothilones A and B by Dixon analysis [5]

References:

1. Gerth K, Bedorf N, Hofle G, et al., & Reichenbach H (1996) Epothilones A and B: Antifungal and cytotoxic compounds from *Sorangium cellulosum* (Myxobacteria). Production, physico-chemical and biological properties. J Antibiot 49:560–563
2. Hoefle G, Bedorf N, Gerth K & Reichenbach H (1993) Patent DE 4138042.
3. Bollag DM, McQueney PA, Zhu J, et al., & Woods CM (1995) Epothilones, a new class of microtubule-stabilizing agents with a taxol-like mechanism of action. Cancer Res 55:2325–2333
4. Hofle G, Bedorf N, Steinmetz H, et al., & Reichenbach H (1996) Epothilone A and B-Novel 16-membered macrolides with cytotoxic activity: Isolation, crystal structure, and conformation in solution. Angew Chem Int Ed Engl 35:1567–1569
5. Kowalski RJ, Giannakakou P & Hamel E (1997) Activities of the microtubule-stabilizing agents epothilones A and B with purified tubulin and in cells resistant to paclitaxel. J Biol Chem 272:2534–2541

Erbstatin

Key Words: [Tyrosine kinase inhibitor]

Structure:

Molecular Formula: $C_9H_9NO_3$
Molecular Weight: 179
Solubility: DMSO, +++ ; H_2O, ± ; MeOH, +++

Discovery:

Erbstatin was isolated as a tyrosine kinase inhibitor, using the membrane fraction of human epidermoid carcinoma cell line A-431 [1], and the structure was determined by X-ray crystallography [2]. From the results of structure-activity relationships on tyrosine kinase inhibitory activity, 2,5-dihydroxycinnamates (2,5-MeC), a stable erbstatin derivative, was developed [3–5]. A variety of erbstatin analogue, tyrophostins, were synthesized and the specific analogues for each tyrosine kinase, e.g., epidermal growth factor receptor or platelet-derived growth factor receptor tyrosine kinase were developed [6,7]. 2,5-MeC and tyrophostins are commercially available.

Biological Studies:

Erbstatin inhibited EGF receptor in a competitive manner against substrate [8]. Several groups have been using erbstatin and its derivatives as a tyrosine kinase specific inhibitor and investigated cellular functions such as the cell cycle [9,10], regulation of phospholipase C [11] and protein kinase C [12], cell adhesion [13] etc. These compounds also showed anti-angiogenic effects [14] and antitumor effects via apoptosis induction [15]. The usage of erbstatin and its analogs were reviewed [16].

Biological Activity:

3.6 µg/ml: IC_{50} for A431 cells growth [1]
5.58 µM: K_i value for EGF receptor [8]
0.15 µg/ml: IC_{50} for EGF receptor [10]

19.8 µM: IC_{50} for protein kinase C [12]
80 ng/egg: ID_{50} for embryonic angiogenesis [14]
10 and 30µg/ml: apoptosis induction in L1210 and Ms-1 cells [15]

References:

1. Umezawa H, Imoto M, Sawa T, et al., & Takeuchi T (1986) Studies on a new epidermal growth factor-receptor kinase inhibitor, erbstatin, produced by MH435-hF3. J Antibiot 39:170–173
2. Nakamura H, Iitaka Y, Imoto M, et al., & Umezawa H (1986) The structure of an epidermal growth factor-receptor kinase inhibitor, erbstatin. J Antibiot 39:314–315
3. Isshiki K, Imoto M, Sawa T, et al., & Tatsuta K (1987) Inhibition of tyrosine protein kinase by synthetic erbstatin analogs. J Antibiot 40:1209–1210
4. Isshiki K, Imoto M, Takeuchi T, & Tatsuta K (1987) Effective synthesis of erbstatin and its analogs. J Antibiot 40:1207–1208
5. Hori T, Kondo T, Tsuji T, & Hiratsuka M (1992) Inhibition of tyrosine kinase and src oncogene functions by stable erbstatin analogues. J Antibiot 45:280–282
6. Gazit A, Yaish P, Gilon C & Levitzki A (1989) Tyrphostins I: synthesis and biological activity of protein tyrosine kinase inhibitor. J Med Chem 32:2344–2352
7. Yaish P, Gazit A, Gilon C & Levitzki A (1988) Blocking of EGF-dependent cell proliferation by EGF receptor kinase inhibitors. Science 242:933–935
8. Imoto M, Umezawa K, Isshiki K, & Umezawa H (1987) Kinetic studies of tyrosine kinase inhibition by erbstatin. J Antibiot 40:1471–1473
9. Umezawa K, Sugata D, Yamashita K, et al., & Shibuya M (1992) Inhibition of epidermal growth factor receptor functions by tyrosine kinase inhibitors in NIH3T3 cells. FEBS Lett 314:289–292
10. Umezawa K, Hori T, Tajima H, et al., & Takeuchi T (1990) Inhibition of epidermal growth factor-induced DNA synthesis by tyrosine kinase inhibitors. FEBS Lett 260:198–200
11. Imoto M, Shimura N, Ui H & Umezawa K (1990) Inhibition of EGF-induced phospholipase C activation in A431 cells by erbstatin, a tyrosine kinase inhibitor. Biochem Biophys Res Commun 173:208–211
12. Bishop WR, Petrin J, Wang L, et al., & Doll RJ (1990) Inhibition of protein kinase C by the tyrosine kinase inhibitor erbstatin. Biochem Pharmacol 40:2129–2135
13. Polanowska-Grabowska R, Geanacopoulos M & Gear AR (1993) Platelet adhesion to collagen via the α 2 β 1 integrin under arterial flow conditions causes rapid tyrosine phosphorylation of pp125[FAK]. Biochem J 296:543–547
14. Oikawa T, Ashino H, Shimamura M, Hasegawa M, Morita I, Murota S, Ishizuka M & Takeuchi T (1993) Inhibition of angiogenesis by erbstatin, an inhibitor of tyrosine kinase. J Antibiot 46:785–790
15. Simizu S, Imoto M, Masuda N, et al., & Umezawa K (1996) Involvement of hydrogen peroxide production in erbstatin-induced apoptosis in human small cell lung carcinoma cells. Cancer Res 56:4978–4982
16. Umezawa K & Imoto M (1991) Use of erbstatin as protein-tyrosine kinase inhibitor. Methods Enzymol 201:379–385

FK409

Key Words: [Nitric oxide releaser][Vasodilator]

Structure:

Molecular Formula: $C_8H_{13}N_3O_4$
Molecular Weight: 215
Solubility: DMSO, ++ ; H_2O, + ; MeOH, ++

Discovery:

FK409, a novel vasodilator with anti-platelet aggregation activity, was isolated from the acid-treated fermentation broth of *Streptomyces griseosporeus*, which was cultured on a medium containing $NaNO_3$ [1]. The structure of FK409 was determined on the basis of spectroscopic and chemical evidence [2,3]. Kita et al. reported that synthesized FR144420 showed a longer duration of effects than FK409 and may be very useful for investigating the in vivo actions of NO [4].

Biological Studies:

FK409 showed potent vasorelaxant effects in isolated conronary artery rings of dogs and these effects were correlated with the elevation of cGMP [5]. FK409 also inhibited contractile responses to norepinephrine, histamine, and 5- hydroxytryptamine in rabbit aorta [6]. Isono et al. found that FK409 activated soluble guanylate cyclase and liberated nitric oxide (NO) spontaneously in rat aortic soluble fraction. They speculated that FK409 shows a vasorelaxant effect via guanylate cyclase activation due to the liberated NO [7].

Several groups investigated the effects of NO using FK409, for example, on acetylcholine release from rat hippocampal slices [8], on renal actions and norepinephrine overflow induced by renal nerve stimulation [9], and in regulation of gastric acid secretion [10].

Biological Activity:

10^{-11}–10^{-5} M: vasorelaxant effects relaxation in isolated coronary artery rings of the dog [5]
10^{-8}–10^{-5} M: inhibition of contractile responses to norepinephrine, histamine, and 5-hydroxytryptamine in rabbit aorta [6]
3.2×10^{-7} M: activation of soluble guanylate cyclase [7]
3 mM: liberates nitric oxide in rat aortic soluble fraction [7]

References:

1. Hino M, Iwami M, Okamoto M, et al., & Imanaka H (1989) FK409, a novel vasodilator isolated from the acid-treated fermentation broth of *Streptomyces griseosporeus*. I. Taxonomy, fermentation, isolation, and physico-chemical and biological characteristics. J Antibiot 42:1578–1583
2. Hino M, Takase S, Itoh Y, et al., & Kohsaka M (1989) Structure and synthesis of FK409, a novel vasodilator isolated from *Streptomyces* as a semi-artificial fermentation product. Chem Pharm Bull 37:2864–2866
3. Hino M, Ando T, Takase S, et al., & Imanaka H (1989) FK409, a novel vasodilator isolated from the acid-treated fermentation broth of *Streptomyces griseosporeus*. II. Structure of FK409 and its precursor FR-900411. J Antibiot 42:1584–1588
4. Kita Y, Ohkubo K, Hirasawa Y, et al., & Yoshida K (1995) FR144420, a novel, slow, nitric oxide-releasing agent. Eur J Pharmacol 275:125–130
5. Yamada H, Yoneyama F, Satoh K & Taira N (1991) Comparison of the effects of the novel vasodilator FK409 with those of nitroglycerin in isolated coronary artery of the dog. Br J Pharmacol 103:1713–1718
6. Shibata S, Satake N, Sato N, et al., & Hester RK (1991) Characteristics of the vasorelaxing action of (3*E*)-4-ethyl-2- hydroximino-5-nitro-3-hexamide FK409, a new vasodilator isolated from microbial sources, in isolated rabbit arteries. J Cardiovasc Pharmacol 17:508–518
7. Isono T, Koibuchi Y, Sato N, et al., & Ohtsuka M (1993) Vasorelaxant mechanism of the new vasodilator, FK409. Eur J Pharmacol 246:205–212
8. Suzuki T, Nakajima K, Fujimoto K, et al., & Kawashima K (1997) Nitric oxide increases stimulation-evoked acetylcholine release from rat hippocampal slices by a cyclic GMP-independent mechanism. Brain Res 760:158–162
9. Tadano K, Matsuo G, Hashimoto T & Matsumura Y (1998) Effects of FK409, a nitric oxide donor, on renal responses to renal nerve stimulation in anesthetized dogs. Eur J Pharmacol 341:191–199
10. Kato S, Kitamura M, Korolkiewicz RP & Takeuchi K (1998) Role of nitric oxide in regulation of gastric acid secretion in rats: effects of NO donors and NO synthase inhibitor. Br J Pharmacol 123:839–846

FK506

Key Words: [Immunosuppressant][PP2B (calcineurin) inhibitor]

Structure:

Molecular Formula: $C_{44}H_{69}NO_{12}$
Molecular Weight: 804
Solubility: DMSO, ++ ; H_2O, ± ; MeOH, ++

Discovery/Isolation:

In the course of the search for immunosuppressive agents, FK506 (tacrolimus®) was isolated from *Streptomyces tsukubaensis* [1,2]. The structure elucidation was also reported [3].

Biological Studies:

FK506 inhibited mixed lymphocyte reaction, cytotoxic T cell generation, and expression of early T cell activation genes [2,4]. Harding et al. and Siekierka et al. purified the FK-506 binding protein, FKBP. FKBP is a cis-trans peptidyl-prolyl isomerase and FK-506 inhibits its activity [5–7]. However, Bierer et al. showed that inhibition of rotamase activity is an insufficient requirement for mediating these effects [8].

Liu et al. identified calcineurin as the binding protein of the FKBP-FK506 complex. They showed that the complexes cyclophilin-Cyclosporin A and FKBP-FK506 (but not cyclophilin, FKBP, FKBP-rapamycin, or FKBP-506BD) competitively bind to and inhibit calcineurin. These results suggest that calcineurin is involved in a common step associated with T cell receptor and IgE receptor signaling pathways and that cyclophilin and FKBP mediate the actions of Cyclosporin A and FK506, respectively, by forming drug-dependent complexes with and altering the activity of calcineurin-calmodulin [9]. Furthermore, they demonstrated that the FKBP-FK506 complex inhibits the phosphatase activity of calcineurin and transcriptional activation by NF-AT, a T cell specific transcription factor that regulates IL-2 gene synthesis in human T cells [10].

Biological Activity:

> 200 mg/kg: LD_{50} for intraperitoneally to BALB/C mice [1]
0.1 nM: IC_{50} for mixed lymphocyte reaction, cytotoxic T cell generation, the production of T cell-derived soluble mediators such as interleukin 2, interleukin 3 and γ-interferon and the expression of the IL-2 receptor [2]

References:

1. Kino T, Hatanaka H, Hashimoto M, et al., & Imanaka H (1987) FK-506, a novel immunosuppressant isolated from a *Streptomyces*. I. Fermentation, isolation, and physicochemical and biological characteristics. J Antibiot 40:1249–1255
2. Kino T, Hatanaka H, Miyata S, et al., & Ochiai T (1987) FK-506, a novel immunosuppressant isolated from a *Streptomyces*. II. Immunosuppressive effect of FK-506 in vitro. J Antibiot 40:1256–1265
3. Tanaka H, Kuroda A, Marusawa H, et al., & Taga T (1987) Structure of FK506: a novel immunosuppressant isolated from *Streptomyces*. J Am Chem Soc 109:5031–5033
4. Tocci MJ, Matkovich DA, Collier KA, et al., & Hutchinson NI (1989) The immunosuppressant FK506 selectively inhibits expression of early T cell activation genes. J Immunol 143:718–726
5. Harding MW, Galat A, Uehling DE & Schreiber SL (1989) A receptor for the immunosuppressant FK506 is a cis-trans peptidyl-prolyl isomerase. Nature 341:758–760
6. Siekierka JJ, Hung SH, Poe M, et al., & Sigal NH (1989) A cytosolic binding protein for the immunosuppressant FK506 has peptidyl-prolyl isomerase activity but is distinct from cyclophilin. Nature 341:755–757
7. Rosen MK, Standaert RF, Galat A, et al., & Schreiber SL (1990) Inhibition of FKBP rotamase activity by immunosuppressant FK506: twisted amide surrogate. Science 248:863–866
8. Bierer BE, Somers PK, Wandless TJ, et al., & Schreiber SL (1990) Probing immunosuppressant action with a nonnatural immunophilin ligand. Science 250:556–559
9. Liu J, J. D. Farmer J, Lane WS, et al., & Schreiber SL (1991) Calcineurin is a common target of cyclophilin-cyclosporin A and FKBP-FK506 complexes. Cell 66:807–815
10. Liu J, Albers MW, Wandless TJ, et al., & Schreiber SL (1992) Inhibition of T cell signaling by immunophilin-ligand complexes correlates with loss of calcineurin phosphatase activity. Biochemistry 31:3896–3901

Fostriecin

Key Words: [Antitumor][Apoptosis][PP1/PP2A inhibitor]

Structure:

Molecular Formula: $C_{19}H_{27}O_9P$
Molecular Weight: 430
Solubility: DMSO, ++ ; H_2O, ± ; MeOH, ++

Discovery/Isolation:

Fostriecin (CI-920) was isolated as an antitumor and antifungal antibiotic from *Streptomyces pulveraceus* subsp. *fostreus* ATCC 31906 [1–3].

Biological and Chemical Studies:

Fostriecin showed antitumor activity to murine P388 lymphocytic and L-1210 lymphoid leukemia [1,4]. Boritzki et al. demonstrated that fostriecin inhibited the catalytic activity of partially purified topoisomerase II from Ehrlich ascites carcinoma [5]. Roberge et al. reported that fostriecin induced premature mitosis and also that it inhibited the PP1 and PP2A more potently [6]. Moreover, fostriecin induced vimentin hyperphosphorylation and intermediate filament reorganization [7]. Recently, Walsh et al. reported that fostriecin is a strong inhibitor of PP2A but a weak inhibitor of PP1 [8].

Cheng et al. reported that fostriecin-mediated G2/M phase growth arrest correlated with abnormal centrosome replication, the formation of aberrant mitotic spindles, and the inhibition of serine/threonine protein phosphatase [9]. Apoptosis inductive activity was reported by several groups [10,11]. Although fostriecin induced G2/M accumulation, apoptotic chromosome condensation in interphase cells did not require p34[cdc2] kinase activity but was associated with H2A/H3 phosphorylation [12].

Biological Activity:

246: T/C values to murine P388 lymphocytic and 207 to L1210 lymphoid leukemia [1]
110 μM: K_i for topoisomerase inhibition, noncompetitive [5]

40 µM: IC_{50} for topoisomerase inhibition [5]
5 µM: G2 arrest in L1210 cells [5]
4 µM and 40 nM: IC_{50} for PP1 and PP2A [6]
131 µM and 3.2 nM: IC_{50} for PP1 and PP2A [8]
1–12 µM: G2/M arrest and antitumor activity in CHO cells [9]

References:

1. Tunac JB, Graham BD & Dobson WE (1983) Novel antitumor agents CI-920, PD 113,270 and PD 113,271. I. Taxonomy, fermentation and biological properties. J Antibiot 36:1595–1600
2. Stampwala SS, Bunge RH, Hurley TR, et al., & French JC (1983) Novel antitumor agents CI-920, PD 113,270 and PD 113,271. II. Isolation and characterization. J Antibiot 36:1601–1605
3. Mamber SW, Okasinski WG, Pinter CD & Tunac JB (1986) Antimycotic effects of the novel antitumor agents fostriecin (CI-920), PD 113,270 and PD 113,271. J Antibiot 39:1467–1472
4. Leopold WR, Shillis JL, Mertus AE, et al., & Jackson RC (1984) Anticancer activity of the structurally novel antibiotic Cl-920 and its analogues. Cancer Res 44:1928–1932
5. Boritzki TJ, Wolfard TS, Besserer JA, et al., & Fry DW (1988) Inhibition of type II topoisomerase by fostriecin. Biochem Pharmacol 37:4063–4068
6. Roberge M, Tudan C, Hung SM, et al., & Anderson H (1994) Antitumor drug fostriecin inhibits the mitotic entry checkpoint and protein phosphatases 1 and 2A. Cancer Res 54:6115–6121
7. Ho DT & Roberge M (1996) The antitumor drug fostriecin induces vimentin hyperphosphorylation and intermediate filament reorganization. Carcinogenesis 17:967–972
8. Walsh AH, Cheng A & Honkanen RE (1997) Fostriecin, an antitumor antibiotic with inhibitory activity against serine/threonine protein phosphatases types 1 (PP1) and 2A (PP2A), is highly selective for PP2A. FEBS Lett 416:230–234
9. Cheng A, Balczon R, Zuo Z, et al., & Honkanen RE (1998) Fostriecin-mediated G2-M-phase growth arrest correlates with abnormal centrosome replication, the formation of aberrant mitotic spindles, and the inhibition of serine/threonine protein phosphatase activity. Cancer Res 58:3611–3619
10. Hotz MA, Traganos F & Darzynkiewicz Z (1992) Changes in nuclear chromatin related to apoptosis or necrosis induced by the DNA topoisomerase II inhibitor fostriecin in MOLT-4 and HL-60 cells are revealed by altered DNA sensitivity to denaturation. Exp Cell Res 201:184–191
11. Gorczyca W, Gong J, Ardelt B, et al., & Darzynkiewicz Z (1993) The cell cycle related differences in susceptibility of HL-60 cells to apoptosis induced by various antitumor agents. Cancer Res 53:3186–3192
12. Guo XW, Thíng JP, Swank RA, et al., & Roberge M (1995) Chromosome condensation induced by fostriecin does not require $p34^{cdc2}$ kinase activity and histone H1 hyperphosphorylation, but is associated with enhanced histone H2A and H3 phosphorylation. EMBO J 14:976–985

FR901228

Key Words: [Antitumor][Histone deacetylase inhibitor]

Structure:

Molecular Formula: $C_{24}H_{36}N_4O_6S_2$
Molecular Weight: 540
Solubility: DMSO, ++ ; H_2O, + ; MeOH, ++

Discovery:

FR901228 was isolated from a broth culture of *Chromobacterium violaceum* as an antitumor bicyclic depsipeptide [1], and its structure was determined by a combination of spectroscopic, chemical evidence and single-crystal X-ray crystallographic analysis [2].

Biological Studies:

FR901228 was isolated as an antitumor compound showing life prolongation of mice bearing murine tumors [3]. FR901228 reversed the transformed morphology of the Ha-*ras* transformants, Ras-1 cells, and inhibited their growth at G0/G1. The reduction of c-*myc* expression was observed in FR901228-treated Ras-1 cells by RNA dot-blot hybridization [4]. It was also reported that FR901228 inhibits c-Myc and Fas ligand expression, then effectively blocks activation-induced apoptosis in T cell hybridomas [5].

FR901228 was reisolated in the course of screening for microbial metabolites that induce transcriptional activation of the SV40 promoter. FR901228 caused arrest of the cell cycle at both G1 and G2/M phases, induction of internucleosomal breakdown of chromatin, and inhibition of intracellular histone deacetylase activity. FR901228 is therefore a new type of histone deacetylase inhibitor, whose chemical structure is unrelated to known inhibitors such as trichostatins and trapoxins [6].

Biological Activity:

0.56–3.2 mg/kg: antitumor activity against human tumor A569 implanted under the kidney capsule of immunosuppressed BDF_1 mice [3]
2.5 ng/ml: morphological reversion of Ras-1 cell, Ha-*ras* transformed NIH/3T3 cell line [4]
10 ng/ml: cell cycle arrest at G1 and G2/M phases in M-8 cells [6]
1.1 nM: IC_{50} against histone deacetylase activity in vitro [6]

References:

1. Ueda H, Nakajima H, Hori Y, et al., & Okuhara M (1994) FR901228, a novel antitumor bicyclic depsipeptide produced by *Chromobacterium violaceum* No. 968. I. Taxonomy, fermentation, isolation, physico-chemical and biological properties, and antitumor activity. J Antibiot 47:301–310
2. Shigematsu N, Ueda H, Takase S, et al., & Tada T (1994) FR901228, a novel antitumor bicyclic depsipeptide produced by *Chromobacterium violaceum* No. 968. II. Structure determination. J Antibiot 47:311–314
3. Ueda H, Manda T, Matsumoto S, et al., & Shimomura K (1994) FR901228, a novel antitumor bicyclic depsipeptide produced by *Chromobacterium violaceum* No. 968. III. Antitumor activities on experimental tumors in mice. J Antibiot 47:315–323
4. Ueda H, Nakajima H, Hori Y, et al., & Okuhara M (1994) Action of FR901228, a novel antitumor bicyclic depsipeptide produced by *Chromobacterium violaceum* no. 968, on Ha-*ras* transformed NIH3T3 cells. Biosci Biotechnol Biochem 58:1579–1583
5. Wang R, Brunner T, Zhang L & Shi Y (1998) Fungal metabolite FR901228 inhibits c-Myc and Fas ligand expression. Oncogene 17:1503–1508
6. Nakajima H, Kim YB, Terano H, et al., & Horinouchi S (1998) FR901228, a potent antitumor antibiotic, is a novel histone deacetylase inhibitor. Exp Cell Res 241:126–133

Fumagillin (AGM-1470)

Key Words: [Angiogenesis inhibitor][Methionine aminopeptidase inhibitor]

Structure:

	Fumagillin	AGM-1470
Molecular Formula:	$C_{26}H_{34}O_7$	$C_{19}H_{28}NO_6Cl$
Molecular Weight:	458	402
Solubility:	DMSO, ++ ; H_2O, - ; MeOH, ++	

Discovery:

Fumagillin was isolated as active concentrates that were capable of inhibiting *Staphylococcus aureus* 209 bacteriophage [1], and found as an extremely potent amebicidal antibiotic [2]. The structure was determined by chemical degradation and X-ray crystallography [3–5] and total synthesis was performed [6].

Biological Studies:

Fumagillin was originally used as an amebicidal antibiotic, but Ingber et al. re-isolated fumagillin as a potent angiogenesis inhibitor [7]. They synthesized several fumagillin derivatives and found that AGM-1470 was the most potent angiogenesis inhibitor. Sin et al. and Griffith et al. showed that fumagillin covalently binds and inhibits the methionine aminopeptidase, MetAP-2 [8,9]. The type 1 enzyme, MetAP1, was not inhibited by the drugs, suggesting that the interaction between the drugs and MetAP2 is highly specific. The reactive ring epoxide in fumagillin was involved in the covalent modification of the MetAP2 active-site histidine, and MetAP2 is irreversibly inhibited [10,11]. Structure of human MetAP2 complexed with fumagillin was analyzed by X-ray crystallography and the crystal structure of free and inhibited human MetAP-2 shows a covalent bond formed between a reactive epoxide of fumagillin and histidine-231 in the active site of MetAP-2 [12].

Biological Activity: (for AGM-1470)

0.037 nM: IC_{50} for growth of bovine aortic endothelial cells [8]
1 nM: IC_{50} for in vitro MetAP2 activity [8]

References:

1. Hanson FR & Eble E (1949) J. Bacteriol. 58:527
2. McGowen MC, Callender ME & Lawlis JF (1951) Fumagillin (H3) a new antibiotic with amoebicidal properties. Science 113:202
3. McCorkindale NJ & Sime JG (1961) The configuration of fumagillin. Proc Chem Soc:331
4. Tarbell DS, Carman RM, Chapman DD, et al., & West RL (1961) The chemistry of fumagillin. J Am Chem Soc 83:3096–113
5. Turner JR & Tarbell DS (1962) The stereochemistry of fumagillin. Proc Natl Acad Sci USA 48:733–735
6. Corey EJ & Snider BB (1972) A total synthesis of (±)-fumagillin. J Am Chem Soc 94:2549–50
7. Ingber D, Fujita T, Kishimoto S, et al., & Folkman J (1990) Synthetic analogues of fumagillin that inhibit angiogenesis and suppress tumour growth. Nature 348:555–7
8. Griffith EC, Su Z, Turk BE, et al., & Liu JO (1997) Methionine aminopeptidase (type 2) is the common target for angiogenesis inhibitors AGM-1470 and ovalicin. Chem Biol 4:461–71
9. Sin N, Meng L, Wang MQ, et al., & Crews CM (1997) The anti-angiogenic agent fumagillin covalently binds and inhibits the methionine aminopeptidase, MetAP-2. Proc Natl Acad Sci USA 94:6099–103
10. Lowther WT, McMillen DA, Orville AM & Matthews BW (1998) The anti-angiogenic agent fumagillin covalently modifies a conserved active-site histidine in the *Escherichia coli* methionine aminopeptidase. Proc Natl Acad Sci USA 95:12153–7
11. Griffith EC, Su Z, Niwayama S, et al., & Liu JO (1998) Molecular recognition of angiogenesis inhibitors fumagillin and ovalicin by methionine aminopeptidase 2. Proc Natl Acad Sci USA 95:15183–15188
12. Liu S, Widom J, Kemp CW, et al., & Clardy J (1998) Structure of human methionine aminopeptidase-2 complexed with fumagillin. Science 282:1324–7

Fumonisin B1

Key Words: [Ceramide synthase inhibitor]

Structure:

Molecular Formula: $C_{34}H_{59}NO_{15}$
Molecular Weight: 722
Solubility: DMSO, ++ ; H_2O, ± ; MeOH, +++

Discovery:

Fumonisins were discovered as mycotoxins produced by *Fusarium moniforme* (Sheldon) and its structures were elucidated [1]. Absolute configuration of fumonisins were determined by the CD exciton chirality method [2].

Biological Studies:

Fumonisins are mycotoxins and naturally contaminated corn. Consumption of grains contaminated with fumonisins causes liver cancer in rats and has been correlated with esophageal cancer in humans. Wang et al. found that fumonisin B1 is a potent inhibitor of ceramide synthase (sphingosine *N*-acyltransferase) in rat liver microsomes [3] and leads to accumulation of sphingoid bases. Schroeder et al. found that fumonisin B1 caused accumulation of sphinganine and sphingosine in Swiss 3T3 cells and then stimulated DNA synthesis. They speculated that carcinogenicity of fumonisins derived from the accumulation of sphingoid bases rather than inhibition of complex sphingolipid biosynthesis per se [4]. Although several groups investigated the roles of ceramide in cell proliferation [5], apoptosis induction [6–9], neuronal outgrowth [10], and cell adhesion [11] by using fumonisin B1, there are several contradictory results. The inhibitory activity of protein phosphatases at high concentrations of fumonisin B1 was also reported [12].

Biological Activity:

0.1 μM: IC_{50} for ceramide synthase in rat liver microsomes [3]
50 μM: Stimulation of DNA synthesis in Swiss 3T3 fibroblast [4]
80–3000 μM: IC_{50} for PP1, PP2A, PP2B, PP2C and PP5 [12]

References:

1. Bezuidenhout SC, Gelderblom WCA, Gorst-Allman CP, et al., & Vleggaar R (1988) Structure elucidation of the fumonisins, mycotoxins from *Fusarium moniliforme*. J Chem Soc Chem Commun 1988:743–745
2. Hartl M & Humpf H-U (1998) Assigning the absolute configuration of fumonisins by the circular dichroism exciton chirality method. Tetrahedron: Asymmetry 9:1549–1556
3. Wang E, Norred WP, Bacon CW, et al., & Merrill AH, Jr. (1991) Inhibition of sphingolipid biosynthesis by fumonisins. Implications for diseases associated with *Fusarium moniliforme*. J Biol Chem 266:14486–14490
4. Schroeder JJ, Crane HM, Xia J, et al., & Merrill AH, Jr. (1994) Disruption of sphingolipid metabolism and stimulation of DNA synthesis by fumonisin B1. A molecular mechanism for carcinogenesis associated with *Fusarium moniliforme*. J Biol Chem 269:3475–3481
5. Lee JY, Leonhardt LG & Obeid LM (1998) Cell-cycle-dependent changes in ceramide levels preceding retinoblastoma protein dephosphorylation in G2/M. Biochem J 334:457–461
6. Boland MP, Foster SJ & O'Neill LA (1997) Daunorubicin activates NFκB and induces κB-dependent gene expression in HL-60 promyelocytic and Jurkat T lymphoma cells. J Biol Chem 272:12952–12960
7. Bose R, Verheij M, Haimovitz-Friedman A, et al., & Kolesnick R (1995) Ceramide synthase mediates daunorubicin-induced apoptosis: an alternative mechanism for generating death signals. Cell 82:405–414
8. Jaffrezou JP, Levade T, Bettaieb A, et al., & Laurent G (1996) Daunorubicin-induced apoptosis: triggering of ceramide generation through sphingomyelin hydrolysis. EMBO J 15:2417–2424
9. Suzuki A, Iwasaki M, Kato M & Wagai N (1997) Sequential operation of ceramide synthesis and ICE cascade in CPT-11-initiated apoptotic death signaling. Exp Cell Res 233:41–47
10. Harel R & Futerman AH (1993) Inhibition of sphingolipid synthesis affects axonal outgrowth in cultured hippocampal neurons. J Biol Chem 268:14476–14481
11. Hidari K, Ichikawa S, Fujita T, et al., & Hirabayashi Y (1996) Complete removal of sphingolipids from the plasma membrane disrupts cell to substratum adhesion of mouse melanoma cells. J Biol Chem 271:14636–14641
12. Fukuda H, Shima H, Vesonder RF, et al., & Nagao M (1996) Inhibition of protein serine/threonine phosphatases by fumonisin B1, a mycotoxin. Biochem Biophys Res Commun 220:160–165

GE3

Key Words: [Antitumor][E2F inhibitor]

Structure:

Molecular Formula: $C_{49}H_{80}N_8O_{14}$
Molecular Weight: 1004
Solubility: DMSO, ++ ; H_2O, + ; MeOH, ++

Discovery:

GE3, a cyclic hexadepsipeptide antibiotic, was isolated as an antitumor drug from the culture broth of *Streptomyces* sp. [1]. The structure of GE3 was determined by spectral studies [2].

Biological Studies:

GE3 exhibited antitumor activity against human pancreatic carcinoma, PSN-1, in vivo. GE3B, a linear peptide form of GE3 showed no antibiotic and cytotoxic activities, suggesting the necessity of the cyclic structure of GE3 for its biological activities [1]. GE3 inhibited cell cycle progression from the G1 to S phase in A431 cells and inhibited cyclin A gene expression in Saos-2 cells. GE3 inhibited the expression of E2F-dependent transcription of cyclin A, cdc2, DNA polymerase α, and c-*myc* but not E2F-independent transcription of c-*fos*. Furthermore, GE3 inhibited the DNA binding of both E2F-1 homodimer and E2F-1/DP-1 heterodimer without interference of DNA binding of both Sp1 and AP-1 in vitro. These data suggested that GE3 exhibited cytotoxicity by acting as an inhibitor of E2F transcriptional factor.

Biological Activity:

1–20 nM: IC_{50} for cytotoxicity against human tumor cell lines [1]
2 mg/kg: antitumor activity against the PSN-1 human pancreatic carcinoma [1]

References:

1. Sakai Y, Yoshida T, Tsujita T, et al., & Mizukami T (1997) GE3, a novel hexadepsipeptide antitumor antibiotic, produced by *Streptomyces* sp. I. Taxonomy, production, isolation, physico-chemical properties, and biological activities. J Antibiot 50:659–664
2. Agatsuma T, Sakai Y, Mizukami T & Saitoh Y (1997) GE3, a novel hexadepsipeptide antitumor antibiotic produced by *Streptomyces* sp. II. Structure determination. J Antibiot 50:704–708

Geldanamycin

Key Words: [Antitumor][Hsp90 inhibitor][Tyrosine kinase inhibitor]

Structure:

Molecular Formula: $C_{29}H_{40}N_2O_9$
Molecular Weight: 560
Solubility: DMSO, ++ ; H_2O, - ; MeOH, ++

Discovery/Isolation/Structure:

Geldanamycin was discovered in the culture filtrates of *Streptomyces hygroscopicus* and showed moderately inhibitory activity in vitro against protozoa, bacteria and fungi [1]. The structure was assigned by spectral analysis and on the basis of its chemical characteristics [2].

Biological Studies:

Geldanamycin is active against L-1210, KB cells, and mammalian cells including cells transformed by viruses and human leukemia cells in vitro and in vivo [1,3].

Uehara et al. found that benzoquinonoid ansamycins, including geldanamycin, reduced the intracellular phosphorylation of p60[src] at a permissive temperature in a rat kidney cell line infected with a temperature-sensitive mutant of Rous sarcoma virus. This effect was reversible and accompanied by morphological changes from the transformed to the normal phenotype [4]. To identify mechanisms, Whitesell et al. prepared a solid phase-immobilized geldanamycin derivative and affinity precipitated the molecular targets. Immobilized geldanamycin bound elements of a major class of heat shock protein (Hsp90) in a stable and pharmacologically specific manner [5]. These results suggest that geldanamycin specifically binds to Hsp90 and disrupts certain multimolecular complexes containing this protein. By using geldanamycin,

Schulte et al. demonstrated that association with Hsp90 is essential for both Raf-1 protein stability and its proper localization in the cell [6], and Thulasiraman and Matts analysed the on the kinetics of chaperone-mediated renaturation [7].

Biological Activity:

0.1 μg/ml: morphological changes from the transformed to the normal phenotype [4]
2 μM: disruption of multimolecular complexes containing Raf-1 protein [6]

References:

1. DeBoer C, Meulman PA, Wnuk RJ & Peterson DH (1970) Geldanamycin, a new antibiotic. J Antibiot 23:442–447
2. Sasaki K, Rinehart Jr. KL, Slomp G, et al., & Olson EC (1970) Geldanamycin. I. Structure assignment. J Am Chem Soc 92:7591–7593
3. Sasaki K, Yasuda H & Onodera K (1979) Growth inhibition of virus transformed cells in vitro and antitumor activity in vivo of geldanamycin and its derivatives. J Antibiot 32:849–851
4. Uehara Y, Hori M, Takeuchi T & Umezawa H (1986) Phenotypic change from transformed to normal induced by benzoquinonoid ansamycins accompanies inactivation of p60^{src} in rat kidney cells infected with Rous sarcoma virus. Mol Cell Biol 6:2198–2206
5. Whitesell L, Mimnaugh EG, Costa BD, et al., & Neckers LM (1994) Inhibition of heat shock protein HSP90-pp60^{v-src} heteroprotein complex formation by benzoquinone ansamycins: Essential role for stress proteins in oncogenic transformation. Proc Natl Acad Sci USA 91:8324–8328
6. Schulte TW, Blagosklonny MV, Ingui C & Neckers L (1995) Disruption of the Raf-1-Hsp90 molecular complex results in destabilization of Raf-1 and loss of Raf-1-Ras association. J Biol Chem 270:24585–24588
7. Thulasiraman V & Matts RL (1996) Effect of geldanamycin on the kinetics of chaperone-mediated renaturation of firefly luciferase in rabbit reticulocyte lysate. Biochemistry 35:13443–13450

Genistein

Key Words: [Tyrosine kinase inhibitor]

Structure:

Molecular Formula: $C_{15}H_{10}O_5$
Molecular Weight: 270
Solubility: DMSO, +++ ; H_2O, - ; MeOH, +++

Discovery:

Genistein, an isoflavone compound, was isolated from subterranean clover (*Trifolium subterraneum* L.) as a main compound responsible for its oestrogenic activity [1]. *Streptomyces xanthophaeus* [2] and *Pseudomonas* sp. [3] were also reported to produce genistein. There are many kinase inhibitors, such as PD98059 which is selective and cell-permeable inhibitor of MAP kinase (MEK), were synthesized as genistein derivatives.

Biological Studies:

Genistein was isolated as an inhibitor of tyrosine phosphorylation from a culture broth of *Pseudomonas* [3]. This compound inhibits the activity of tyrosine kinases, such as epidermal growth factor receptor and pp60[src], but scarcely inhibits the activity of serine and threonine kinases such as cAMP-dependent protein kinase [4]. The inhibition was competitive with respect to ATP and noncompetitive to a phosphate acceptor, histone H2B. Several groups used genistein as a tyrosine kinase-specific inhibitor for analyses of cellular functions. Use and specificity of genistein was reviewed [5]. However, it should be remembered that genistein is not such a specific inhibitor for tyrosine kinases [6].

Biological Activity:

2.6 µM: IC_{50} for autophosphorylation activity of EGF receptor [4]
25.9 µM: IC_{50} for phosphorylation of casein by pp60src [4]

References:

1. Bradbury RB & White DE (1951) The chemistry of subterranean clover. I. Isolation of formononetin and genistein. J Chem Soc 1951:3447–3449
2. Hazato T, Naganawa H, Kumagai M, et al., & Umezawa H (1979) β-Galactosidase-inhibiting new isoflavonoids produced by actinomycetes. J Antibiot 32:217–222
3. Ogawara H, Akiyama T, Ishida J, et al., & Suzuki K (1986) A specific inhibitor for tyrosine protein kinase from *Pseudomonas*. J Antibiot 39:606–608
4. Akiyama T, Ishida J, Nakagawa S, et al., & Fukami Y (1987) Genistein, a specific inhibitor of tyrosine-specific protein kinases. J Biol Chem 262:5592–5595
5. Akiyama T & Ogawara H (1991) Use and specificity of genistein as inhibitor of protein -tyrosine kinases. Methods Enzymol 201:362–370
6. Osada H, Magae J, Watanabe C & Isono K (1988) Rapid screening method for inhibitors of protein kinase C. J Antibiot 41:925–931

Herbimycin A

Key Words: [Antitumor][Tyrosine kinase inhibitor]

Structure:

Molecular Formula: $C_{30}H_{42}N_2O_9$
Molecular Weight: 574
Solubility: DMSO, ++ ; H_2O, ± ; MeOH, +

Discovery/Isolation/Structure:

Herbimycin, has potent herbicidal activity against most mono- and dicotyledonous plants, and was isolated from the fermentation broth of *Streptomyces hygroscopicus* [1]. The structure and biosynthetic pathway were determined.

Biological Studies:

Uehara et al. and Murakami et al. found that benzoquinonoid ansamycins, including herbimycin, reduce the intracellular phosphorylation of p60[src] at a permissive temperature in a rat kidney cell line infected with a temperature-sensitive mutant of Rous sarcoma virus. This effect was reversible and accompanied by morphological changes from the transformed to the normal phenotype [3,4]. To identify the mechanisms, Whitesell et al. prepared a solid phase-immobilized geldanamycin derivative and affinity precipitated the molecular targets. Immobilized geldanamycin bound elements of a major class of heat shock protein (HSP90) in a stable and pharmacologically specific manner. The formation of the complex was inhibited by herbimycin as well as by geldanamycin [5]. These results suggest that benzoquinonoid ansamycins, geldanamycin and herbimycin, specifically bind to Hsp90 and disrupt certain multi-molecular complexes containing this protein.

Yamashita et al. reported anti-angiogenesis activity of herbimycin A in a three assay system, the proliferation of cultured capillary endothelial cells, angiogenesis in chick chorioallantoic membrane, and angiogenesis induced by crude tumor angiogenesis factor in rabbit cornea [6].

Biological Activity:

0.5 µg/ml: morphological changes from the transformed to the normal phenotype [4]

0.013 µg/ml: IC_{50} for growth of endothelial cells [6]

0.01 µg/pellet: suppression of capillary growth in TAF assay [6]

References:

1. Omura S, Iwai Y, Takahashi Y, et al., & Ikai T (1979) Herbimycin, a new antibiotic produced by a strain of *Streptomyces*. J Antibiot 32:255–261
2. Omura S, Nakagawa A & Sadakane N (1979) Structure of herbimycin, a new ansamycin antibiotic. Tetrahedron Lett 44:4323–4326
3. Murakami Y, Mizuno S, Hori M & Uehara Y (1988) Reversal of transformed phenotypes by herbimycin A in src oncogene expressed rat fibroblasts. Cancer Res 48:1587–1590
4. Uehara Y, Hori M, Takeuchi T & Umezawa H (1986) Phenotypic change from transformed to normal induced by benzoquinonoid ansamycins accompanies inactivation of p60[src] in rat kidney cells infected with Rous sarcoma virus. Mol Cell Biol 6:2198–2206
5. Whitesell L, Mimnaugh EG, Costa BD, et al., & Neckers LM (1994) Inhibition of heat shock protein HSP90-pp60[v-src] heteroprotein complex formation by benzoquinone ansamycins: Essential role for stress proteins in oncogenic transformation. Proc Natl Acad Sci USA 91:8324–8328
6. Yamashita T, Sakai M, Kawai Y, et al., & Takahashi K (1989) A new activity of herbimycin A: Inhibition of angiogenesis. J Antibiot 42:1015–1017

Inostamycin

Key Words: [Antitumor][CDP-DG:inositol transferase inhibitor]

Structure:

Molecular Formula: $C_{38}H_{68}O_{11}$
Molecular Weight: 700
Solubility: DMSO, + ; H_2O, - ; MeOH, +++

Discovery/Isolation/Structure:

Inostamycin was found as an inhibitor of phosphatidylinositol turnover from the culture broth of a *Streptomyces* strain. Spectroscopic and crystallographic analysis revealed that it has a novel polyether structure [1]. Inostamycin related compounds, inostamycins B and C, were also isolated [2].

Biological Studies:

Imoto et al. found that inostamycin inhibited cellular inositol phosphate formation only when it was added at the same time as labeled inositol, and that it inhibited in vitro CDP-DG:inositol transferase activity of the A431 cell membrane [3].

Imoto et al. demonstrated that PI synthesis is involved in the regulation of S phase induction by using inostamycin [4]. This early G1 arrest might be due to the inhibition of cyclin D1 and E expression [5]. Inostamycin not only decreased cyclin D1 expression, but induced apoptosis in human small cell lung carcinoma Ms-1 cells [6].

Kawada et al. found that inostamycin reverses multidrug resistance in KB cells and could inhibit P-glycoprotein irreversibly by binding to plasma membranes irreversibly through phosphatidylethanolamine [7–9].

Biological Activity:

$0.5\,\mu g/ml$: IC_{50} for phosphatidylinositol turnover inhibition in cultured A431 cells [1]
$0.02\,\mu g/ml$: IC_{50} for in vitro CDP-DG:inositol transferase activity of the A431 cell membrane [3]
$1\,\mu g/ml$: G1 arrest induction in NRK cells [4]
$1\,\mu g/ml$: inhibit active [^{3}H]vinblastine efflux from multidrug-resistant KB-C4 cells [7]

References:

1. Imoto M, Umezawa K, Takahashi Y, et al., & Takeuchi T (1990) Isolation and structure determination of inostamycin, a novel inhibitor of phosphatidylinositol turnover. J Nat Prod 53:825–829
2. Odai H, Shindo K, Odagawa A, et al., & Takeuchi T (1994) Inostamycins B and C, new polyether antibiotics. J Antibiot 47:939–41
3. Imoto M, Taniguchi Y & Umezawa K (1992) Inhibition of CDP-DG: inositol transferase by inostamycin. J Biochem 112:299–302
4. Imoto M, T TM, Deguchi A & Umezawa K (1994) Involvement of phosphatidylinositol synthesis in the regulation of S phase induction. Exp Cell Res 215:228–233
5. Deguchi A, Imoto M & Umezawa K (1996) Inhibition of G1 cyclin expression in normal rat kidney cells by inostamycin, a phosphatidylinositol synthesis inhibitor. J Biochem 120:1118–22
6. Imoto M, Tanabe K, Simizu S, et al., & Umezawa K (1998) Inhibition of cyclin D1 expression and induction of apoptosis by inostamycin in small cell lung carcinoma cells. Jpn J Cancer Res 89:315–22
7. Kawada M & Umezawa K (1991) Long-lasting accumulation of vinblastine in inostamycin-treated multidrug-resistant KB cells. Jpn J Cancer Res 82:1160–4
8. Kawada M, Sumi S, Umezawa K, et al., & Seto H (1992) Circumvention of multidrug resistance in human carcinoma KB cells by polyether antibiotics. J Antibiot 45:556–62
9. Kawada M & Umezawa K (1995) Inostamycin, an inhibitor of P-glycoprotein function, interacts specifically with phosphatidylethanolamine. Jpn J Cancer Res 86:873–8

K-252a

Key Words: [Cell cycle inhibitor][Kinase inhibitor]

Structure:

Molecular Formula: $C_{27}H_{21}N_3O_5$
Molecular Weight: 467
Solubility: DMSO, +++ ; H_2O, - ; MeOH, +

Discovery/Isolation:

K-252a was isolated from the culture broth of *Nocardiopsis* sp., which exhibits an extremely potent inhibitory activity on protein kinase C [1]. The structure of K-252a was determined by spectral studies and chemical conversion [2].

Biological and Chemical Studies:s

Koizumi et al. found that K-252a selectively inhibits the actions of NGF, but not FGF or dibutyryl cAMP, on PC12 cells [3]. The specific effect on NGF-dependent neuronal differentiation is thought to be caused by selective inhibition of tyrosine kinase activity of the NGF receptor, gp140^{rk} [4–7]. It is interesting that K-252a showed neurotrophic effects on SH-SY5Y cells, which is a human neuroblastoma cell line lacking NGF receptor [8].

Usui et al. found that K-252a caused DNA re-replication in rat diploid fibroblasts 3Y1 without an intervening mitosis, producing tetraploid cells and suggested that a putative protein kinase(s) sensitive to K-252a plays an important role in the mechanism for preventing over-replication after the completion of previous DNA synthesis [9].

Biological Activity:

32.9 nM: IC_{50} for Protein kinase C [1]
200 nM: prevention of neurite generation initiated by NGF in PC12 cells [3]
20 nM: K_i value for chicken gizzard myosin light chain kinase in a competitive manner with respect to ATP [10]
1.7 nM: IC_{50} for phosphorylase kinase [11]

References:

1. Kase H, Iwahashi K & Matsuda Y (1986) K-252a, a potent inhibitor of protein kinase C from microbial origin. J Antibiot 39:1059–1065
2. Yasuzawa T, Iida T, Yoshida M, et al., & Sano H (1986) The structures of the novel protein kinase C inhibitors K-252a, b, c and d. J Antibiot 39:1072–1078
3. Koizumi S, Contreras ML, Matsuda Y, et al., & Guroff G (1988) K-252a: A specific inhibitor of the action of nerve growth factor on PC 12 cells. J Neurosci 8:715–721
4. Berg MM, Sternberg DW, Parada LF & Chao MV (1992) K-252a inhibits nerve growth factor-induced trk proto-oncogene tyrosine phosphorylation and kinase activity. J Biol Chem 267:13–16
5. Muroya K, Hashimoto Y, Hattori S & Nakamura S (1992) Specific inhibition of NGF receptor tyrosine kinase activity by K-252a. Biochim Biophys Acta 1135:353–356
6. Nye SH, Squinto SP, Glass DJ, et al., & Yancopoulos GD (1992) K-252a and staurosporine selectively block autophosphorylation of neurotrophin receptors and neurotrophin-mediated responses. Mol Biol Cell 3:677–686
7. Ohmichi M, Decker SJ, Pang L & Saltiel AR (1992) Inhibition of the cellular actions of nerve growth factor by staurosporine and K252A results from the attenuation of the activity of the trk tyrosine kinase. Biochemistry 31:4034–4039
8. Maroney AC, Lipfert L, Forbes ME, et al., & Dionne CA (1995) K-252a induces tyrosine phosphorylation of the focal adhesion kinase and neurite outgrowth in human neuroblastoma SH-SY5Y cells. J Neurochem 64:540–549
9. Usui T, Yoshida M, Abe K, et al., & Beppu T (1991) Uncoupled cell cycle without mitosis induced by a protein kinase inhibitor, K-252a. J Cell Biol 115:1275–1282
10. Nakanishi S, Yamada K, Kase H, et al., & Nonomura Y (1988) K-252a, a novel microbial product, inhibits smooth muscle myosin light chain kinase. J Biol Chem 263:6215–6219
11. Elliott LH, Wilkinson SE, Sedgwick AD, et al., & Nixon JS (1990) K252a is a potent and selective inhibitor of phosphorylase kinase. Biochem Biophys Res Commun 171:148–154

Kaitocephalin

Key Words: [Glutamate receptor antagonist]

Structure:

Molecular Formula: $C_{18}H_{21}N_3O_9Cl_2$
Molecular Weight: 493
Solubility: DMSO, ++ ; H_2O, + ; MeOH, +

Discovery:

Kaitocephalin was isolated in the course of screening for substances that protect chick primary telencephalic neurons from kainate toxicity [1]. It consists of a pyrrolidine moiety with tricarboxylic acids and a dichlorohydroxybenzoate substructure.

Biological Studies and Activity:

Kaitocephalin was isolated from *Eupenicillium shearii* as a glutamate receptor antagonist which protects chick telencephalic neurons as well as rat hippocampal neurons from kainate toxicity at 500 µM with EC_{50} values of 0.68 µM and 2.4 µM, respectively. Kaitocephalin also protected chick primary telencephalic and rat hippocampal neurons from AMPA/cyclothiazide (500 µM/50 µM) toxicity with EC_{50} values of 0.6 and 0.4 µM, respectively.

References:

1. Shin-ya K, Kim J-S, Furihata K, et al., & Seto H (1997) Structure of kaitocephalin, a novel glutamate receptor antagonist produced by *Eupenicillium shearii.* Tetrahedron Lett 38:7079–7082

Kifunensine (FR900494)

Key Words: [Immunomodulator]

Structure:

Molecular Formula: $C_8H_{12}N_2O_6$
Molecular Weight: 232
Solubility: DMSO, + ; H_2O, ++ ; MeOH, +

Discovery:

Kifunensine was isolated as an immunoactive substance produced by an actino-mycete [1] and its structure was determined by chemical and physical evidence and by X-ray crystal analysis [2]. Kifunensine was re-isolated as a compound that potentiates the expression of class II MHC molecules on the cell surface of macrophages, which are antigen-presenting cells.

Biological Studies:

Kifunensine was isolated and exhibited a competitive action against an immunosuppressive factor produced in the serum of tumor bearing mice and has the capacity to restore the depression of lymphocytes [1]. However, Kayakiri et al. found that kifunensine is a weak inhibitor of jack bean α-mannosidase [2], Elbein et al. found kifunensine is a very potent inhibitor of mannosidase I, but not mannosidase II [3]. Kifunensine caused a complete shift in the structure of the N-linked oligosaccharides from complex chains to $Man_9(GlcNAc)_2$ structures in cultured cells. Ahrens reported that target cell N-glycosides influence the NK-target interaction mediated by adhesion molecules such as ICAM-1 by using kifunensine and other N-glycan processing inhibitors that promote accumulation of high mannose-type glycosides [4]. However kifunensine showed interferon-γ-like immunosuppressive effects in a mice system, but the mechanism is unknown.

Biological Activity:

1.2×10^{-4} M: IC_{50} for jack bean α-mannosidase [2]
$2\text{-}5 \times 10^{-8}$ M: IC_{50} for mannosidase I [3]
1 μg/ml: a complete shift in the structure of the N-linked oligosaccharides from complex chains to $Man_9(GlcNAc)_2$ structures in cell culture [3]

References:

1. Iwami M, Nakayama O, Terano H, et al., & Imanaka H (1987) A new immunomodulator, FR-900494: taxonomy, fermentation, isolation, and physico-chemical and biological characteristics. J Antibiot 40:612–622
2. Kayakiri H, Takase S, Shibata T, et al., & Koda S (1989) Structure of kifunensine, a new immunomodulator isolated from an actinomycete. J Org Chem 54:4015–4016
3. Elbein AD, Tropea JE, Mitchell M & Kaushal GP (1990) Kifunensine, a potent inhibitor of the glycoprotein processing mannosidase I. J Biol Chem 265:15599–5605
4. Ahrens PB (1993) Role of target cell glycoproteins in sensitivity to natural killer cell lysis. J Biol Chem 268:385–391

Lactacystin

Key Words: [Cell cycle inhibitor] [Neurite outgrowth inducer][Proteasome inhibitor]

Structure:

Molecular Formula: $C_{15}H_{24}N_2O_7S$
Molecular Weight: 376
Solubility: DMSO, ++ ; H_2O, - ; MeOH, ++

Discovery/Isolation:

During the course of screening for microbial metabolites which induce differentiation of the mouse neuroblastoma cell line Neuro 2A, a novel compound designated lactacystin was isolated from the cultured broth of a *Streptomyces* sp. OM-6519 [1]. The structure was elucidated by NMR spectroscopic and X-ray crystallographic analyses [2].

Biological Studies:

A neurite-like structure generation and transient increase of intracellular cAMP level were observed in lactacystin-treated Neuro 2A cells [1]. Fenteany et al. reported that lactacystin inhibited progression of the M phase of Neuro 2A cells and MG-63 osteosarcoma cells beyond the G1 phase of the cell cycle [3]. Furthermore, Katagiri et al. reported that lactacystin arrested the cell cycle at both G0/G1 and G2 phases in Neuro 2A cells [4].

Fenteany et al. identified the 20S proteasome as a lactacystin specific cellular target. The ability of lactacystin analogs to inhibit cell cycle progression and induce neurite outgrowth correlated with their ability to inhibit the proteasome. Lactacystin appeared to modify covalently the highly conserved amino-terminal threonine of the mammalian proteasome subunit X. They suggested that this threonine residue has a catalytic role [3].

Soldatenkov and Dritschilo found that exposure of Ewing's sarcoma cells to lactacystin resulted in accumulation of ubiquitinated proteins and apoptotic pathways, and suggested that functional disorder of the ubiquitin-proteasome system plays an important role in apoptosis [5].

Application:

Several groups used lactacystin for the analysis of proteasomal degradation. Mori et al. suggest that the degradation process of the ligand-stimulated platelet-derived growth factor β-receptor involves the ubiquitin-proteasome proteolytic pathway [6]. Jensen et al. also suggested that CFTR and presumably other intrinsic membrane proteins are substrates for proteasomal degradation during their maturation within the ER [7]. Cui et al. used lactacystin to study the role of the proteasome in the activation-induced cell death of T cells [8].

Biological Activity:

1.3 µM: a neurite-like structure in Neuro 2A cells [1]
1.3 µM: cell cycle arrest at both G0/G1 and G2 phases in Neuro 2A cells [4]
5.0 µM: accumulation of ubiquitinated proteins and induction of apoptosis in Ewingís sarcoma cells [5]

References:

1. Omura S, Fujimoto T, Otoguro K, et al., & Sasaki Y (1991) Lactacystin, a novel microbial metabolite, induces neuritogenesis of neuroblastoma cells. J Antibiot 44:113–116
2. Omura S, Matsuzaki K, Fujimoto T, et al., & Nakagawa A (1991) Structure of lactacystin, a new microbial metabolite which induces differentiation of neuroblastoma cells. J Antibiot 44:117–118
3. Fenteany G, Standaert RF, Reichard GA, et al., & Schreiber SL (1994) A β-lactone related to lactacystin induces neurite outgrowth in a neuroblastoma cell line and inhibits cell cycle progression in an osteosarcoma cell line. Proc Natl Acad Sci USA 91:3358–3362
4. Katagiri M, Hayashi M, Matsuzaki K, et al., & Omura S (1995) The neuritogenesis inducer lactacystin arrests cell cycle at both G0/G1 and G2 phases in Neuro 2a cells. J Antibiot 48:344–346
5. Soldatenkov VA & Dritschilo A (1997) Apoptosis of Ewing's sarcoma cells is accompanied by accumulation of ubiquitinated proteins. Cancer Res 57:3881–3885
6. Mori S, Tanaka K, Omura S & Saito Y (1995) Degradation process of ligand-stimulated platelet-derived growth factor β-receptor involves ubiquitin-proteasome proteolytic pathway. J Biol Chem 270:29447–29452
7. Jensen TJ, Loo MA, Pind S, et al., & Riordan JR (1995) Multiple proteolytic systems, including the proteasome, contribute to CFTR processing. Cell 83:129–135
8. Cui H, Matusi K, Omura S, et al., & Ju S(1997) Proteasome regulation of activation-induced T cell death. Proc Natl Acad Sci USA 94:7515–7520

Leptomycin
(Leptomycins/Anguinomycins/Kazusamycins/
Leptolstatin/Leptofuranins)

Key Words: [Antitumor][Apoptos][Cell cycle inhibitor]

Structure:

		R1	R2	R3	R4	MW
Leptomycin A	$(C_{32}H_{46}O_6)$;	CH_3	CH_3	CH_3	COOH	526
Leptomycin B	$(C_{33}H_{48}O_6)$;	CH_3	CH_2CH_3	CH_3	COOH	540
Anguinomycin A	$(C_{31}H_{44}O_6)$;	H	CH_3	CH_3	COOH	512
Anguinomycin B	$(C_{32}H_{46}O_6)$;	H	CH_2CH_3	CH_3	COOH	526
Anguinomycin C	$(C_{31}H_{46}O_4)$;	H	CH_3	CH_3	CH_3	483
Anguinomycin D	$(C_{32}H_{48}O_4)$;	H	CH_2CH_3	CH_3	CH_3	497
Kazusamycin A	$(C_{33}H_{48}O_7)$;	CH_3	CH_2CH_3	CH_2OH	COOH	556
Kazusamycin B	$(C_{32}H_{46}O_7)$;	CH_3	CH_3	CH_2OH	COOH	542
Leptolstatin	$(C_{31}H_{46}O_5)$;	H	CH_3	CH_3	CH_2OH	498

		R1	R2	MW
Leptofuranin A	$(C_{32}H_{48}O_5)$;	CH_2OH	H	512
Leptofuranin B	$(C_{33}H_{50}O_5)$;	CH_2OH	CH_3	526
Leptofuranin C	$(C_{32}H_{46}O_5)$;	CHO	H	511
Leptofuranin D	$(C_{33}H_{48}O_5)$;	CHO	CH_3	525

Solubility: DMSO, +; H_2O, -; MeOH, +++

Discovery/Isolation:

Leptomycin A and B (LMA and LMB) were isolated as strong antifungal antibiotics [1]. A low concentration of LMB caused inhibition of cell division, producing elongated cells with morphologically altered nuclei and several cell plates in *Schizosaccharomyces pombe* [2]. Anguinomycins were isolated as antitumor antibiotics [3, 4]. Kazusamycin A and B were also isolated as antitumor antibiotics. These antibiotics did not possess antibacterial activity against Gram-positive and Gram-negative bacteria, but showed strong cytotoxic activity against HeLa cells, L1210 and P388 in vitro [5, 6]. Leptolstatin was isolated as a gap phase-specific inhibitor of the mammalian cell cycle [7, 8]. Leptofuranin A, B, C and D were isolated as antitumor antibiotics against pRB-inactivated cells. The leptofuranins arrested the growth of normal cells and induced apoptotic cell death against tumor cells and cells transformed with the adenovirus E1A gene [9].

Chemical Studies/Structure:

These compounds have the structure characteristics of an unsaturated, branched-chain fatty acid with a terminal δ-lactone ring [3–5, 7, 10, 11]. The leptofuranins are novel leptomycin-related substances containing a tetrahydrofuran ring. Leptofuranins C and D were revealed to be in tautomeric isomerism and their relative stereo-chemistries were analyzed [12].

Application:

Leptomycins

Leptomycin B reversibly blocked both the G1 and G2 phases in mammalian cells and *S. pombe*. The treated mammalian cells were introduced into a resting state, and after removal of LMB, proliferative tetraploid cells were produced from the cells which had been arrested by LMB at the G2 phase as a result of DNA replication without passage through the M phase [13]. Nishi et al. revealed that the molecular target of leptomycin B was CRM1 which is required for maintaining higher order chromosome structures, but the function was unknown [14]. The function of crm1 was revealed by another approach, i.e., the discovery that leptomycin B inhibited the nuclear export of proteins which contain nuclear export signals (NESs). Leptomycin B is a potent and specific inhibitor of the NES-dependent nuclear export of proteins and the binding of p110 to NES is inhibited by LMB. p110 was found to be CRM1. Thus, the CRM1 protein could act as a NES receptor involved in nuclear protein export [15–18].

Anguinomycins

Anguinomycin A and B were highly cytotoxic to murine P388 leukemia cells and displayed potent antitumor activity in mice [3]. The anguinomycins induced growth arrest against normal cells and induced cell death against transformed cells in which pRB was inactivated [4].

Kazusamycins

Kazusamycin A and B were tested as antitumor drugs in mice with several types of schedules. Kazusamycins B and A, and an analog of B seemed to show no significant differences in their effectiveness. The effective dose range and toxicity were markedly dependent on the tumor lines tested and the regimen used [19].

Takamiya et al. examined the effect on the cell cycle of L1210 and antitumor activity. Kazusamycin B arrested L1210 cells at the G1 phase and induced apoptosis accompanied with abnormal condensation of nuclei coincided with the appearance of unidentified population [21].

Biological Activity:

Leptomycins

12–250 ng/ml: MIC for *Schizosaccharomyces, Mucor* sp. [1]
1–200 ng/ml: reversible cell cycle arrest at G1/G2 in 3Y1 cells [13]
400 nM: inhibition of nuclear export in vitro [17]

Anguinomycins

0.1–0.2 ng/ml: IC_{50}s for cytotoxicity to murine P388 leukemia cells [3]

Kazusamycins

1 ng/ml: IC_{50}s for growth of tumor cells in vitro [19,20]

Leptolstatin

1–200 ng/ml: reversible cell cycle arrest at G1/G2 in 3Y1 cells [8].

Leptofuranins

100 ng/ml: apoptosis induction in HeLa cells [9].

References:

1. Hamamoto T, Gunji S, Tsuji H, & Beppu T (1983) Leptomycins A and B, new antifungal antibiotics. I. Taxonomy of the producing strain and their fermentation, purification and characterization. J Antibiot 36: 639–645.
2. Hamamoto T, Uozumi T, & Beppu T (1985) Leptomycins A and B, new antifungal antibiotics. III. Mode of action of leptomycin B on *Schizosaccharomyces pombe*. J Antibiot 38: 1573–1580.
3. Hayakawa Y, Adachi K, & Komeshima N (1987) New antitumor antibiotics, anguinomycins A and B. J Antibiot 40: 1349–1352.
4. Hayakawa Y, Sohda K, Shin-ya K, et al., & Seto H (1995) Anguinomycins C and D, new antitumor antibiotics with selective cytotoxicity against transformed cells. J Antibiot 48: 954–961.

5. Funaishi K, Kawamura K, Sugiura Y, et al., & Komiyama K (1987) Kazusamycin B, a novel antitumor antibiotic. J Antibiot 40: 778–785.
6. Umezawa I, Komiyama K, Oka H, et al., & Takano S (1984) A new antitumor antibiotic, kazusamycin. J Antibiot 37: 706–711.
7. Abe K, Yoshida M, Naoki H, et al., & Beppu T (1993) Leptolstatin from *Streptomyces* sp. SAM1595, a new gap phase-specific inhibitor of the mammalian cell cycle. II. Physicochemical properties and structure. J Antibiot 46: 735–40.
8. Abe K, Yoshida M, Horinouchi S, & Beppu T (1993) Leptolstatin from *Streptomyces* sp. SAM1595, a new gap phase-specific inhibitor of the mammalian cell cycle. I. Screening, taxonomy, purification and biological activities. J Antibiot 46: 728–34.
9. Hayakawa Y, Sohda K, Furihata K, et al., & Seto H (1996) Studies on new antitumor antibiotics, leptofuranins A, B, C and D. I. Taxonomy, fermentation, isolation and biological activities. J Antibiot 49: 974–979.
10. Hamamoto T, Seto H, & Beppu T (1983) Leptomycins A and B, new antifungal antibiotics. II. Structure elucidation. J Antibiot 36: 646–650.
11. Komiyama K, Okada K, Oka H, et al., & Umezawa I (1985) Structural study of a new antitumor antibiotic, kazusamycin. J Antibiot 38: 220–229.
12. Hayakawa Y, Sohda K, & Seto H (1996) Studies on new antitumor antibiotics, leptofuranins A, B, C and D. II. Physicochemical properties and structure elucidation. J Antibiot 49: 980–984.
13. Yoshida M, Nishikawa M, Nishi K, et al., & Beppu T (1990) Effects of leptomycin B on the cell cycle of fibroblasts and fission yeast cells. Exp Cell Res 187: 150–156.
14. Nishi K, Yoshida M, Fujiwara D, et al., & Beppu T (1994) Leptomycin B targets a regulatory cascade of crm1, a fission yeast nuclear protein, involved in control of higher order chromosome structure and gene expression. J Biol Chem 269: 6320–6324.
15. Fukuda M, Asano S, Nakamura T, et al., & Nishida E (1997) CRM1 is responsible for intracellular transport mediated by the nuclear export signal. Nature 390: 308–311.
16. Kudo N, Khochbin S, Nishi K, et al., & Horinouchi S (1997) Molecular cloning and cell cycle-dependent expression of mammalian CRM1, a protein involved in nuclear export of proteins. J Biol Chem 272: 29742–29751.
17. Fornerod M, Ohno M, Yoshida M, & Mattaj IW (1997) CRM1 is an export receptor for leucine-rich nuclear export signals. Cell 90: 1051–1060.
18. Ossareh-Nazari B, Bachelerie F, & Dargemont C (1997) Evidence for a role of CRM1 in signal-mediated nuclear protein export. Science 278: 141–144.
19. Yoshida E, Komiyama K, Naito K, et al., & Umezawa I (1987) Antitumor effect of kazusamycin B on experimental tumors. J Antibiot 40: 1596–1604.
20. Takamiya K, Yoshida E, Takahashi T, et al., & Umezawa I (1988) The effect of kazusamycin B on the cell cycle and morphology of cultured L1210 cells. J Antibiot 41: 1854–1861.

Lovastatin
(Compactin, Pravastin)

Lovastatin is the common name of monacolin K and mevinolin.
Compactin and ML-236B are the same compound.

Key Words: [Apoptosis][HMG-CoA reductase inhibitor][Cell cycle inhibitor]

Structure:

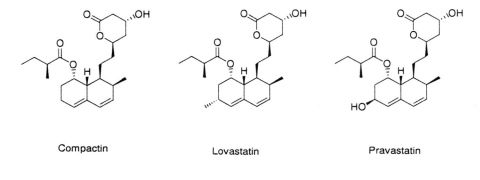

Compactin   Lovastatin   Pravastatin

	Compactin	Lovastatin	Pravastatin
Molecular Formula:	$C_{23}H_{34}O_5$	$C_{24}H_{36}O_5$	$C_{23}H_{34}O_6$
Molecular Weight:	390	404	406
Solubility:	DMSO, ++ ; H_2O, ++ ; MeOH, ++		

Discovery/Isolation:

Compactin was isolated from cultures of *Penicillium brevicompactum* as an antifungal metabolite and its structure was determined by a combination of spectroscopic, chemical, and X-ray crystallographic methods [1]. Endo et al. isolated the same (ML-236B) and related (ML-236A and C) compounds as new inhibitors of cholesterogenesis from *P. citrinum* [2].

Monacolin K was isolated from *Monascus* sp. [3] and mevinolin, the same compound, was isolated from *Asperugillus terreus* [4] as a hypo-cholesterolemic agent.

Pravastatin (Mevalotin®; 3β-hydroxy compactin) was first found as a minor urinary metabolite of compactin in dogs, and Serizawa et al. isolated several microorganisms which are able to hydroxylate compactin at the 3β-position [5]. Matsuoka et al. investigated the hydroxylation mechanism in *Streptomyces carbophilus* in detail [6].

Biological and Chemical Studies:

Alberts et al. found that mevinolin was a potent competitive inhibitor of hydroxymethylglutaryl-coenzyme A (HMG-CoA) reductase. They showed that treatment of dogs for 3 weeks with mevinolin at 8 mg/kg per day resulted in a 29.3% lowering of plasma cholesterol [4]. Pravastatin has a distinct advantage in inhibiting the HMG-CoA reductase in a more tissue-selective manner for clinical usage [7].

Several groups reported that compactin inhibited cell proliferation in the G1 phase and cell cycle progression was restored by the addition of cholesterol or mevalonic acid [8–12]. Quesney-Huneeus found a dual role of mevalonate in the cell cycle, i.e., the cholesterol requirement is limited to the early and mid-G1 phases, whereas the isopentenyl effect is required at the late G1-S interphase of the cell cycle [9]. Jakobisiak et al. investigated the effects of lovastatin on the cell cycle progression of the human bladder carcinoma T24 cell line expressing activated $p21^{ras}$ and found that lovastatin altered the intracellular location of $p21^{ras}$ without changing the cellular content [13].

Faust and Dice prepared antibodies that recognize isopentenyladenosine (i6A), a modified nucleoside derived from mevalonic acid. In immunoblot assays, affinity-purified anti-i6A antibodies specifically bound to a 26 kDa protein (i6A26) in Chinese hamster ovary cells and the expression of i6A26 correlated with cellular proliferation and growth. They speculated that i6A26 contains isopentenyladenine moieties and mediates isoprenoid regulation of DNA synthesis [14].

It was reported that mevinolin stimulated neurite outgrowth and acetylcholinesterase activity in Neuro-2A murine neuroblastoma cells [15]. Apoptosis induction activity was also reported [16].

Biological Activity:

The effects of 6 HMG-CoA reductase inhibitors: pravastatin, lovastatin, simvastatin, atorvastatin, fluvastatin and cerivastatin were analyzed in cultured human smooth muscle cells, fibroblasts, endothelial cells and myoblasts [17].

Compactin

2.6×10^{-8} M: IC_{50} for cholesterol synthesis to digitonin-precipitable sterols in a rat liver enzyme system [2]
0.25 to 2.5 µM: The inhibition of DNA synthesis by compactin could be completely prevented by adding monkey low density lipoproteins, cholesterol, or mevalonic acid [12].

Lovastatin/Monacolin K/Mevinolin

0.6 nM: K_i for HMG-CoA reductase, competitive [4]
46 µg/kg: IC_{50} for orally administered in the rat in an acute assay
100 nM: apoptosis in human malignant glioma cell lines A172 and U87-MG [13]

References:

1. Brown AG, Smale TC, King TJ, et al., & Thompson RH (1976) Crystal and molecular structure of compactin, a new antifungal metabolite from *Penicillium brevicompactum*. J Chem Soc Perkin Trans I:1165–1170

2. Endo A, Kuroda M & Tsujita Y (1976) ML-236A, ML-236B, and ML-236C, a new inhibitors of cholesterogenesis produced by *Penicillium citrinum*. J Antibiot 29:1346–1348

3. Endo A (1979) Monacolin K, a new hypo-cholesterolemic agent produced by a *Monascus* species. J Antibiot 32:852–854

4. Alberts AW, Chen J, Kuron G, et al., & Springer J (1980) Mevinolin: A highly potent competitive inhibitor of hydroxymethylglutaryl-coenzyme A reductase and a cholesterol-lowering agent. Proc Natl Acad Sci USA 77:3957–3961

5. Serizawa N, Serizawa S, Nakagawa K, et al., & Terahara A (1983) Microbial hydroxylation of ML-236B (Compactin) studies on microorganisms capable of 3b-hydroxylation of ML-236B. J Antibiot 36:887–891

6. Matsuoka T, Miyakoshi S, Tanzawa K, et al., & Serizawa N (1989) Purification and characterization of cytochrome P-450sca from *Streptomyces carbophilus*. ML-236B (compactin) induces a cytochrome P-450sca in *Streptomyces carbophilus* that hydroxylates ML-236B to pravastatin sodium (CS-514), a tissue-selective inhibitor of 3-hydroxy- 3-methylglutaryl-coenzyme-A reductase. Eur J Biochem 184:707–13

7. Tsujita Y, Kuroda M, Shimada Y, et al., & Fujii S (1986) CS-514, a competitive inhibitor of 3-hydroxy-3-methylglutaryl coenzyme A reductase: tissue-selective inhibition of sterol synthesis and hypolipidemic effect on various animal species. Biochim Biophys Acta 877:50–60

8. Keyomarsi K, Sandoval L, Band V & Pardee AB (1991) Synchronization of tumor and normal cells from G1 to multiple cell cycles by lovastatin. Cancer Res 51:3602–3609

9. Quesney-Huneeus V, Galick HA, Siperstein MD, et al., & Nelson JA (1983) The dual role of mevalonate in the cell cycle. J Biol Chem 258:378–385

10. Doyle JW & Kandutsch AA (1988) Requirement for mevalonate in cycling cells: Quantitative and temporal aspects. J Cell Physiol 137:133–140

11. Cornell RB & Horwitz AF (1980) Apparent coordination of the biosynthesis of lipids in cultured cells: its relationship to the regulation of the membrane sterol:phospholipid ratio and cell cycling. J Cell Biol 86:810–819

12. Habenicht AJ, Glomset JA & Ross R (1980) Relation of cholesterol and mevalonic acid to the cell cycle in smooth muscle and swiss 3T3 cells stimulated to divide by platelet-derived growth factor. J Biol Chem 255:5134–5140

13. Jakobisiak M, Bruno S, Skierski JS & Darzynkiewicz Z (1991) Cell cycle-specific effects of lovastatin. Proc Natl Acad Sci USA 88:3628–3632

14. Faust JR & Dice JF (1991) Evidence for isopentenyladenine modification on a cell cycle-regulated protein. J Biol Chem 25:9961–9970

15. Maltese WA & Sherida KM (1985) Differentiation of neuroblastoma cells induced by an inhibitor of mevalonate synthesis: relation of neurite outgrowth and acetylcholinesterase activity to changes in cell proliferation and blocked isoprenoid synthesis. J Cell Physiol 125:540–558

16. Jones KD, Couldwell WT, Hinton DR, et al., & RE REL (1994) Lovastatin induces growth inhibition and apoptosis in human malignant glioma cells. Biochem Biophys Res Commun 205:1681–1687

17. Negre-Aminou P, van Vliet AK, van Erck M, et al., & Cohen LH (1997) Inhibition of proliferation of human smooth muscle cells by various HMG-CoA reductase inhibitors; Comparison with other human cell types. Biochim Biophys Acta 1345:259–268

Macrosphelide A

Key Words: [Anti-adherent]

Structure:

Molecular Formula: $C_{16}H_{22}O_8$
Molecular Weight: 342
Solubility: DMSO, ++ ; H_2O, + ; MeOH, ++

Discovery:

Macrosphelide A and B were isolated as anti-adherent compounds [1] and the structure was determined to be a 16-membered macrolide antibiotic possessing three ester bonds in the ring structure [2]. Macrosphelide C and D were also isolated [3].

Biological Studies:

Macrosphelide A and B dose-dependently inhibited the adhesion of HL-60 cells to a LPS-activated HUVEC monolayer [1].

Biological Activity:

3.5 μM: IC_{50} of macrosphelide A against the adhesion of HL-60 cells to LPS-activated HUVEC monolayer [1]
36 μM: IC_{50} of macrosphelide B against the adhesion of HL-60 cells to LPS-activated HUVEC monolayer [1]

References:

1. Hayashi M, Kim YP, Hiraoka H, et al., & Omura S (1995) Macrosphelide, a novel inhibitor of cell-cell adhesion molecule. I. Taxonomy, fermentation, isolation and biological activities. J Antibiot 48:1435–1439
2. Takamatsu S, Kim YP, Hayashi M, et al., & Omura S (1996) Macrosphelide, a novel inhibitor of cell-cell adhesion molecule. II. Physiochemical properties and structural elucidation. J Antibiot 49:95–98
3. Takamatsu S, Hiraoka H, Kim YP, et al., & Omura S (1997) Macrosphelides C and D, novel inhibitors of cell adhesion. J Antibiot 50:878–880

Madindoline

Key Words: [Inhibitor of IL-6 activity]

Structure:

Molecular Formula: $C_{22}H_{27}NO_4$
Molecular Weight: 369
Solubility: DMSO, +++ ; H_2O, ± ; MeOH, +++

Discovery:

Hayashi et al. isolated madindolines A and B from *Streptomyces* sp. as inhibitors of IL-6 activity; they are 3a-hydroxy-indoline with diketocyclopentene at the N position [1,2]. Madindoline B appears to be a stereoisomer of madindoline A.

Biological Studies:

Madindolines A and B exhibited dose-dependent inhibition of MH60 cells, an IL-6-dependent cell line, in the presence of 0.1 U/ml IL-6 [2]. These compounds did not inhibit the growth of cell lines which are not IL-6 dependent and the growth inhibition of the MH60 cell line was reversed by addition of excess 0.4 U/ml of IL-6 to the culture medium.

These compounds did not show any microbial activity at a concentration of 1 mg/ml.

Biological Activity:

8 µM: IC_{50} of madindoline A for MH60 cells growth [2]
30 µM: IC_{50} of madindoline B for MH60 cells growth [2]

References:

1. Takamatsu S, Kim YP, Enomoto A, et al., & Omura S (1997) Madindolines, novel inhibitors of IL-6 activity from *Streptomyces* sp. K93-0711. II. Physico-chemical properties and structural elucidation. J Antibiot 50:1069–1072
2. Hayashi M, Kim YP, Takamatsu S, et al., & Omura S (1996) Madindoline, a novel inhibitor of IL-6 activity from *Streptomyces* sp. K93-0711. I. Taxonomy, fermentation, isolation and biological activities. J Antibiot 49:1091–1095

Manoalide

Key Words: [Phospholipase A2 inhibitor]

Structure:

Molecular Formula: $C_{25}H_{36}O_5$
Molecular Weight: 416
Solubility: DMSO, +++ ; H_2O, +++ ; MeOH, +++

Discovery:

Manoalide was isolated from a sponge, *Luffariella variabilis*, and the structure was determined by spectral analysis and chemical transformation [1]. Kobayashi et al. isolated several of the manoalide family as cytotoxic sesterterpenes from the marine sponge, *Hyrtios erecta*, and the absolute configurations were determined [2].

Biological Studies:

Manoalide antagonizes phorbol-induced inflammation but not that induced by arachidonic acid. Lombardo and Dennis showed that manoalide is a potent irreversible inhibitor of the cobra venom phospholipase A2 with lysine residues in the enzyme [3]. Manoalide also inhibits phosphoinositide-specific phospholipase C (PI-PLC) [4] and the Ca^{2+} channel [5]. It was also reported that manoalide inhibited 5-lipoxygenase (5-LO) activity by an indirect mechanism [6,7].

Several groups reported the inhibition mechanism of manoalide on phospholipase A2 and suggested that lysine residue(s) of phospholipase A2 were modified [8–11].

Recently, its ability to induce apoptosis was reported [12], and its derivatives, manoalide 25-acetals, exhibited in vivo antitumor activity and inhibition of the DNA-relaxing activity of mouse DNA topoisomerase I and the DNA-unknotting activity of calf thymus DNA topoisomerase II [2].

Biological Activity:

2 µM: IC_{50} for cobra venom phospholipase A2 [3]
3–5 µM: IC_{50} for both cytosolic and purified PI-PLC [4]
0.3 µM: IC_{50} for 5-lipoxygenase activity in RBL- cells [7]

References:

1. Dilip de Silva E & Scheuer PJ (1980) Manoalide, an antibiotic sesterterpenoid from the marine sponge *Luffariella variabilis* (Polejaeff). Tetrahedron Lett 21:1611–1614
2. Kobayashi M, Okamoto T, Hayashi K, et al., & Kitagawa I (1994) Marine natural products. XXXII. Absolute configurations of C-4 of the manoalide family, biologically active sesterterpenes from the marine sponge *Hyrtios erecta*. Chem Pharm Bull 42:265–270
3. Lombardo D & Dennis EA (1985) Cobra venom phospholipase A2 inhibition by manoalide. A novel type of phospholipase inhibitor. J Biol Chem 260:7234–7240
4. Bennett CF, Mong S, Wu HL, et al., & Crooke ST (1987) Inhibition of phosphoinositide-specific phospholipase C by manoalide. Mol Pharmacol 32:587–593
5. Wheeler LA, Sachs G, De Vries G, et al., & Muallem S (1987) Manoalide, a natural sesterterpenoid that inhibits calcium channels. J Biol Chem 262:6531–6538
6. Cabre F, Carabaza A, Suesa N, et al., & Carganico G (1996) Effect of manoalide on human 5-lipoxygenase activity. Inflamm Res 45:218–223
7. De Vries GW, Amdahl L, Mobasser A, et al., & Wheeler LA (1988) Preferential inhibition of 5-lipoxygenase activity by manoalide. Biochem Pharmacol 37:2899–2905
8. Glaser KB, Vedvick TS & Jacobs RS (1988) Inactivation of phospholipase A2 by manoalide. Localization of the manoalide binding site on bee venom phospholipase A2. Biochem Pharmacol 37:3639–3646
9. Bianco ID, Kelley MJ, Crowl RM & Dennis EA (1995) Identification of two specific lysines responsible for the inhibition of phospholipase A2 by manoalide. Biochim Biophys Acta 1250:197–203
10. Fujii S, Tahara Y, Toyomoto M, et al., & Hayashi K (1995) Chemical modification and inactivation of phospholipases A2 by a manoalide analogue. Biochem J 308:297–304
11. Ortiz AR, Pisabarro MT & Gago F (1993) Molecular model of the interaction of bee venom phospholipase A2 with manoalide. J Med Chem 36:1866–1879
12. Miao JY, Kaji K, Hayashi H & Araki S (1997) Inhibitors of phospholipase promote apoptosis of human endothelial cells. J Biochem 121:612–618

Manumycins

Key Words: [Farnesyltransferase inhibitors]

Structure: show manumycin A

Molecular Formula: $C_{31}H_{38}N_2O_7$
Molecular Weight: 550
Solubility: DMSO, ++ ; H_2O - ; MeOH, +++

Discovery:

Manumycin was isolated as an antibiotic against Gram-positive bacteria and fungi and furthermore, an inhibition of the developmental processes of some insects [1]. Its absolute structure was determined by chemical degradation and CD spectra [2]. Many manumycin group antibiotics have been isolated.

Biological Studies:

Manumycins were originally isolated as antibiotics and antitumor compounds. Hara et al. reisolated manumycin in the course of screening of inhibitors of farnesyltransferase. Kinetic analyses of the inhibition suggested that manumycin acts as a competitive inhibitor of farnesyltransferase with respect to farnesyl pyrophosphate, and acts as a noncompetitive inhibitor of farnesyltransferase with respect to the farnesyl acceptor, the Ras protein [3]. Manumycin showed significant activity to inhibit the growth of the Ki-*ras*-point mutated human pancreatic cancer cell line, MIA PaCa-2 in mice [4].

Several groups investigated the roles of small-G proteins in neurite formation [5], and signal transduction [6,7].

Tanaka et al. isolated the manumycin-related compounds, concluding that manumycin A and B as interleukin-1β converting enzyme (ICE) inhibitors [8].

Biological Activity:

5 µM: IC_{50} for farnesyltransferase [3]
1.2 µM: K_i value for farnesyltransferase with respect to farnesyl pyrophosphate [3]
0.65 µM: IC_{50} of manumycin B for ICE [8]

References:

1. Zeeck A, Schroder K, Frobel K, et al., & Thiericke R (1987) The structure of manumycin. I. Characterization, structure elucidation and biological activity. J Antibiot 40:1530–1540
2. Thiericke R, Stellwaag M, Zeeck A & Snatzke G (1987) The structure of manumycin. III. Absolute configuration and conformational studies. J Antibiot 40:1549–1554
3. Hara M, Akasaka K, Akinaga S, et al., & Tamanoi F (1993) Identification of Ras farnesyltransferase inhibitors by microbial screening. Proc Natl Acad Sci USA 90:2281–2285
4. Ito T, Kawata S, Tamura S, et al., & Matasuzawa Y (1996) Suppression of human pancreatic cancer growth in BALB/c nude mice by manumycin, a farnesyl:protein transferase inhibitor. Jpn J Cancer Res 87:113–116
5. Hiwasa T, Kondo K, Hishiki T, et al., & Nakagawara A (1997) GDNF-induced neurite formation was stimulated by protein kinase inhibitors and suppressed by Ras inhibitors. Neurosci Lett 238:115–118
6. Lee-Kwon W, Park D & Bernier M (1998) Involvement of the Ras/extracellular signal-regulated kinase signalling pathway in the regulation of ERCC-1 mRNA levels by insulin. Biochem J 331:591–597
7. Lee-Kwon W, Park D, Baskar PV, et al., & Bernier M (1998) Antiapoptotic signaling by the insulin receptor in Chinese hamster ovary cells. Biochemistry 37:15747–15757
8. Tanaka T, Tsukuda E, Uosaki Y & Matsuda Y (1996) EI-1511-3, -5 and EI-1625-2, novel interleukin-1β converting enzyme inhibitors produced by *Streptomyces* sp. E-1511 and E-1625. III. Biochemical properties of EI-1511-3, -5 and EI-1625-2. J Antibiot 49:1085–1090

Matlystatins

Key Words: [Type IV collagenase inhibitors]

Structure: shows matlystatin A

Molecular Formula: $C_{27}H_{47}N_5O_8S$
Molecular Weight: 601
Solubility: DMSO, ++ ; H_2O, + ; MeOH, ++

Discovery:

Matlystatins were isolated in the course of a screening for inhibitors of type IV collagenases [1], and their structures were determined by a systematic application of homo- and heteronuclear 2D NMR and FAB-MS/MS techniques [2]. Their structures were characterized by the presence of piperazic acid and hydroxamic acid moieties, structural motifs often seen in protease inhibitors. The absolute configuration of matlystatin B was unambiguously determined by total synthesis and the structure-activity relationships of its stereoisomers were investigated [3]. Tamaki et al. synthesized and investigated the structure-activity relationships of gelatinase inhibitors derived from matlystatins [4].

Biological Studies:

Matlystatin A inhibits type IV collagenases while 7- to 11-fold greater concentrations are required to inhibit thermolysin and aminopeptidase M. The inhibition is reversible and competitive with respect to gelatin. It inhibits the invasion of basement membrane Matrigel by human fibrosarcoma HT1080 dose-dependently [5,6].

Biological Activity:

0.3 µM: IC_{50} of matlystatin A for 92 kDa type IV collagenases [6]
0.56 µM: IC_{50} for matlystatin A for 72 kDa type IV collagenases [6]

21.6 μM: IC_{50} for matlystatin A for invasion of basement membrane Matrigel by human fibrosarcoma HT1080 [6]

References:

1. Ogita T, Sato A, Enokita R, et al., & Tanzawa K (1992) Matlystatins, new inhibitors of typeIV collagenases from *Actinomadura atramentaria*. I. Taxonomy, fermentation, isolation, and physico-chemical properties of matlystatin-group compounds. J Antibiot 45:1723–1732
2. Haruyama H, Ohkuma Y, Nagaki H, et al., & Kinoshita T (1994) Matlystatins, new inhibitors of type IV collagenases from Actinomadura atramentaria. III. Structure elucidation of matlystatins A to F. J Antibiot 47:1473–1480
3. Tamaki K, Kurihara S, Oikawa T, et al., & Sugimura Y (1994) Matlystatins, new inhibitors of type IV collagenases from *Actinomadura atramentaria*. IV. Synthesis and structure-activity relationships of matlystatin B and its stereoisomers. J Antibiot 47:1481–1492
4. Tamaki K, Tanzawa K, Kurihara S, et al., & Sugimura Y (1995) Synthesis and structure-activity relationships of gelatinase inhibitors derived from matlystatins. Chem Pharm Bull 43:1883–1893
5. Fujii H, Nakajima M, Aoyagi T & Tsuruo T (1996) Inhibition of tumor cell invasion and matrix degradation by aminopeptidase inhibitors. Biol Pharm Bull 19:6–10
6. Tanzawa K, Ishii M, Ogita T & Shimada K (1992) Matlystatins, new inhibitors of type IV collagenases from *Actinomadura atramentaria*. II. Biological activities. J Antibiot 45:1733–1737

Maytansine/Ansamitocin P-3

Key Words: [Antitumor][Cell cycle inhibitor][Microtubule inhibitor]

Structure:

	Maytansine	Ansamitocin P-3
Molecular Formula:	$C_{34}H_{46}N_3O_{10}Cl$	$C_{32}H_{43}N_2O_9Cl$
Molecular Weight:	692	635
Solubility:	DMSO, ++ ; H_2O, + ; MeOH, ++	

Discovery/Isolation:

First *ansa* compound isolated from a plant. Kupchan et al. isolated maytansine from various *Maytenus* species as an antitumor agent that significantly inhibits mouse P388 leukemia, L-1210 leukemia, Lewis lung carcinoma, and B16 melanoma. They elucidated the maytansine structure [1].

Ansamitocins were isolated from *Nocardia* sp as a new group of ansamycin antibiotics with potent antitumor activity. Structures of ansamitocins were found to be similar to maytansine and related maytansinoids obtained from plant sources [2–4]. A mutant strain of *Nocardia* which produces large amount of ansamitocins was isolated by treatment with ethidium bromide [5,6].

Biological and Chemical Studies:

Maytansine irreversibly inhibits cell division in eggs of sea urchins and clams. Maytansine does not affect formation of the mitotic organizing center but does inhibit in vitro polymerization of tubulin [7]. Maytansine binding to crude rat brain tubulin is a temperature- and ionic strength-dependent process. Maytansine competitively inhibits vincristine binding and the binding of both drugs is reversible. Both drugs appear to share a common binding site, although an additional site specific for maytansine seems to be present [8,9]. Fellous et al. speculated that two of the main groups of microtubule-associated proteins, Tau and MAP2, interact with tubulin by a strikingly different mechanism, as judged from the results of maytansine inhibition kinetics to Tau- and MAP2-catalyzed tubulin assembly [10].

Ansamitocins also show antimicrobial activities as well as inhibition of microtubule assembly [2,5].

Biological Activity of Maytansine:

6×10^{-8} M: irreversible inhibition of cell division of egg urchins and clams [7]

1.5×10^6 M^{-1}: K_A for binding maytansine to rat brain tubulin [9]

10×10^{-6} M: K_i for competitive inhibition of vincristine binding to tubulin [9]

$0.5 - 2 \mu$M: inhibition of tau- and MAP2-catalyzed tubulin assembly [10]

References:

1. Kupchan SM, Komoda Y, Court WA, et al., & Bryan RF (1972) Maytansine, a novel antileukemic *ansa* macrolide from *Maytenus ovatus*. J Am Chem Soc 94:1354–1356
2. Higashide E, Asai M, Ootsu K, et al., & Yoneda M (1977) Ansamitocin, a group of novel maytansinoid antibiotics with antitumour properties from *Nocardia*. Nature 270:721–722
3. Asai M, Mizuta E, Izawa M, et al., & Kishi T (1979) Isolation, chemical characterization and structure of ansamitocin, a new antitumor ansamycin antibiotic. Tetrahedron 35:1079–1085
4. Tanida S, Hasegawa T, Hatano K, et al., & Yoneda M (1980) Ansamitocins, maytansinoid antitumor antibiotics producing organism, fermentation, and antimicrobial activities. J Antibiot 33:192–198
5. Tanida S, Izawa M & Hasegawa T (1981) Ansamitocin analogs from a mutant strain of *Nocardia*. I. Isolation of the mutant, fermentation and antimicrobial properties. J Antibiot 34:489–495
6. Izawa M, Tanida S & Asai M (1981) Ansamitocin analogs from a mutant strain of *Nocardia*. II. Isolation and structure. J Antibiot 34:496–506
7. Remillard S, Rebhun LI, Howie GA & Kupchan SM (1975) Antimitotic activity of the potent tumor inhibitor maytansine. Science 189:1002–1005
8. Bhattacharyya B & Wolff J (1977) Maytansine binding to the vinblastine sites of tubulin. FEBS Lett 75:159–162
9. Mandelbaum-Shavit F, Wolpert-DeFilippes MK & Johns DG (1976) Binding of maytansine to rat brain tubulin. Biochem Biophys Res Commun 72:47–54
10. Fellous A, Luduena RF, Prasad V, et al., & Smith PT (1985) Effects of Tau and MAP2 on the interaction of maytansine with tubulin: inhibitory effect of maytansine on vinblastine-induced aggregation of tubulin. Cancer Res 45:5004–5010

Mitomycin C

Key Words: [Alkylating reagent][Antitumor][Apoptosis]

Structure:

Molecular Formula: $C_{15}H_{18}N_4O_5$
Molecular Weight: 334
Solubility: DMSO, ++ ; H_2O, ++ ; MeOH, ++

Discovery/Isolation/Structure:

Mitomycin, an antibiotic which is active against bacteria as well as Ehrlich carcinoma and Yoshida sarcoma, was isolated from *Streptomyces* sp. [1,2]. The structures of mitomycins A, B and C were reported [3–5]. Stevens et al. studied a series of degradation products from mitomycin C converted by acid hydrolysis [6]. The structure of *N*-brosylmitomycin A, a crystalline derivative, was determined using X-ray crystallography [7]. The absolute configuration of mitomycin C was determined by referring to that of *N*-(*p*-bromobenzoyl)mitomycin C [8,9]. Absolute configurations of mitomycin A and B were determined by the Bijvoet difference method [10].

Biological Studies:

Iyer and Szybalski investigated the molecular mechanism of mitomycin action. They found that mitomycins behave as bifunctional "alkylating" agents upon chemical or enzymatic reduction and that activated mitomycins reacted in vitro with purified DNA, linking its complementary strands. A high content of guanine and cytosine favors this crosslinking reaction, which is the basis of the lethal effect in vivo [11,12]. Tomasz and Lipman studied alkylation reactions of mitomycin C and UTP at acid pH [13], and Hashimoto et al. investigated the alkylation of GTP by activated mitomycin C [14].

Apoptosis was also induced with mitomycin C. Xu et al. reported that cells expressing wild-type p53 underwent extensive apoptosis after exposure to mitomycin C, whereas cells carrying mutated p53 responded weakly [15].

Biological Activity:

$0.5\,\mu g/ml$: IC_{90} for antibacterial activity against *Bacillus subtilis* [11]
$30\,\mu M$: apoptosis induction in normal human mammary epithelial cell 70N [15]

References:

1. Hata T, Sano Y, Sugawara R, et al., & Hoshi T (1956) Mitomycin, a new antibiotic from streptomyces. I. J Antibiot Ser A 9:141–146
2. Sugawara R & Hata T (1956) Mitomycin, a new antibiotic from *Streptomyces*. II. Description of the strain. J Antibiot Ser A 9:147–151
3. Tulinsky A (1962) The structure of mitomycin A. J Am Chem Soc 84:3188–3190
4. Webb JS, Cosulich DB, Mowat JH, et al., & Lancaster JE (1962) The structures of mitomycins A, B and C and porfiromycin. Part I. J Am Chem Soc 84:3185–3187
5. Webb JS, Cosulich DB, Mowat JH, et al., & Lancaster JE (1962) The structures of mitomycins A, B and C and porfiromycin. Part II. J Am Chem Soc 84:3187–3188
6. Stevens CL, Taylor KG, Munk ME, et al., & Uzu K (1964) Chemistry and structure of mitomycin C. J Med Chem 8:1–10
7. Tulinsky A & van den Hende JH (1967) The crystal and molecular structure of N-brosylmitomycin A. J Am Chem Soc 89:2905–2911
8. Ogawa K, Nomura A, Fujiwara T & Tomita K (1979) Crystal and molecular structure of mitomycin C, an anticancer antibiotic. Bull Chem Soc Japan 52:2334–2338
9. Shirahata K & Hirayama N (1983) Reviseid absolute configuration of mitomycin C. X-ray analysis of 1-*N*-(*p*-bromobenzoyl) mitomycin C. J Am Chem Soc 105:7199–7200
10. Hirayama N & Shirahata K (1987) Structural studies of mitomycins. I. Absolute configurations of mitomycin A and B. Acta Cryst B43:555–559
11. Iyer VN & Szybalski W (1963) A molecular mechanism of mitomycin action: Linking of complementary DNA strands. Microbiology 50:355–362
12. Iyer VN & Szybalski W (1964) Mitomycins and porfiromycin: Chemical mechanism of activation and cross-linking of DNA. Science 145:55–58
13. Tomasz M & Lipman R (1979) Alkylation reactions of mitomycin C at acid pH. J Am Chem Soc 101:20:6063–6067
14. Hashimoto Y, Shudo K & Okamoto T (1980) Alkylation of 5'-guanylic acid by activated mitomycin C. Chem Pharm Bull 28(3):1961–1963
15. Xu C, Meikrantz W, Schlegel R & Sager R (1995) The human papilloma virus 16E6 gene sensitizes human mammary epithelial cells to apoptosis induced by DNA damage. Proc Natl Acad Sci USA 92:7829–7833

Myriocin
(ISP-1/Thermozymocidin)

Key Words: [Antifungal][Immunosuppressant][Serine palmitoyltransferase inhibitor]

Structure:

Molecular Formula: $C_{21}H_{39}NO_6$
Molecular Weight: 401
Solubility: DMSO, +++ ; H_2O, ± ; MeOH, +++

Discovery:

Myriocin was isolated as an antifungal compound from *Myriococcum albomyces* [1]. Fujita et al. reisolated myriocin as a potent immunosuppressant [2]. The stereochemistry was confirmed [3] and total synthesis was performed by Yoshikawa et al. [4].

Biological Studies:

The structure of myriocin is homologous to sphingosine. Miyake et al. found that myriocin inhibits serine palmitoyltransferase which catalyzes the first step of sphingolipid biosynthesis. The growth inhibition of CTLL-2 induced by myriocin was completely abolished by the addition of sphingosines or sphingosine-1-phosphate, but not by sphingomyelin or glycosphingolipids. These results suggest that myriocin suppresses T cell proliferation by the modulation of sphingolipid metabolism [5]. Recently, Chen et al. synthesized myriocin derivatives and purified two specific myriocin-binding proteins by affinity chromatography. These two proteins were identified as murine LCB1 and LCB2, mammalian homologs of two yeast proteins that have been genetically linked to sphingolipid biosynthesis [6]

Several groups investigated the roles of sphingolipid metabolism in apoptosis [7] and ER-Golgi transport of GPI-anchored proteins [8,9].

Biological Activity:

1 µg/ml: MIC against *Candida albicans* [1]
10-18 nM: IC_{50} of mouse allogenic mixed lymphocyte reaction [2]
15 nM: IC_{50} of CTLL-2 cells growth inhibition [5]
0.28 nM: IC_{50} of serine palmitoyltransferase inhibition [5]
47 nM: Apoptosis induction in CTLL-2 cells [7]

References:

1. Kluepfel D, Bagli J, Baker H, et al., & Kudelski A (1972) Myriocin, a new antifungal antibiotic from *Myriococcum albomyces*. J Antibiot 25:109–115
2. Fujita T, Inoue K, Yamamoto S, et al., & Okumoto T (1994) Fungal metabolites. Part 11. A potent immunosuppressive activity found in Isaria sinclairii metabolite. J Antibiot 47:208–215
3. Bagli JF, Kluepfel D & St-Jacques M (1973) Elucidation of structure and stereochemistry of myriocin. A novel antifungal antibiotic. J Org Chem 38:1253–1260
4. Yoshikawa M, Yokokawa Y, Okuno Y & Murakami N (1994) Total synthesis of a novel immunosuppressant, myriocin (thermozymocidin, ISP-1), and Z-myriocin. Chem Pharm Bull 42:994–996
5. Miyake Y, Kozutsumi Y, Nakamura S, et al., & Kawasaki T (1995) Serine palmitoyltransferase is the primary target of a sphingosine-like immunosuppressant, ISP-1/myriocin. Biochem Biophys Res Commun 211:396–403
6. Chen JK, Lane WS & Schreiber SL (1999) The identification of myriocin-binding proteins. Chem Biol 6:221–35
7. Nakamura S, Kozutsumi Y, Sun Y, et al., & Kawasaki T (1996) Dual roles of sphingolipids in signaling of the escape from and onset of apoptosis in a mouse cytotoxic T-cell line, CTLL-2. J Biol Chem 271:1255–1257
8. Reggiori F, Canivenc-Gansel E & Conzelmann A (1997) Lipid remodeling leads to the introduction and exchange of defined ceramides on GPI proteins in the ER and Golgi of *Saccharomyces cerevisiae*. EMBO J 16:3506–18
9. Horvath A, Sutterlin C, Manning-Krieg U, et al., & Riezman H (1994) Ceramide synthesis enhances transport of GPI-anchored proteins to the Golgi apparatus in yeast. EMBO J 13:3687–95

Neocarzinostatin

Key Words: [Antitumor][Kinase inhibitor]

Structure: here show the chromophore

Molecular Formula: $C_{35}H_{33}NO_{12}$
Molecular Weight: 659
Solubility: DMSO, ++ ; H_2O, + ; MeOH, ++

Discovery:

Neocarzinostatin (NCS) is an acidic polypeptide antitumor antibiotic purified from the culture filtrate of *Streptomyces carzinostaticus* [1] and which consisted of chromophore (MW. 659) and NCS-apoprotein (MW. approx. 11,000). Primary structure of the NCS-apoprotein shows considerable homology with the other antitumor antibiotic proteins macromomycin and actinoxanthin [2,3]. Determination of the full chemical structure of the NCS-chromophore as an enediyne compound was performed by Edo et al. [4]. The crystal structure of neocarzinostatin was resolved [5].

Biological Studies:

Neocarzinostatin shows antitumor activity by inducing DNA strand scission [6]. The chromophore of NCS is an active compound and has been identified as a molecule responsible for DNA degradation in vitro and in vivo [7,8]. Several groups studied the DNA scission mechanism of neocarzinostatin in detail [9–12].

It was reported that neocarzinostatin also inhibits protein kinases such as microtubule-associated protein kinase [13], protein kinase A [14], and casein kinase II [15].

Biological Activity:

0.3 µg/ml: MIC for HeLa cells [1]
0.8–3.2 mg/kg/day injection: 100% survival of S-180 sarcoma bearing mice [1]
0.5–5.0 mg/ml: DNA strand scission in HeLa cell [6]
60 nM: ID_{50} for casein kinase II [15]

References:

1. Ishida N, Miyazaki K, Kumagai K & Rikimaru M (1965) Neocarzinostatin, an antitumor antibiotic of high molecular weight. J Antibiot Ser. A 18:68–76
2. Meienhofer J, Maeda H, Glaser CB, et al., & Kuromizu K (1972) Primary structure of neocarzinostatin, an antitumor protein. Science 178:875–876
3. Gibson BW, Herlihy WC, Samy TS, et al., & Biemann K (1984) A revised primary structure for neocarzinostatin based on fast atom bombardment and gas chromatographic-mass spectrometry. J Biol Chem 259:10801–10806
4. Edo K, Mizugaki N, Koide Y, et al., & Ishida N (1985) The chemical structure of antitumor polypeptide antibiotic neocarzinostatin chromophore. Tetrahedron Lett 26:331–334
5. Kim KH, Kwon BM, Myers AG & Rees DC (1993) Crystal structure of neocarzinostatin, an antitumor protein-chromophore complex. Science 262:1042–1046
6. Beerman TA & Goldberg IH (1974) DNA strand scission by the antitumor protein neocarzinostatin. Biochem Biophys Res Commun 59:1254–1261
7. Kappen LS, Napier MA & Goldberg IH (1980) Roles of chromophore and apo-protein in neocarzinostatin action. Proc Natl Acad Sci U S A 77:1970–1974
8. Ohtsuki K & Ishida N (1980) The biological effect of a nonprotein component removed from neocarzinostatin. J Antibiot 33:744–750
9. Chin DH & Goldberg IH (1986) Generation of superoxide free radical by neocarzinostatin and its possible role in DNA damage. Biochemistry 25:1009–1015
10. Kappen LS, Ellenberger TE & Goldberg IH (1987) Mechanism and base specificity of DNA Breakage in intact cells by neocarzinostatin. Biochemistry 26:384–390
11. Kappen LS & Goldberg IH (1993) Site-specific cleavage at a DNA bulge by neocarzinostatin chromophore via a novel mechanism. Biochemistry 32:13138–3145
12. Kappen LS & Goldberg IH (1997) Characterization of a covalent monoadduct of neocarzinostatin chromophore at a DNA bulge. Biochemistry 36:14861–4867
13. Ohtsuki K, Koike T, Sato T, et al., & Satake N (1980) Neocarzinostatin (NCS) and microtubule-associated kinase. J Antibiot 33:1590–1593
14. Ohtsuki K, Sato T, Koike T, et al., & Ishida N (1981) The inhibitory mechanism of in vitro protein phosphorylation by a nonprotein chromophore removed from neocarzinostatin. Biochim Biophys Acta 673:147–156
15. Tanoue S, Karino A, Nayuki Y & Ohtsuki K (1998) Neocarzinostatin-chromophore: a potent inhibitor of casein kinase II in vitro. J Antibiot 51:95–98

Okadaic Acid

Key Words: [Antitumor][PP1/PP2A inhibitor]

Structure:

Molecular Formula:　$C_{44}H_{68}O_{13}$
Molecular Weight:　805
Solubility:　　　　　DMSO, ++ ;H_2O, ± ; MeOH, +++

Discovery/Isolation:

Tachibana et al. isolated okadaic acid, which is implicated as the causative agent of diarrhetic shellfish poisoning, independently from two sponges *Halichondria* (syn *Reniera*) *okadai* Kadota, a black sponge, commonly found along the Pacific coast of Japan, and *H. melanodocia*, a Caribbean sponge collected in the Florida Keys [1].

Biological Studies:

Shibata et al. found that okadaic acid caused a long-lasting tonic contraction in human umbilical arteries and indicated that okadaic acid has a unique contractile action on smooth muscle which shows a strong resistance to Ca^{2+} deficiency [2]. Bialojan and Takai investigated the inhibitory effect of okadaic acid on PP1, PP2A, PP2B and PP2C protein phosphatases as well as on a polycation-modulated (PCM) phosphatase. PP2A was most potently inhibited. PP2B was inhibited to a lesser extent [3]. Haystead et al. report that okadaic acid rapidly stimulated protein phosphorylation in intact cells, and behaves like a specific protein phosphatase inhibitor in a variety of metabolic processes. Their results indicate that PP1 and PP2A are the dominant protein phosphatases acting on a wide range of phosphoproteins in vivo [4].

　　Boe et al. reported that okadaic acid induced morphological changes typical of apoptosis in mammalian cells [5].

Biological Activity:

192 µg/kg: LC_{50} *i.p.* mice [1]
5 ng/ml: IC_{80} for growth inhibition of KB cells [1]
1×10^{-7}-1×10^{-4} M: a long-lasting tonic contraction in human umbilical arteries [2]
1 nM, 0.1–0.5 µM, and 4–5 µM: ID_{50} for PP2A, PP1 and PP2B [3]
0.1 to 1 µM: morphological alterations induction in freshly isolated rat hepatocytes in suspension or in primary culture, the human mammary carcinoma cell line MCF-7, the human neuroblastoma cell line SK-N-SH, rat pituitary adenoma GH3 cells, and rat promyelocytic IPC-81 cells [5]

References:

1. Tachibana K, Seheuer PJ, Tsukitani Y, et al., & Schmitz FJ (1981) Okadaic acid, a cytotoxic polyether from two marine sponges of the genus halichondria. J Am Chem Soc 103:2469–2471
2. Shibata S, Ishida Y, Kitano H, et al., & Kikuchi H (1982) Contractile effects of okadaic acid, a novel ionophore-like substance from black sponge, on isolated smooth muscles under the condition of Ca deficiency. J Pharmacol Exp Therapeutics 1982:135–143
3. Bialojan C & Takai A (1988) Inhibitory effect of a marine-sponge toxin, okadaic acid, on protein phosphatases. Specificity and kinetics. Biochem J 256:283–290
4. Haystead TAJ, Sim ATR, Carling D, et al., & Hardie DG (1989) Effects of the tumor promoter okadaic acid on intracellular protein phosphorylation and metabolism. Nature 337:78–81
5. Boe R, Gjertsen BT, Vintermyr OK, et al., & Doskeland SO (1991) The protein phosphatase inhibitor okadaic acid induces morphological changes typical of apoptosis in mammalian cells. Exp Cell Res 195:237–246

Oxanosine

Key Words: [Antitumor]

Structure:

Molecular Formula: $C_{10}H_{12}N_4O_6$
Molecular Weight: 284
Solubility: DMSO, ± ; H_2O, ++ ; MeOH, +

Discovery/Isolation:

Oxanosine was isolated from the culture broth of *Streptomyces capreolus* as a weak antibacterial antibiotic [1].

The structure was determined to be 5-amino-3-β-d-ribofuranosyl-3*H*-imidazo[4,5-d][1,3]oxazin-7-one by X-ray crystallographic analysis [2]. Total synthesis of oxanosine was reported by Yagisawa et al., and 2'-deoxyoxanosine was synthesized by chemical modification [3]. The antibacterial activities of 2'-deoxyoxanosine were enhanced remarkably in comparison with oxanosine. 2'-deoxyoxanosine had also a stronger activity in inhibiting the growth of L-1210 in vitro than oxanosine [4].

Biological and Chemical Studies:

The antibacterial activity was antagonized by addition of guanine, guanosine and 5'-guanylic acid. Oxanosine suppressed the growth of L-1210 leukemia in mice and this inhibition was also reversed by guanylic acid [5]. Oxanosine was confirmed to be a competitive inhibitor of GMP synthetase [6]. Oxanosine-5'-monophosphate was a potent nearly competitive inhibitor, with respect to IMP, of IMP dehydrogenase [7]. Uehara et al. found that oxanosine inhibited cell growth in vitro, as well as nucleic acid synthesis in mammalian cells. The inhibition of cell growth and nucleic acid synthesis was reversed by guanosine, GMP, and to a lesser extent by adenosine and inosine. Oxanosine also inhibits the conversion of [^{14}C]hypoxanthine to guanine nucleotides in cells [7].

Itoh et al. reported that oxanosine altered the transformed morphology of rat kidney cells integrating a K-ras*rs* gene into "normal" morphology at permissive temperatures. The cells under these conditions had lower levels of guanine nucleotides, and unstable and less palmitylated K-*ras* gene product [8].

Biological Activity:

$12.5\,\mu$g/ml: MIC for *Escherichia coli* K-12 on peptone agar [1]

200 mg/kg mice (*i.p.*): no toxicity[5]

0.53 and $0.15\,\mu$g/ml: IC_{50s} of oxanosine and 2'-deoxyoxanosine for antitumor activity to L1210[4]

7.4×10^{-4} M: K_i for GMP synthetase inhibition; competitive inhibition [6]

References:

1. Shimada N, Yagisawa N, Naganawa H, et al., & Umezawa H (1981) Oxanosine, a novel nucleoside from *Actinomycetes*. J Antibiot 34:1216–1218
2. Nakamura H, Yagisawa N, Shimada N, et al., & Iitaka Y (1981) The X-ray structure determination of oxanosine. J Antibiot 34:1219–1221
3. Yagisawa N, Kato K, Shimada N, et al., & Umezawa H (1983) A facile total synthesis of oxanosine, an novel nucleoside antibiotic. Tetrohedron Lett 24:931–932
4. Kato K, Yagisawa N, Shimada N, et al., & Umezawa H (1984) Chemical modification of oxanosine. I. Synthesis and biological properties of 2'-deoxyoxanosine. J Antibiot 37:941–942
5. Yagisawa N, Shimada N, Naganawa H, et al., & Iitaka Y (1981) Oxanosine, a novel nucleoside from *Actinomycetes*. Nuc Acids Res (Symposium Series) 10:55–58
6. Yagisawa N, Shimada N, Takita T, et al., & Umezawa H (1982) Mode of action of oxanosine, a novel nucleoside antibiotic. J Antibiot 35:755–759
7. Uehara Y, Hasegawa M, Hori M & Umezawa H (1985) Increased sensitivity to oxanosine, a novel nucleoside antibiotic, of rat kidney cells upon expression of the integrated viral *src* gene. Cancer Res 45:5230–5234
8. Itoh O, Kuroiwa S, Atsumi S, et al., & Hori M (1989) Induction by the guanosine analogue oxanosine of reversion toward the normal phenotype of K-*ras*-transformed rat kidney cells. Cancer Res 49:996–1000

Oxetanocin A and G

Key Words: [Antivirus]

Structure:

	Oxetanocin A	Oxetanocin G
Molecular Formula:	$C_{10}H_{13}N_5O_3$	$C_{10}H_{13}N_5O_4$
Molecular Weight:	251	267
Solubility:	DMSO, + ; H_2O, + ; MeOH, +	

Discovery/Isolation/Chemical Studies:

Oxetanocin was isolated from the culture broth of *Bacillus megaterium* as an anti-DNA virus and antibacterial antibiotic [1]. The structure of oxetanocin was found to be 9-[(2R,3R,4S)-3,4-bis(hydroxymethyl)-2-oxetanyl]adenine by X-ray crystallographic analysis [2]. Derivatives of oxetanocin, oxetanocins H, X and G, and 2-aminooxetanocin A were synthesized and antivirus activity of these compounds were examined [3].

Biological Studies:

Oxetanocin inhibits RNA virus infectivity such as that of HIV as well as DNA viruses [4,5]. Nishiyama et al. reported that oxetanocin G (OTX-G) was very potent and selective in inhibiting the replication of human cytomegalovirus and herpes simplex virus in vitro. The mode of action of OXT-G is by impairing viral DNA synthesis [6]. Carbocyclic OTX-G was also a most active antiviral compound but the mode of action is different from that of OXT-G because the antiviral activity was only partially reversed even by the addition of 100-fold excess deoxyguanosine [7]. Nagahata et al. found that chemically synthesized OXT-GTP inhibited the HBV endogenous DNA polymerase reaction and was incorporated into HBV DNA strands at a low efficiency compared with the incorporation of dGTP. A synthetic primer-template study revealed that OXT-GTP was incorporated into DNA strands at a low efficiency and that further extension of the DNA strand by using the 2' position of the incorporated OXT-G could take place [8].

Biological Activity of Oxetanocin G:

1.0 μg/ml: the median effective concentration for HCMV strain AD169 [6]
3.5 μg/ml: the median effective concentration for herpes simplex virus type 2 strain 186 [6]
> 1,000 μM: Cytotoxicity [8]
1,000 μM: the concentration at which OXT-G did not inhibit cellular DNA synthesis or viral RNA synthesis [8]

References:

1. Shimada N, Hasegawa S, Harada T, et al., & Takita T (1986) Oxetanocin, a novel nucleoside from bacteria. J Antibiot 39:1623–1625
2. Nakamura H, Hasegawa S, Shimada N, et al., & Iitaka Y (1986) The X-ray structure determination of oxetanocin. J Antibiot 39:1626–1629
3. Shimada N, Hasegawa S, Saito S, et al., & Takita T (1987) Derivatives of oxetanocin: oxetanocins H, X and G, and 2-aminooxetanocin A. J Antibiot 40:1788–1790
4. Hoshino H, Shimizu N, Shimada N, et al., & Takeuchi T (1987) Inhibition of infectivity of human immunodeficiency virus by oxetanocin. J Antibiot 40:1077–1078
5. Seki J-i, Shimada N, Takahashi K, et al., & Takeuchi T (1989) Inhibition of infectivity of human immunodeficiency virus by a novel nucleoside, oxetanocin, and related compounds. Antimicrob Agents Chemother 33:773–775
6. Nishiyama Y, Yamamoto N, Takahashi K & Shimada N (1988) Selective inhibition of human cytomegalovirus replication by a novel nucleoside, oxetanocin G. Antimicrob Agents Chemother 32:1053–1056
7. Nishiyama Y, Yamamoto N, Yamada Y, et al., & Takahashi K (1989) Anti-herpes virus activity of carbocyclic oxetanocin G in vitro. J Antibiot 42:1854–1859
8. Nagahata T, Kitagawa M & Matsubara K (1994) Effect of oxetanocin G, a novel nucleoside analog, on DNA synthesis by hepatitis B virus virions. Antimicrob Agents Chemother 38:707–712

Paclitaxel

Key Words: [Antitumor][Microtubule inhibitor]

Structure:

Molecular Formula: $C_{47}H_{51}NO_{14}$
Molecular Weight: 853
Solubility: DMSO, ++ ; H_2O, + ; MeOH, +++

Discovery/Isolation:

Paclitaxel (Taxol®) was isolated from the stem bark of the western yew, *Taxus brevifolia*. Paclitaxel has potent antileukemic and tumor inhibitory properties [1].

Biological Studies:

Schiff et al. found that paclitaxel completely inhibits division of exponentially grow- ing HeLa cells without effects on DNA, RNA or protein synthesis. They showed that paclitaxel acts as a promoter of calf brain microtubule assembly in vitro, in contrast to plant products such as colchicine and podophyllotoxin, which inhibit assembly. Paclitaxel decreases the critical concentration of tubulin required for assembly and microtubules polymerised in the presence of paclitaxel are resistant to depolymerisation by cold and $CaCl_2$ [2]. Kumar investigated the effects on microtubule assembly and found that there was a marked difference in the kinetics of tubulin polymerized in the presence of both paclitaxel and MAP2 as compared to that obtained with either of them alone [3]. The molecular antiproliferative effects of paclitaxel was studied in human K562 leukemia, and in MCF-7 breast and OVCAR-5 ovarian carcinoma cell cultures. Olah et al. observed the apoptosis inductive activity of paclitaxel [4].

Biological Activity:

0.25 μM: inhibition of division on HeLa cells [2]
2 – 100 nM: both a differentiation program and apoptosis on K562 leukemia cells [4]

References:

1. Wani MC, Taylor HL, Wall ME, et al., & McPhail AT (1971) Plant antitumor agents. VI. The isolation and structure of taxol, a novel antileukemic and antitumor agent from *Taxus brevifolia*. J Am Chem Soc 93:2325–2327
2. Schiff PB, Fant J & Horwitz SB (1979) Promotion of microtubule assembly in vitro by taxol. Nature 227:665–667
3. Kumar N (1981) Taxol-induced polymerization of purified tubulin. Mechanism of action. J Biol Chem 256:10435–10441
4. Olah E, Csokay B, Prajda N, et al., & Weber G (1996) Molecular mechanisms in the antiproliferative action of taxol and tiazofurin. Anticancer Res 16:2469–2478

Phosmidosine

Key Words: [Cell cycle inhibitor][Protein synthesis inhibitor]

Structure:

Molecular Formula: $C_{16}H_{24}N_7O_8P$
Molecular Weight: 473
Solubility: DMSO, ++ ; H_2O, ++ ; MeOH, ++

Discovery:

Phosmidosine was isolated as an antifungal proline-containing nucleotide antibiotic by Uramoto et al. [1] and derivatives, phosmidosines B and C, were isolated as detransforming compounds by Matsuura et al. [2].

Biological Studies:

Phosmidosine inhibited spore formation of *Botrytis cinerea* and showed morpho-logical reversion activity in v-*src*-transformed-NRK cells [1,2]. Matsuura et al. and Kakeya et al. reported that phosmidosine suppressed S-phase entry and arrested cell cycle progression at the G1 phase [2,3]. Phosmidosine inhibits protein synthesis in situ (unpublished data) and irreversibly inhibits the cell cycle progression at the G1 phase without affecting the G2-M transition. Phosmidosine acts at an earlier point in the G1 phase compared to mimosine or aphidicolin, well-known cell cycle blockers at the G1/S boundary [3].

Biological Activity:

0.25 μg/ml: MIC of spore formation of *Botrytis cinerea* [1]
4 μg/ml: Morphology reversion activity of v-*src*-NRK cells [2]
10 μM: inhibition of serum-induced G1 phase entry of WI-38 cells [3]

References:

1. Uramoto M, Kim CJ, Shin-ya K, et al., & McCloskey JA (1991) Isolation and characterization of phosmidosine. A new antifungal nucleotide antibiotic. J Antibiot 44:375–381
2. Matsuura N, Onose R & Osada H (1996) Morphology reversion activity of phosmidosine
   and phosmidosine B, a newly isolated derivative, on *src* transformed NRK cells. J
   Antibiot 49:361–365
3. Kakeya H, Onose R, Liu PC, et al., & Osada H (1998) Inhibition of cyclin D1 expression
   and phosphorylation of retinoblastoma protein by phosmidosine, a nucleotide antibiotic. Cancer Res 58:704–710

Pironetin

Key Words: [Antitumor activity][Cell cycle inhibitor][Microtubule binder]

Structure:

Molecular Formula: $C_{19}H_{32}O_4$
Molecular Weight: 325
Solubility: DMSO, +++ ; H_2O, - ; MeOH, ++

Discovery/Isolation:

Pironetin was originally isolated as a plant growth regulator from the culture broth of *Streptomyces* sp. by Kobayashi et al. [1]. The same compound, named PA-48153C, was isolated as an immunosuppressant by Yoshida et al. (Lactone with immunosuppressive activity. European Patent EP 560 389 A1 December 3, 1993). The structure of pironetin was determined to be (5R,6R)-5-ethyl-5,6-dihydro-6-[(E)-(2R,3S,4R,5S)-2-hydroxy-4-methoxy-3,5-dimethyl-7-nonenyl]-2H-pyran-2-one by FAB-MS, ^{1}H and ^{13}C NMR, COSY, COLOC, DEPT, IR, X-ray crystallographic analyses and an adapted Mosher's method [2]. A demethyl analog of pironetin, NK10958P, was also isolated [3].

Biological and Chemical Studies:

Kobayashi et al. and Yoshida et al. isolated pironetin as a plant growth regulator and an immunosuppressant, respectively. Yasui et al. reported the total synthesis of pironetin and prepared pironetin derivatives by partial synthesis from natural product. They found that derivatives retain the inhibitory activity on the responses of both T and B cells to mitogens and that the C-8 hexanoate showed potent suppressive effects on mitogen responses with less cytotoxicity to EL4 cells and was selected for in vivo evaluation [4].

Kondoh et al. reported that pironetin arrested the cell cycle progression at the M-phase and showed an antitumor activities in the mice model [5]. They also showed that pironetin inhibits the microtubule assembly both in vitro and in situ. As pironetin inhibited the binding of vinblastine but not colchicine to tubulin, it is thought that pironetin binds to tubulin directly and that inhibits microtubule assembly [6].

Biological Activity:

23% inhibition on the growth of rice plants without any loss of crop yield at 10 g/a on 9 days before heading [1]
10–20 ng/ml: completely inhibited the cell proliferation of 3Y1 cells [5]
5–25 ng/ml: antiproliferative effects in several tumor cell lines [5]
T/C, 128%: the intraperitoneal administration of 6.3 mg/kg pironetin over a 5-day period showed a moderate antitumor effect in CDF_1-SLC mice bearing P388 leukemia cells [5]

References:

1. Kobayashi S, Tsuchiya K, Harada T, et al., & Kobayashi K (1994) Pironetin, a novel plant growth regulator produced by *Streptomyces* sp. NK10958. I. Taxonomy, production, isolation and preliminary characterization. J Antibiot 47:697–702
2. Kobayashi S, Tsuchiya K, Kurokawa T, et al., & Iitaka Y (1994) Pironetin, a novel plant growth regulator produced by *Streptomyces* sp. NK10958. II. Structural elucidation. J Antibiot 47:703–707
3. Tsuchiya K, Kobayashi S, Nishikiori T, et al., & Tatsuta K (1997) NK10958P, a novel plant growth regulator produced by *Streptomyces* sp. J Antibiot 50:259–260
4. Yasui K, Tamura Y, Nakatani T, et al., & Ohtani M (1996) Chemical modification of PA-48153C, a novel immunosuppressant isolated from *Streptomyces prunicolor* PA-48153. J Antibiot 49:173–180
5. Kondoh M, Usui T, Kobayashi S, et al., & Osada H (1998) Cell cycle arrest and antitumor activity of pironetin and its derivatives. Cancer Lett 126:29–32
6. Kondoh M, Usui T, Nishikiori T, et al., & Osada H (1999) Apoptosis induction via microtubule disassembly by an antitumor compound, pironetin. Biochem J 340:411–416

Poststatin

Key Words: [Prolyl endopeptidase inhibitor]

Structure:

Molecular Formula: $C_{26}H_{47}N_5O_7$
Molecular Weight: 541
Solubility: DMSO, ± ; H_2O, +++ ; MeOH, +

Discovery:

Poststatin was discovered as a new inhibitor of prolyl endopeptidase in the fermentation broth of *Streptomyces viridochromogenes* [1]. The structure was defined as l-
valyl-l-valyl-3-amino-2- oxovaleryl-d-leucyl-l-valine by analysis of spectral properties and chemical studies of poststatin and its derivatives [2]. The absolute configuration of poststatin was also determined [3]. Total synthesis of poststatin was achieved
by both liquid phase and solid phase methods [4].

Biological Studies:

It is considered that prolyl endopeptidase plays an important role in biological regulation because its activity increased with the progress of the erythematosus-like
syndrome in model mice NZB/WF1, and was significantly high in the occipital lobe of
Alzheimer patients. To investigate the roles of prolyl endopeptidase, Aoyagi et al.
screened its inhibitor and found a potent and specific inhibitor, poststatin [1].
Poststatin is also an effective inhibitor of kinin-degrading enzyme in rat urine but not
in rat plasma [5]. Several derivatives were synthesized and their structure-activity
relationships were investigated [6–9]. The α-keto group of postine in poststatin
plays the most important role in the inhibitory mechanism [2].

Biological Activity:

0.03 µg/ml: IC_{50} for prolyl endopeptidase [1]
5.6×10^{-8} M: K_i values for prolyl endopeptidase [1]
4.0×10^{-4} M: K_m values for prolyl endopeptidase [1]

References:

1. Aoyagi T, Nagai M, Ogawa K, et al., & Takeuchi T (1991) Poststatin, a new inhibitor of prolyl endopeptidase, produced by *Streptomyces viridochromogenes* MH534-30F3. I. Taxonomy, production, isolation, physico-chemical properties and biological activities. J Antibiot 44:949–955
2. Nagai M, Ogawa K, Muraoka Y, et al., & Takeuchi T (1991) Poststatin, a new inhibitor of prolyl endopeptidase, produced by *Streptomyces viridochromogenes* MH534-30F3. II. Structure determination and inhibitory activities. J Antibiot 44:956–961
3. Tsuda M, Muraoka Y, Nagai M, et al., & Takeuchi T (1996) Poststatin, a new inhibitor of prolyl endopeptidase. III. Optical resolution of 3-amino-2-hydroxyvaleric acid and absolute configuration of poststatin. J Antibiot 49:281–286
4. Tsuda M, Muraoka Y, Nagai M, et al., & Aoyagi T (1996) Poststatin, a new inhibitor of prolyl endopeptidase. IV. The chemical synthesis of poststain. J Antibiot 49:287–291
5. Majima M, Shima C, Saito M, et al., & Aoyagi T (1993) Poststatin, a novel inhibitor of bradykinin-degrading enzymes in rat urine. Eur J Pharmacol 232:181–190
6. Tsuda M, Muraoka Y, Nagai M, et al., & Takeuchi T (1996) Poststatin, a new inhibitor of prolyl endopeptidase. VIII. Endopeptidase inhibitory activity of non-peptidyl poststatin analogues. J Antibiot 49:1022–1030
7. Tsuda M, Muraoka Y, Nagai M, et al., & Takeuchi T (1996) Poststatin, a new inhibitor of prolyl endopeptidase. VII. *N*-cycloalkylamide analogues. J Antibiot 49:909–920
8. Tsuda M, Muraoka Y, Someno T, et al., & Takeuchi T (1996) Poststatin, a new inhibitor of prolyl endopeptidase. VI. Endopeptidase inhibitory activity of poststatin analogues containing pyrrolidine ring. J Antibiot 49:900–908
9. Tsuda M, Muraoka Y, Nagai M, et al., & Takeuchi T (1996) Poststatin, a new inhibitor of prolyl endopeptidase. V. Endopeptidase inhibitory activity of poststatin analogues. J Antibiot 49:890–899

Prodigiosin 25-C

Key Words: [Immunomodulator][V-ATPase inhibitor]

Structure:

Molecular Formula: $C_{25}H_{35}N_3O$
Molecular Weight: 393
Solubility: DMSO, +++ ; H_2O, - ; MeOH, ++

Discovery:

Prodigiosins were originally isolated as pigments from *Serratia marcescens* and *Streptomyces* sp. Prodigiosin 25-C was re-isolated as an immunomodulating substance showing inhibition of cytotoxic T cell induction in a mixed lymphocyte reaction [1].

Biological Studies:

Prodigiosin 25-C completely inhibited the induction of cytotoxic T cells in vitro and in vivo [2], and moderately prolonged survival of MHC-mismatched skin grafts [3]. Kataoka et al. found that prodigiosin 25-C inhibited proton pump activity, but did not affect ATPase activity in rat liver lysosomes. These results indicate that prodigiosin 25-C raises the pH of acidic compartments through inhibition of the proton pump activity of vacuolar type H^+-ATPase [4]. Prodigiosin 25-C completely abrogated the perforin activity without affecting perforin content, then inhibited the activity of cytotoxic T cells [5].

Sato et al. reported that prodigiosins (prodigiosin, metacycloprodigiosin, and prodigiosin 25-C) inhibited the acidification activity of H^+-ATPase chloride dependently, but not membrane potential formation by promotion of H^+/Cl^- symport across vesicular membranes [6].

Suppressive activity of Prodigiosin 25-C and metacycloprodigiosin on osteoclast bone resorption was also reported [7].

Biological Activity:

Prodigiosin 25-C

30 nM: IC_{50} values of proton pump activity in rat liver lysosomes [4]
> 1 mM: inhibition of ATPase activity in rat liver lysosomes [4]

Prodigiosins

30–120 pmol/mg of protein: IC_{50} values of intralysosomal pH through inhibition of lysosomal acidification driven by V-ATPase [8]

References:

1. Nakamura A, Nagai K, Ando K & Tamura G (1986) Selective suppression by prodigiosin of the mitogenic response of murine splenocytes. J Antibiot 39:1155–1159
2. Nakamura A, Magae J, Tsuji RF, et al., & Nagai K (1989) Suppression of cytotoxic T cell induction in vivo by prodigiosin 25-C. Transplantation 47:1013–1016
3. Tsuji RF, Magae J, Yamashita M, et al., & Yamasaki M (1992) Immunomodulating properties of prodigiosin 25-C, an antibiotic which preferentially suppresses induction of cytotoxic T cells. J Antibiot 45:1295–1302
4. Kataoka T, Muroi M, Ohkuma S, et al., & Nagai K (1995) Prodigiosin 25-C uncouples vacuolar type H^+-ATPase, inhibits vacuolar acidification and affects glycoprotein processing. FEBS Lett 359:53–59
5. Togashi K, Kataoka T & Nagai K (1997) Characterization of a series of vacuolar type H^+-ATPase inhibitors on CTL-mediated cytotoxicity. Immunol Lett 55:139–144
6. Sato T, Konno H, Tanaka Y, et al., & Ohkuma S (1998) Prodigiosins as a new group of H^+/Cl^- symporters that uncouple proton translocators. J Biol Chem 273:21455–21462
7. Woo JT, Ohba Y, Tagami K, et al., & Nagai K (1997) Prodigiosin 25-C and metacycloprodigiosin suppress the bone resorption by osteoclasts. Biosci Biotechnol Biochem 61:400–402
8. Ohkuma S, Sato T, Okamoto M, et al., & Wasserman HH (1998) Prodigiosins uncouple lysosomal vacuolar-type ATPase through promotion of H^+/Cl^- symport. Biochem J 334:731–741

Pyrrolomycin B

Key Words: [Antimicrobial][Immunomodulator]

Structure:

Molecular Formula: $C_{11}H_6N_2O_3Cl_4$
Molecular Weight: 354
Solubility: DMSO, + ; H_2O, ++ ; MeOH, ++

Discovery:

Pyrrolomycin A and B were isolated as antimicrobial substances [1] and their structures were determined by spectroscopic and X-ray crystallographic analysis [2,3]. Several derivatives, strongly active against Gram-positive bacteria and fungi were isolated [4–6].

Biological Studies:

Pyrrolomycins were isolated as anti Gram-positive bacteria and antifungal antibiotics. Umezawa et al. found that pyrrolomycin B enhanced humoral immune response in mice, mitogenesis in spleen cell culture in combination with concanavalin A, and phagocytosis by peritoneal macrophages after in vivo administration to mice [7]. These results suggests that pyrrolomycin B is an immunopotentiator.

Masuda et al. reisolated pyrrolomycin group antibiotics as inhibitors of substance P-induced release of myeloperoxidase from human polymorphonuclear leukocytes [8].

Biological Activity:

0.1–0.39 µg/ml: MIC values for Gram-positive bacteria [1]
16 µg/mouse: 40% enhancement of antibody formation against SRBC when intraperitoneally injected in mice at the time of immunization [7]
16 µg/mouse: 50% enhancement of phagocytosis by peritoneal macrophages in mice [7]
0.1 µg/ml: 40% inhibition of substance P-induced release of myeloperoxidase [8]

References:

1. Ezaki N, Shomura T, Koyama M, et al., & Niida T (1981) New chlorinated nitro-pyrrole antibiotics, pyrrolomycin A and B (SF- 2080 A and B). J Antibiot 34:1363–1365
2. Kaneda M, Nakamura S, Ezaki N & Iitaka Y (1981) Structure of pyrrolomycin B, a chlorinated nitro-pyrrole antibiotic. J Antibiot 34:1366–1368
3. Koyama M, Kodama Y, Tsuruoka T, et al., & Inouye S (1981) Structure and synthesis of pyrrolomycin A, a chlorinated nitro-pyrrole antibiotic. J Antibiot 34:1569–1576
4. Ezaki N, Koyama M, Shomura T, et al., & Inouye S (1983) Pyrrolomycins C, D and E, new members of pyrrolomycins. J Antibiot 36:1263–1267
5. Ezaki N, Koyama M, Kodama Y, et al., & Sakai S (1983) Pyrrolomycins F1, F2a, F2b and F3, new metabolites produced by the addition of bromide to the fermentation. J Antibiot 36:1431–1438
6. Koyama M, Ezaki N, Tsuruoka T & Inouye S (1983) Structural studies on pyrrolomycins C, D and E. J Antibiot 36:1483–1489
7. Umezawa K, Ishizuka M, Sawa T & Takeuchie T (1984) Enhancement of mouse immune system by pyrrolomycin B. J Antibiot 37:1253–1256
8. Masuda K, Suzuki K, Ishida-Okawara A, et al., & Koyama M (1991) Pyrrolomycin group antibiotics inhibit substance P-induced release of myeloperoxidase from human polymorphonuclear leukocytes. J Antibiot 44:533–540

Radicicol

Key Words: [Angiogenisis inhibitor][Antifungal][Detransforming agent]
[Tyrosine kinase inhibitor]

Structure:

Molecular Formula: $C_{18}H_{17}O_6Cl$
Molecular Weight: 364
Solubility: DMSO, ++ ; H_2O, + ; MeOH, +++

Discovery/Isolation:

Radicicol was isolated as an antifungal substance from *Monosporuim bonorden* [1].
The compound was assigned the molecular formula $C_{17}H_{16}O_7$ on the basis of preliminary analytical data, but McCapra et al. revised the formula to $C_{18}H_{17}O_6Cl$ [2].

Biological Studies:

Kwon et al. found that radicicol induced the reversal of transformed phenotypes of v-
src-transformed fibroblasts and showed specific inhibition of p60^{v-src} both in vivo
and in vitro [3]. The down-regulation and dephosphorylation of RB [4], differentiation, and cell cycle arrest at G1/G2 phases [5] in HL60 cells were reported. Interestingly, radicicol induced reversal of the transformed phenotype of not only *src*-, but
ras- and *mos*-transformed cells [5–7]. Kwon et al. showed that the morphological
change by radicicol required de novo mRNA and protein synthesis [6], and that one
of the proteins involved in suppression of morphological transformation was gelsolin
[8]. Pillay et al. also found that the phosphorylation of Sam68 (Src-associated in
mitosis 68 kDa), which is a protein that associates with and is tyrosine phosphorylated by Src in a mitosis-specific manner, was inhibited by radicicol and the cell cycle
was blocked in mitosis [9]. Furthermore, Soga et al. reported that the level of Raf
kinase was significantly decreased in radicicol-treated cells [10].
     Recently, the target molecule of radicicol was elucidated by Sharma et al. They
synthesized a biotinylated derivative of radicicol and used as a probe to visualize
cellular proteins that interact with radicicol. The most prominent cellular protein was
determined as HSP90. Taken together with other studies, they suggested that the
anti-transformation effects of radicicol may be mediated by the association of radicicol

with HSP90 and the consequent dissociation of the Raf/HSP90 complex leading to the attenuation of the Ras/MAP kinase signal transduction pathway [11].

Anti-angiogenic action and inhibition of both proliferation of vascular endothelial cells and plasminogen activator production by these cells were also reported [13,14].

Biological Activity:

$0.1\,\mu g/ml$: IC_{50} for transphosphorylation activities of purified p60^{v-src} [3]
25 ng/ml: reversal of the transformed phenotype of ras-transformed NIH3T3 cells [6]
200 ng/egg: ID_{50} for inhibition embryonic angiogenesis [13]

References:

1. Delmotte P & Delmotte-Plaquee J (1953) A new antifungal substance of fungal origin. Nature 171:344
2. McCapra F, Scott AI, Delmotte P, et al., & Bhacca NS (1964) The constitution of monorden, an antibiotic with tranquilising action. Tetrahedron Lett 15:869–875
3. Kwon HJ, Yoshida M, Fukui Y, et al., & Beppu T (1992) Potent and specific inhibition of p60^{v-src} protein kinase both in vivo and in vitro by radicicol. Cancer Res 52:6926–6930
4. Yen A, Soong S, Kwon HJ, et al., & Varvayanis S (1994) Enhanced cell differentiation when RB is hypophosphorylated and down- regulated by radicicol, a SRC-kinase inhibitor. Exp Cell Res 214:163–171
5. Shimada Y, Ogawa T, Sato A, et al., & Tsujita Y (1995) Induction of differentiation of HL-60 cells by the anti-fungal antibiotic, radicicol. J Antibiot 48:824–830
6. Kwon HJ, Yoshida M, Muroya K, et al., & Horinouchi S (1995) Morphology of ras-transformed cells becomes apparently normal again with tyrosine kinase inhibitors without a decrease in the Ras-GTP complex. J Biochem 118:221–228
7. Zhao JF, Nakano H & Sharma S (1995) Suppression of RAS and MOS transformation by radicicol. Oncogene 11:161–173
8. Kwon HJ, Yoshida M, Nagaoka R, et al., & Horinouchi S (1997) Suppression of morphological transformation by radicicol is accompanied by enhanced gelsolin expression. Oncogene 15:2625–2631
9. Pillay I, Nakano H & Sharma SV (1996) Radicicol inhibits tyrosine phosphorylation of the mitotic Src substrate Sam68 and retards subsequent exit from mitosis of Src-transformed cells. Cell Growth Differ 7:1487–1499
10. Soga S, Kozawa T, Narumi H, et al., & Mizukami T (1997) Radicicol leads to selective depletion of raf kinase and disrupts K-ras-activated aberrant signaling pathway. J Biol Chem 273:822–828
11. Sharma SV, Agatsuma T & Nakano H (1998) Targeting of the protein chaperone, HSP90, by the transformation suppressing agent, radicicol. Oncogene 16:2639–2645
12. Oikawa T, Ito H, Ashino H, et al., & Murota S-i (1993) Radicicol, a microbial cell differentiation modulator, inhibits in vivo angiogenesis. Eur J Pharmacol 241:221–227
13. Onozawa C, Shimamura M, Iwasaki S & Oikawa T (1997) Inhibition of angiogenesis by rhizoxin, a microbial metabolites containing two epoxide groups. Jpn J Cancer Res 88:1125–1129

Rapamycin

Key Words: [Antifungal][Immunosuppressant]

Structure:

Molecular Formula: $C_{51}H_{79}NO_{13}$
Molecular Weight: 914
Solubility: DMSO, ++ ; H_2O, + ; MeOH, ++

Discovery/Isolation:

Rapamycin was isolated as an antifungal antibiotic produced by *Streptomyces hygroscopicus* [1,2]

Biological Studies:

Rapamycin inhibited the immune response in rats and the immunosuppressant activity of rapamycin appears to be related to inhibition of the lymphatic system [3]. Interestingly, FK506, which is structurally related to rapamycin, showed immunosuppressive properties by a different mechanism. Remarkably, these two drugs inhibit each other's actions, raising the possibility that both act by means of a common immunophilin, FKBP [4]. Heitman et al. found that *Saccharomyces cerevisiae* treated with rapamycin was irreversibly arrested in the G1 phase and isolated TOR1 and TOR2, that participate in rapamycin toxicity [5]. Brown et al. and Sabatini et al. identified a mammalian protein targeted by the G1-arresting rapamycin-receptor complex, FRAP, that is highly related to TOR1/2 [6,7]. The rapamycin/FKBP complex selectively inhibits the $p70^{S6k}$ activation cascade, however, the kinase activity of FRAP alone is not sufficient for control of $p70^{S6k}$ [8–10].

Recently, several groups showed that rapamycin inhibits cap-dependent, but not cap-independent, translation. This inhibition is related to dephosphorylation and consequent activation of 4E-BP1/PHAS-I, a protein identified as a repressor of the cap-binding protein, eIF-4E. TOR phosphorylated PHAS-I in vitro, and these modifications inhibited the binding of PHAS-I to eIF-4E. These studies define a role for TOR in translational control and offer further insights into the mechanism whereby rapamycin inhibits G1-phase progression in mammalian cells [11,12].

Biological Activity:

0.2 nM: K_d for rapamycin to FKBP [5]
0.05–0.2 nM: inhibition of IL-2-induced S phase entry of T cells [10]

References:

1. Vezina C, Kudelski A & Sehgal SN (1975) Rapamycin (AY-22,989), a new antifungal antibiotic. I. Taxonomy of the producing Streptomycete and isolation of the active principle. J Antibiot 28:721–726
2. Sehgal SG, Baker H & Vezina C (1975) Rapamycin (AY-22,989), a new antifungal antibiotic. II. Fermentation, isolation and characterization. J Antibiot 28:727–732
3. Martel RR, Klicius J & Galet S (1977) Inhibition of the immune response by rapamycin, a new antifungal antibiotic. Can J Physiol Pharmacol 55:48–51
4. Bierer BE, Mattila PS, Standaert RF, et al., & Schreiber SL (1990) Two distinct signal transmission pathways in T lymphocytes are inhibited by complexes formed between an immunophilin and either FK506 or rapamycin. Proc Natl Acad Sci USA 87:9231–9235
5. Heitman J, Movva NR & Hall MN (1991) Targets for cell cycle arrest by the immuno-suppressant rapamycin in yeast. Science 253:905–9
6. Brown EJ, Albers MW, Shin TB, et al., & Schreiber SL (1994) A mammalian protein targeted by G1-arresting rapamycin-receptor complex. Nature 369:756–758
7. Sabatini DM, Erdjument-Bromage H, Lui M, et al., & Snyder SH (1994) RAFT1: a mammalian protein that binds to FKBP12 in a rapamycin- dependent fashion and is homologous to yeast TORs. Cell 78:35–43
8. Kuo CJ, Chung J, Fiorentino DF, et al., & Crabtree GR (1992) Rapamycin selectively inhibits interleukin-2 activation of p70 S6 kinase. Nature 358:70–73
9. Chung J, Kuo CJ, Crabtree GR & Blenis J (1992) Rapamycin-FKBP specifically blocks growth-dependent activation of and signaling by the 70 kd S6 protein kinases. Cell 69:1227–1236
10. Brown EJ, Beal PA, Kelth CT, et al., & Schreiber SL (1995) Control of p70 S6 kinase by kinase activity of FRAP in vivo. Nature 377:441–446
11. Beretta L, Gingras AC, Svitkin YV, et al., & Sonenberg N (1996) Rapamycin blocks the phosphorylation of 4E-BP1 and inhibits cap-dependent initiation of translation. EMBO J 15:658–664
12. Brunn GJ, Hudson CC, Sekulic A, et al., & Abraham RT (1997) Phosphorylation of the translational repressor PHAS-I by the mammalian target of rapamycin. Science 277:99–101

Reveromycin A

Key Words: [Detransforming agent][Protein synthesis inhibitor]

Structure:

Molecular Formula: $C_{36}H_{52}O_{11}$
Molecular Weight: 660
Solubility: DMSO, ++ ; H_2O, + ; MeOH, ++

Discovery/Isolation:

Reveromycin A was isolated as inhibitor of mitogenic activity induced by EGF in a mouse epidermal keratinocyte [1,2].

Biological Studies:

Reveromycin A showed inhibitory activity against EGF-stimulated mitogen response in Balb/MK cells and exhibited morphological reversion of src^{ts}-NRK cells, antiproliferative activity against human tumor cell lines and antifungal activity. In vitro studies revealed that reveromycin A is a selective inhibitor of protein synthesis in eukaryotic cells [3]. Reveromycin A showed little antitumor effect against three murine tumors tested, but showed a strong antitumor effect against a human ovarian carcinoma BG-1, which is known to be a TGF-α-secreting and estrogen receptor-expressing cell line. In BG-1 cells, Reveromycin A inhibited cell proliferation induced by TGF-α, but did not inhibit the proliferation induced by 17β-estradiol [4].

Cui et al. cloned and characterized a *Saccharomyces cerevisiae* gene *YRS1* that complements the phenotype of the mutant sensitive to the anionic drug reveromycin A. The *YRS1*, which is identical to *YOR1*, encodes a protein with extensive homology to the human multidrug resistance-associated protein. They showed that Yrs1 is a multispecific organic anion transporter important for tolerance against toxic environmental organic anions [5].

Biological Activity:

1 μM: IC_{50} for inhibition of protein synthesis in src^{ts}-NRK cells [3]
40 nM: IC_{50} for inhibition of protein synthesis in rabbit reticulocyte lysate [3]
30–300 nM: inhibition of cell proliferation induced by TGF-α in BG-1 cells [4]
1 μg/ml: growth inhibition of *Saccharomyces cerevisiae* on YPD plate adjusted at pH4.5 [5]

References:

1. Osada H, Koshino H, Isono K, et al., & Kawanishi G (1991) Reveromycin A, a new antibiotic which inhibits the mitogenic activity of epidermal growth factor. J Antibiot 44:259–261
2. Takahashi H, Osada H, Koshino H, et al., & Isono K (1992) Reveromycins, new inhibitors of eukaryotic cell growth. I. Producing organism, fermentation, isolation and physicochemical properties. J Antibiot 45:1409–1413
3. Takahashi H, Osada H, Koshino H, et al., & Isono K (1992) Reveromycins, new inhibitors of eukaryotic cell growth. II. Biological activities. J Antibiot 45:1414–1419
4. Takahashi H, Yamashita Y, Takaoka H, et al., & Osada H (1997) Inhibitory action of reveromycin A on TGF-α-dependent growth of ovarian carcinoma BG-1 in vitro and in vivo. Oncol Res 9:7–11
5. Cui Z, Hirata D, Tsuchiya E, et al., & Miyakawa T (1996) The multidrug resistance-associated protein (MRP) subfamily (Yrs1/Yor1) of *Saccharomyces cerevisiae* is important for the tolerance to a broad range of organic anions. J Biol Chem 271:14712–14716

Rhizoxin

Key Words: [Antifungal][Anti-angiogenic activity][Antitumor]
[Microtubule inhibitor]

Structure:

Molecular Formula: $C_{35}H_{47}NO_9$
Molecular Weight: 625
Solubility: DMSO, ++ ; H_2O, ± ; MeOH, ++

Discovery/Isolation:

Rhizoxin was isolated as a toxin produced by *Rhizopus chinensis,* the causal agent of
rice seedling blight [1]. The skeletal structure was determined by detailed NMR spectroscopic and X-ray analysis [1,2].

Biological Studies:

Takahashi et al. found that rhizoxin inhibits tubulin polymerization in vitro [3]. They
further studied the binding of rhizoxin to porcine brain tubulin and suggested that
rhizoxin binds to the maytansine-binding site and that the binding sites of rhizoxin
and vinblastine are not the same [4]. Sullivan et al. found that rhizoxin acts like
maytansine in that it completely prevents formation of an intrachain cross-link in β-
tubulin by *N,N*-ethylenebis(iodoacetamide), whereas vinblastine only partially inhibits this [5]. The binding site of rhizoxin was investigated by analysis of sensitive
mutants of *Aspergillus nidulans* and it was found that rhizoxin resistance has a
common basis in both naturally occurring species and experimentally selected mutants in the substitution of Ile or Val for Asn-100 in β-tubulin [6,7].

Rhizoxin is effective against B16 melanoma inoculated *i.p.* or *s.c.* Rhizoxin showed effective antitumor activity against human and murine tumor cells resistant to vincristine and adriamycin in vitro and in vivo [8]. Onozawa et al. showed that rhizoxin is a novel inhibitor of angiogenesis, and that it has potential as a new therapeutic agent for cancer [9].

Biological Activity:

10 ng/ml: abnormal swelling of rice seedling roots [1]

1.0×10^{-5} M: inhibition of in vitro tubulin polymerization [3]

1.7×10^{-7} M: K_d value for binding to porcine brain tubulin [4]

2.5 μM: IC_{50} for prevention of an intrachain crosslink in β-tubulin [5]

2 ng/egg: ID_{50} for inhibition of embryonic angiogenesis [9]

2 mg/kg *i.p*: suppression of neovascularization induced by M5076 mouse tumor cells in a mouse dorsal air sac assay system [9]

References:

1. Iwasaki S, Kobayashi H, Furukawa J, et al., & Noda T (1984) Studies on macrocyclic lactone antibiotics. VII. Structure of a phytotoxin "rhizoxin" produced by *Rhizopus chinensis.* J Antibiot 37:354–362
2. Iwasaki S, Namikoshi M, Kobayashi H, et al., & Sato Z (1986) Studies on macrocyclic lactone antibiotics. VIII. Absolute structures of rhizoxin and a related compound. J Antibiot 39:424–429
3. Takahashi M, Iwasaki S, Kobayashi H, et al., & Nagano H (1987) Studies on macrocyclic lactone antibiotics. XI. Anti-mitotic and anti-tubulin activity of new antitumor antibiotics, rhizoxin and its homologues. J Antibiot 40:66–72
4. Takahashi M, Iwasaki S, Kobayashi H, et al., & Sato Y (1987) Rhizoxin binding to tubulin at the maytansine-binding site. Biochim Biophys Acta 926:215–223
5. Sullivan AS, Prasad V, Roach MC, et al., & Luduena RF (1990) Interaction of rhizoxin with bovine brain tubulin. Cancer Res 50:4277–4280
6. Takahashi M, Matsumoto S, Iwasaki S & Yahara I (1990) Molecular basis for determining the sensitivity of eucaryotes to the antimitotic drug rhizoxin. Mol Gen Genet 222:169–175
7. Takahashi M, Kobayashi H & Iwasaki S (1989) Rhizoxin resistant mutants with an altered β-tubulin gene in *Aspergillus nidulans.* Mol Gen Genet 220:53–59
8. Tsuruo T, Oh-hara T, Iida H, et al., & Arakawa M (1986) Rhizoxin, a macrocyclic lactone antibiotic, as a new antitumor agent against human and murine tumor cells and their vincristine-resistant sublines. Cancer Res 46:381–385
9. Onozawa C, Shimamura M, Iwasaki S & Oikawa T (1997) Inhibition of angiogenesis by rhizoxin, a microbial metabolites containing two epoxide groups. Jpn J Cancer Res 88:1125–1129

RK-682

Key Words: [Protein tyrosine phosphatase inhibitor]

Structure:

Molecular Formula: $C_{21}H_{36}O_5$
Molecular Weight: 368
Solubility: DMSO, +++ ; H_2O, - ; MeOH, +

Discovery/Isolation:

3-hexadecanoyl-5-hydroxymethyltetronic acid (RK-682, TAN-1364A) was isolated from *Streptomyces* sp. [1]. Total synthesis was performed by Sodeoka et al [2].
    This compound is now commercially available as a phosphatase inhibitor, RK-682.

Biological Studies:

Originally, RK-682 was isolated as a phospholipase A_2 inhibitor and then two groups reisolated the same compound as a HIV protease inhibitor [3,4] and VHR inhibitor, a dual-specificity protein phosphatase [5]. RK-682 inhibited protein phosphatase activity of VHR in a competitive manner in vitro. Fujii et al. used RK-682 for investigation of the mechanism of ATP-induced long-term potentiation involving extracellular phosphorylation of membrane proteins in guinea-pig hippocampal CA1 neurons [6], and also extracellular phosphorylation of the membrane protein modified theta burst-induced long-term potentiation in CA1 neurons of guinea-pig hippocampal slices [7].

Biological Activity:

16 µM: IC_{50} for phospholipase A_2 in vitro [1]
84 µM: IC_{50} for HIV-1 protease in vitro [4]
11.5 µM: IC_{50} for VHR in vitro

References:

1. Shinagawa S (1993) Jpn Kokai Tokkyo Koho JP 05–43568
2. Sodeoka M, Sampe R, Kagamizono T & Osada H (1996) Asymmetric synthesis of RK-682 and its analogs, and evaluation of their protein phosphatase inhibitory activities. Tetrahedron Lett 37:8775–8778
3. Roggo BE, Hug P, Moss S, et al., & Peter HH (1994) 3-Alkanoyl-5-hydroxymethyl tetronic acid homologues: new inhibitors of HIV-1 protease. II. Structure determination. J Antibiot 47:143–147
4. Roggo BE, Petersen F, Delmendo R, et al., & Roesel J (1994) 3-Alkanoyl-5-hydroxymethyl tetronic acid homologues and resistomycin: new inhibitors of HIV-1 protease. I. Fermentation, isolation and biological activity. J Antibiot 47:136–142
5. Hamaguchi T, Sudo T & Osada H (1995) RK-682, a potent inhibitor of tyrosine phosphatase, arrested the mammalian cell cycle progression at G1phase. FEBS Lett 372:54–58
6. Fujii S, Kato H, Furuse H, et al., & Kuroda Y (1995) The mechanism of ATP-induced long-term potentiation involves extracellular phosphorylation of membrane proteins in guinea-pig hippocampal CA1 neurons. Neurosci Lett 187:130–132
7. Fujii S, Ito K, Osada H, et al., & Kato H (1995) Extracellular phosphorylation of membrane protein modifies theta burst- induced long-term potentiation in CA1 neurons of guinea-pig hippocampal slices. Neurosci Lett 187:133–136

Saintopin

Key Words: [Topoisomerase I and II inhibitor]

Structure:

Molecular Formula: $C_{18}H_{10}O_7$
Molecular Weight: 338
Solubility: DMSO, +++ ; H_2O, - ; MeOH, +

Discovery:

Saintopin was isolated from *Paecilomyces* as an antitumor antibiotic with topoisomerase II dependent DNA cleavage activity [1].

Biological Studies:

Saintopin induced both topoisomerase I and II mediated DNA cleavage through the mechanism of stabilizing the reversible enzyme-DNA "cleavable complex". Consistent with the cleavable complex formation with both topoisomerases, saintopin inhibited catalytic activities of both topoisomerases I and II [2]. It was also revealed that saintopin is a weak DNA intercalator like *m*-AMSA by DNA unwinding assay using T4 DNA ligase. Leteurtre et al. found that a camptothecin-resistant topoisomerase I with a mutation ($Asn^{722} \rightarrow Ser$) next to the catalytic tyrosine (Tyr^{723}) was cross-resistant to saintopin [3]. They suggested that both saintopin and camptothecin interact with the protein near the catalytic tyrosine and proposed the "drug-stacking" model that topoisomerase inhibitors bind, possibly through hydrogen bonding and/or stacking, with one of the bases flanking the DNA termini and within the enzyme catalytic pocket, most likely by stacking with the catalytic tyrosine.

Biological Activity:

$0.35 \mu g/ml$: IC_{50} for HeLa S3 cells [1]
25 mg/kg: statistically significant increase in life span to 30% in murine leukemia P388 harboring mice [1]

References:

1. Yamashita Y, Saitoh Y, Ando K, et al., & Nakano H (1990) Saintopin, a new antitumor antibiotic with topoisomerase II dependent DNA cleavage activity, from *Paecilomyces*. J Antibiot 43:1344–1346
2. Yamashita Y, Kawada S, Fujii N & Nakano H (1991) Induction of mammalian DNA topoisomerase I and II mediated DNA cleavage by saintopin, a new antitumor agent from fungus. Biochemistry 30:5838–5845
3. Leteurtre F, Fujimori A, Tanizawa A, et al., & Pommier Y (1994) Saintopin, a dual inhibitor of DNA topoisomerases I and II, as a probe for drug-enzyme interactions. J Biol Chem 269:28702–28707

Scyphostatin

Key Words: [Membrane-bound neutral sphingomyelinase inhibitor]

Structure:

Molecular Formula: $C_{29}H_{43}NO_5$
Molecular Weight: 485
Solubility: DMSO, ++ ; H_2O, - ; MeOH, +

Discovery:

Scyphostatin was isolated as a membrane-bound neutral sphingomyelinase (N-SMase) inhibitor from a mycelial extract of *Dasyscyphus mollissimus* [1]. Scyphostatin is composed of a fatty acid and an amino alcohol substituted by a highly oxygenated cyclohexenone moiety.

Biological Studies:

Ceramide is an intracellular lipid second messenger and takes part in the regulation of cell proliferation, differentiation, and apoptosis in a wide variety of cell types. In particular, sphingomyelin breakdown by N-SMase appears downstream of signaling events of inflammatory cytokines including TNFα and IL-1β, and this ceramide generation has been reported to mediate prostaglandin production and cytokine gene expression. Scyphostatin was isolated as a potent N-SMase inhibitor [1]. The structural similarity between scyphostatin and ceramide represented by the *N*-acylamino alcohol moiety suggests that scyphostatin may exhibit inhibitory activity as a substrate or product analogue of the enzymatic reaction.

Biological Activity:

1.0 μM: IC_{50} for N-SMase activity [1]
49.3 μM: IC_{50} for acidic SMase activity [1]

References:

1. Tanaka M, Nara F, Suzuki-Konagai K, et al., & Ogita T (1997) Structural elucidation of scyphostatin, an inhibitor of membrane-bound neutral sphingomyelinase. J Am Chem Soc 119:7871–7872

Spergualin

Key Words: [Immunosuppressant]

Structure:

Molecular Formula: $C_{17}H_{37}N_7O_4$
Molecular Weight: 403
Solubility: DMSO, +++ ; H_2O, ± ; MeOH, ++

Discovery:

Spergualin was isolated as an antitumor antibiotic from *Bacillus laterosporus*[1] and its structure was determined by several spectral analysis and chemical modification[2]. Spergualin and 15-deoxyspergualin, a derivative which shows stronger inhibition against mouse leukemia L-1210 than spergualin, were synthesized[3,4].

Biological Studies:

Spergualin was isolated as an antitumor antibiotic, and shows immunosuppressive effects [5,6], antitumor activity [7–9] and angiogenesis inhibition [10]. The immunosuppressive activity of spergualin may be caused by direct inhibition of antibody production through blocking the transcriptional activation of kappa L chain expression during certain stages of B cell development and inhibits the antibody production [11,12].

Nadler et al. identified an intracellular 15-deoxyspergualin binding protein as Hsc70, the constitutive or cognate member of the Hsp70 protein family [13]. Nadeau et al. estimated the kinetic constant of 15-deoxyspergualin with Hsc70 and Hsp90, and validated binding of these novel immunosuppressant agents to these molecular chaperones, at concentrations in the range of pharmacologically active doses [14]. The members of the Hsp70 family of heat shock proteins are important for many cellular processes, including immune responses, and this finding suggests that heat shock proteins may represent a class of immunosuppressant binding proteins, or immunophilins.

Biological Activity:

6.25 mg/kg/day: antitumor activity against L-1210 in vivo, T/C 732% [1]
150 mg/kg: LD_{50} by intraperitoneal injection [1]
4 μM: K_d values for 15-deoxyspergualin binding to Hsc70 [14]
5 μM: K_d values for 15-deoxyspergualin binding to Hsp90 [14]

References:

1. Takeuchi T, Iinuma H, Kunimoto S, et al., & Umezawa H (1981) A new antitumor antibiotic, spergualin: isolation and antitumor activity. J Antibiot 34:1619–1621
2. Umezawa H, Kondo S, Iinuma H, et al., & Takeuchi T (1981) Structure of an antitumor antibiotic, spergualin. J Antibiot 34:1622–1624
3. Kondo S, Iwasawa H, Ikeda D, et al., & Umezawa H (1981) The total synthesis of spergualin, an antitumor antibiotic. J Antibiot 34:1625–1627
4. Iwasawa H, Kondo S, Ikeda D, et al., & Umezawa H (1982) Synthesis of (-)-15-deoxyspergualin and (-)-spergualin-15-phosphate. J Antibiot (Tokyo) 35:1665–9
5. Umezawa H, Ishizuka M, Takeuchi T, Abe F, Nemoto K, Shibuya K & Nakamura T (1985) Suppression of tissue graft rejection by spergualin. J Antibiot 38:283–284
6. Nemoto K, Hayashi M, Ito J, et al., & Umezawa H (1987) Effect of spergualin in autoimmune disease mice. J Antibiot 40:1448–1451
7. Umezawa H, Nishikawa K, Shibasaki C, et al., & Takeuchi T (1987) Involvement of cytotoxic T-lymphocytes in the antitumor activity of spergualin against L1210 cells. Cancer Res 47:3062–3065
8. Nishikawa K, Shibasaki C, Takahashi K, et al., & Umezawa H (1986) Antitumor activity of spergualin, a novel antitumor antibiotic. J Antibiot 39:1461–1466
9. Hiratsuka M, Kuramochi H, Takahashi K, et al., & Oshimura M (1991) Cytostatic effect of deoxyspergualin on a murine leukemia cell line L1210. Jpn J Cancer Res 82:1065–1068
10. Oikawa T, Shimamura M, Ashino-Fuse H, et al., & Takeuchi T (1991) Inhibition of angiogenesis by 15-deoxyspergualin. J Antibiot 44:1033–1035
11. Fujii H, Takada T, Nemoto K, et al., & Takeuchi T (1990) Deoxyspergualin directly suppresses antibody formation in vivo and in vitro. J Antibiot 43:213–219
12. Tepper MA, Nadler SG, Esselstyn JM & Sterbenz KG (1995) Deoxyspergualin inhibits kappa light chain expression in 70Z/3 pre-B cells by blocking lipopolysaccharide-induced NF-κB activation. J Immunol 155:2427–2436
13. Nadler SG, Tepper MA, Schacter B & Mazzucco CE (1992) Interaction of the immunosuppressant deoxyspergualin with a member of the Hsp70 family of heat shock proteins. Science 258:484–486
14. Nadeau K, Nadler SG, Saulnier M, et al., & Walsh CT (1994) Quantitation of the interaction of the immunosuppressant deoxyspergualin and analogs with Hsc70 and Hsp90. Biochemistry 33:2561–2567

Spicamycin (KRN5500)

Key Words: [Antitumor][Protein synthesis inhibitor]

Spicamycin is a nucleoside antibiotic containing fatty acids with a variety of chain lengths (C12–C18). Here, the structure, molecular formula and molecular weight of one of the synthetic analogs, KRN5500 (6-[4-Deoxy-4-(2*E*,4*E*)-tetradecadienoylglycyl]amino-l-glycero-β- l-mannoheptopyranosyl]amino-9H-purine) is shown.

Structure:

Molecular Formula: $C_{28}H_{43}N_7O_7$
Molecular Weight: 589
Solubility: DMSO, ++ ; H_2O, + ; MeOH, ++

Discovery/Isolation:

Spicamycin was isolated as a potent inducer of differentiation of human promyelocytic leukemia cells, HL-60. This compound was a mixture of closely related compounds and was difficult to separate by chromatographic procedures [1].

Biological Studies:

Spicamycin, a nucleoside antibiotic containing fatty acids with a variety of chain lengths (C12–C18), showed potent antitumor activity against human gastric cancer SC-9 and human breast cancer MX-1 in a xenograft model. Kamishohara et al. and Sakai et al. made several semi-synthetic spicamycin analogues (SPMs) which differed in the chain length of the fatty acid moiety, and examined their structure-antitumor activity relationship [2–4]. KRN5500 was semi-synthesized in an attempt to increase the therapeutic effects of spicamycin analogues. Lee et al. examined the cytotoxicity of KRN5500 in small and non-small cell lung cancer cell lines and showed that it was active against a wide range of lung cancer cell lines [5]. Kamishohara et al. investigated the mode of action using KRN5500 and found that KRN5500 inhibited protein synthesis in P388 cells, but was ineffective in rabbit reticulocyte lysates. Interestingly, 4'-N-glycyl spicamycin amino nucleoside (SAN-Gly), which is an intracellular metabolite of KRN5500, exhibited a marked inhibitory effect on protein synthesis in this cell-free system but showed a weaker cytotoxicity and a much lower intracellular incorporation than those of KRN5500. These results suggested the intracellular conversion of KRN5500 to SAN-Gly to exert an antitumor effect [6].

Biological Activity:

2.5 – 640 ng/ml: induction of phagocytic activity of HL60 cells [1]
140–154% T/C: against P388 leukemia in mice when administered ($i.p.$) [1]
1.5 µM: IC_{50} for inhibitory effect on protein synthesis in P388 cells [6]
>170 µM: IC_{50} for protein synthesis inhibition in rabbit reticulocyte lysates [6]
2.3 µM: IC_{50} for inhibitory effect of SAN-Gly on protein synthesis in rabbit reticulocyte lysates [6]

References:

1. Hayakawa Y, Nakagawa M, Kawai H, et al., & Otake N (1983) Studies on the differentiation inducers of myeloid leukemic cells. III. Spicamycin, a new inducer of differentiation of HL-60 human promyelocytic leukemia cells. J Antibiot 36:934–937
2. Kamishohara M, Kawai H, Odagawa A, et al., & Otake N (1993) Structure-antitumor activity relationship of semi-synthetic spicamycin analogues. J Antibiot 46:1439–1446
3. Sakai T, Kawai H, Kamishohara M, et al., & Otake N (1995) Synthesis and antitumor activities of glycine-exchanged analogs of spicamycin. J Antibiot 48:504–508
4. Sakai T, Kawai H, Kamishohara M, et al., & Otake N (1995) Structure-antitumor activity relationship of semi-synthetic spicamycin derivatives. J Antibiot 48:1467–1480
5. Lee YS, Nishio K, Ogasawara H, et al., & Saijo N (1995) In vitro cytotoxicity of a novel antitumor antibiotic, spicamycin derivative, in human lung cancer cell lines. Cancer Res 55:1075–1079
6. Kamishohara M, Kawai H, Sakai T, et al., & Otake N (1994) Antitumor activity of a spicamycin derivative, KRN5500, and its active metabolite in tumor cells. Oncol Res 6:383–390

Squalestatins (Zaragozic acids)

Key Words: [Squalene synthase inhibitors]

Structure: shows squalestatin I/zaragozic acid A

Molecular Formula: $C_{35}H_{46}O_{14}$
Molecular Weight: 690
Solubility: DMSO, ++; H_2O, ±; MeOH, ++

Discovery/Isolation:

Squalestatins 1–3 [1] and zaragozic acids A–C [2] were isolated as inhibitors of squalene synthase, which is a key enzyme in cholesterol biosynthesis. The structures were determined by a combination of spectroscopic, X-ray crystallographic and chemical methods; these natural products incorporate the highly functionalized bicyclic core, [1S-(1α,3α,4β,5α,6α,7β)]-4,6,7-trihydroxy-2,8-dioxabicyclo-[3.2.1]octane-3,4,5-tricar-boxylic acid [3]. Squalestatin I is identical to zaragozic acid A.

Biological and Chemical Studies:

Squalestatins/zaragozic acids are potent competitive inhibitors of rat liver squalene synthase with respect to farnesyl pyrophosphate [2,4,5]. These compounds also inhibit farnesyl-protein transferase competitively with respect to farnesyl pyrophos-phate [6]. Lindsey and Harwood investigated the inhibition mechanism of squalestatin I in detail and suggested that inhibition of squalene synthetase by squalestatin I occurs through a mechanism-based irreversible inactivation process, presumably with formation of a covalent squalestatin I-squalene synthetase adduct that is cata-lytically incompetent [7].

Several groups used the squalestatins for investigation of regulation of gene expression involved in isoprenoid biosynthesis [8–10] and utilization of farnesol in both protein isoprenylation and cholesterol biosynthesis [11].

Biological Activity: Squalestatin I

12 ± 5 nM: IC_{50} for squalene synthase in vitro [4]
39 nM: IC_{50} for cholesterol biosynthesis from [^{14}C]acetate by isolated rat hepatocytes [4]
78 pM: K_i for rat liver squalene synthase [2]
6 µg: IC_{50} for the incorporation of [^3H]mevalonate into cholesterol in HepG2 [2]
200 µg/kg: IC_{50} for acute hepatic cholesterol synthesis in mouse [2]

References:

1. Dawson MJ, Farthing JE, Marshall PS, et al., & Hayes MV (1992) The squalestatins, novel inhibitors of squalene synthase produced by a species of *Phoma*. I. Taxonomy, fermentation, isolation, physico-chemical properties and biological activity. J Antibiot 45:639–647
2. Bergstrom JD, Kurtz MM, Rew DJ, et al., & Alberts AW (1993) Zaragozic acids: a family of fungal metabolites that are picomolar competitive inhibitors of squalene synthase. Proc Natl Acad Sci USA 90:80–84
3. Sidebottom PJ, Highcock RM, Lane SJ, et al., & Watson NS (1992) The squalestatins, novel inhibitors of squalene synthase produced by a species of *Phoma*. II. Structure elucidation. J Antibiot 45:648–658
4. Baxter A, Fitzgerald BJ, Hutson JL, et al., & Wright C (1992) Squalestatin 1, a potent inhibitor of squalene synthase, which lowers serum cholesterol in vivo. J Biol Chem 267:11705–708
5. Hasumi K, Tachikawa K, Sakai K, et al., & Endo A (1993) Competitive inhibition of squalene synthetase by squalestatin 1. J Antibiot 46:689–691
6. Gibbs JB, Pompliano DL, Mosser SD, et al., & Oliff A (1993) Selective inhibition of farnesyl-protein transferase blocks ras processing in vivo. J Biol Chem 268:7617–7620
7. Lindsey S & Harwood HJ, Jr. (1995) Inhibition of mammalian squalene synthetase activity by zaragozic acid A is a result of competitive inhibition followed by mechanism-based irreversible inactivation. J Biol Chem 270:9083–9096
8. Ness GC, Zhao Z & Keller RK (1994) Effect of squalene synthase inhibition on the expression of hepatic cholesterol biosynthetic enzymes, LDL receptor, and cholesterol 7a hydroxylase. Arch Biochem Biophys 311:277–285
9. Correll CC & Edwards PA (1994) Mevalonic acid-dependent degradation of 3-hydroxy-3-methylglutaryl coenzyme A reductase in vivo and in vitro. J Biol Chem 269:633–638
10. Ness GC, Eales S, Lopez D & Zhao Z (1994) Regulation of 3-hydroxy-3-methylglutaryl coenzyme A reductase gene expression by sterols and nonsterols in rat liver. Arch Biochem Biophys 308:420–425
11. Crick DC, Andres DA & Waechter CJ (1995) Farnesol is utilized for protein isoprenylation and the biosynthesis of cholesterol in mammalian cells. Biochem Biophys Res Commun 211:590–599

Staurosporine

Key Words: [Cell cycle inhibitor][Kinase inhibitor]

Structure:

Molecular Formula: $C_{28}H_{26}N_4O_3$
Molecular Weight: 466
Solubility: DMSO, +++ ; H_2O, - ; MeOH, +

Discovery/Isolation:

Staurosporine was originally discovered as an antifungal antibiotic [1] and its structure was elucidated to be an indolo [2,3-a] carbazole derivative [2,3].

Biological and Chemical Studies:

There are good reviews about the characteristics, biological activities and usage of
staurosporine [4,5]. Soon after discovery, it was shown that staurosporine possesses
various biological activities, including induction of differentiation [6], apoptosis induction [7], cell cycle inhibition [8] and so on. It is important to use this compound as
a potent but nonspecific protein kinase inhibitor. After Tamaoki et al. reported that
staurosporine is a potent inhibitor for protein kinase C in 1986 [9], they also reported
that many protein kinases other than protein kinase C are strongly inhibited by
staurosporine [10,11]. As staurosporine inhibits the proteolytically generated catalytic domain of protein kinase C [9], the limited selectivity for different protein kinases
is attributable to its interaction with the essential region of the catalytic domain of
protein kinases that share a homologous region.

Biological Activity:

71 pmol/egg: ID_{50} for embryonic angiogenesis [12]
1–10 ng/ml; reversible cell cycle arrest of rat 3Y1 fibroblasts at the early G1 phase [8]
100 ng/ml; reversible cell cycle arrest of rat 3Y1 fibroblasts at G2 [8]

References:

1. Omura S, Iwai Y, Hirano A, et al., & Masuma R (1977) A new alkaloid AM-2282 of *Streptomyces* origin. Taxonomy, fermentation, isolation and preliminary characterization. J Antibiot 30:275–282
2. Furusaki A, Hahiba N, Matsumoto T, et al., & Omura S (1982) The crystal and molecular structure of staurosporine, a new alkaloid from a *Streptomyces* strain. Bull Chem Soc Jpn 55:3681–3685
3. Furusaki A, Hashiba N, Matsumoto T, et al., & Omura S (1978) X-ray crystal structure of staurosporine: a new alkaloid from a *Streptomyces* strain. J Chem Soc Chem Commun 1978:800–801
4. Tamaoki T (1991) Use and specificity of staurosporine, UCN-01, and calphostin C as protein kinase inhibitors. Methods Enzymol 201:340–347
5. Omura S, Sasaki Y, Iwai Y & Takeshima H (1995) Staurosporine, a potentially important gift from a microorganism. J Antibiot 48:535–548
6. Morioka H, Ishihara M, Shibai H & Suzuki T (1985) Staurosporine-induced differentiation in a human neuroblastoma cell line, NB-1. Agric Biol Chem 49:1959–1963
7. Falcieri E, Martelli AM, Bareggi R, *et al.* & Cocco L (1993) The protein kinase inhibitor staurosporine induces morphological changes typical of apoptosis in MOLT-4 cells without concomitant DNA fragmentation. Biochem Biophys Res Commun 193:19–25
8. Abe K, Yoshida M, Usui T, et al., & Beppu T (1991) Highly synchronous culture of fibroblasts from G2 block caused by staurosporine, a potent inhibitor of protein kinases. Exp Cell Res 192:122–127
9. Tamaoki T, Nomoto H, Takahashi I, et al., & Tomita F (1986) Staurosporine, a potent inhibitor of phospholipid/Ca++-dependent protein kinase. Biochem Biophys Res Commun 135:397–402
10. Nakano H, Kobayashi E, Takahashi I, et al., & Iba H (1987) Staurosporine inhibits tyrosine-specific protein kinase activity of Rous sarcoma virus transforming protein p60. J Antibiot 40:706–708
11. Fujita-Yamaguchi Y & Kathuria S (1988) Characterization of receptor tyrosine-specific protein kinases by the use of inhibitors. Staurosporine is a 100-times more potent inhibitor of insulin receptor than IGF-I receptor. Biochem Biophys Res Commun 157:955–962
12. Oikawa T, Shimamura M, Ashino H, et al., & Murota S (1992) Inhibition of angiogenesis by staurosporine, a potent protein kinase inhibitor. J Antibiot 45:1155–1160

Stevastelin B

Key Words: [Dual-specificity protein phosphatase inhibitor][Immunosuppressant]

Structure:

Molecular Formula: $C_{34}H_{61}N_3O_9$
Molecular Weight: 655
Solubility: DMSO, +++ ; H_2O, ± ; MeOH, +++

Discovery:

Stevastelins A, A3, B, B3, C3, D3, and E3 were isolated as immunosuppressants from *Penicillium* sp [1,2]. Structures of stevastelin A, B, and B3 were determined by spectroscopic and chemical studies [3] and the absolute stereochemistry of stevastelin B was determined by synthetic methods [4].

Biological Studies:

Stevastelins inhibits both OKT-3-stimulated human T cell proliferation and ConA-stimulated mouse spleen lymphocytes. As stevastelin B and B3 showed potent activity, the substitution of the -OH function of threonine or serine affected stevastelin activities against T cell blastogenesis [2]. Humaguchi et al. have systematically synthesized stevastelin analogues and reported that stevastelin B inhibited gene expression is dependent on IL-2 or IL-6 promoters in situ, but that it had no inhibitory activity against any protein phosphatases in vitro. In contrast, stevastelin A, which is a sulforylated derivative of stevastelin B, inhibited the phosphatase activity of a dual-specificity phosphatase, VHR, in vitro, but it had no inhibitory activity against gene expression in situ [5].

Biological Activity:

1.8 µg/ml: IC_{50} of stevastelin B inhibition against OKT-3-stimulated human T cell proliferation [2]

3–5 µg/ml: IC_{50} of stevastelin B inhibition against IL-2 or IL-6 dependent β-galactosidase gene expression [5]

2.7 µM: IC_{50} of stevastelin A against VHR protein phosphatase activity [5]

References:

1. Morino T, Shimada K, Masuda A, et al., & Saito S (1996) Stevastelin A3, D3 and E3, novel congeners from a high producing mutant of *Penicillium* sp. J Antibiot 49:1049–1051
2. Morino T, Masuda A, Yamada M, et al., & Shimada N (1994) Stevastelins, novel immunosuppressants produced by *Penicillium*. J Antibiot 47:1341–1343
3. Morino T, Shimada K, Masuda A, et al., & Saito S (1996) Structural determination of stevastelins, novel depsipeptides from *Penicillium* sp. J Antibiot 49:564–568
4. Shimada K, Morino T, Masuda A, et al., & Saito S (1996) Absolute structural determination of stevastelin B. J Antibiot 49:569–574
5. Hamaguchi T, Masuda A, Morino T & Osada H (1997) Stevastelins, a novel group of immunosuppressants, inhibit dual-specificity protein phosphatases. Chem Biol 4:279–286

Swainsonine

Key Words: [α-Mannosidase inhibitor][Immunomodulator]

Structure:

Molecular Formula: $C_8H_{15}NO_3$
Molecular Weight: 173
Solubility: DMSO, +++ ; H_2O, +++ ; MeOH, +++

Discovery:

Swainsonine was isolated as an α-mannosidase inhibitor from *Swainsona canescens* [1]. Hino et al. re-isolated swainsonine as an immunomodulator, which enhances the mouse immune system in vitro [2].

Biological Studies:

Swainsonine inhibited mannosidase in a liver particulate enzyme preparation [3] and in vivo [4]. Interestingly, swainsonine inhibits not only α-mannosidases, but also degradation of endocytosed glycoproteins in isolated rat liver and mannose glyco-protein uptake by macrophages [5,6]. These effects were thought to be indirect via inhibition of α-mannosidases.

Swainsonine has potential as an immunomodulator for the treatment of immunocompromised hosts [2,7]. The addition of swainsonin into the drinking water of mice reduced the incidence of lung colonization by melanoma cells [8]. The inhibition of metastasis was completely abrogated by depletion of NK cells [9]. Immunomodulation activity of swainsonine may be caused by changes in the glycosylation state of the tumor cell, which can markedly influence its recognition by allogeneic lymphocytes, and further, that different T cell populations differ in their response to such changes [10]. Other immunomodulative activities, such as inhibition of the production of immunoglobulin [11] and enhancement of IL-2/IL-2 receptor production, were reported [12].

Biological Activity:

0.1–1.0 μM: inhibition of the mannosidase that releases [³H]mannose from a high mannose glycopeptide in a liver particulate enzyme preparation [3]

0.1–10 μg/ml: enhancement of the expression of ConA receptor in spleen cells [2]

10 μg/ml: enhancement of IL-2 receptor expression, IL-2 production, and proliferation in ConA-stimulated lymphocyte [12]

100 ng/ml: IC_{50} for pulmonary colonization by 24 hr administration of swainsonine-supplemented drinking water [9]

References:

1. Colegate SM, Dorling PR & Huxtable CR (1979) A spectroscopic investigation of swainsonine: an α-mannosidase inhibitor isolated from *Swainsona canescens*. Aust J Chem 32:2257–2264
2. Hino M, Nakayama O, Tsurumi Y, et al., & Imanaka H (1985) Studies of an immunomodulator, swainsonine. I. Enhancement of immune response by swainsonine in vitro. J Antibiot 38:926–935
3. Elbein AD, Solf R, Dorling PR & Vosbeck K (1981) Swainsonine: an inhibitor of glyco-protein processing. Proc Natl Acad Sci USA 78:7393–7397
4. Abraham DJ, Sidebothom R, Winchester BG, et al., & Dell A (1983) Swainsonine affects the processing of glycoproteins in vivo. FEBS Lett 163:110–113
5. Chung KN, Shepherd VL & Stahl PD (1984) Swainsonine and castanospermine blockade of mannose glycoprotein uptake by macrophages. Apparent inhibition of receptor-mediated endocytosis by endogenous ligands. J Biol Chem 259:14637–14641
6. Winkler JR & Segal HL (1984) Inhibition by swainsonine of the degradation of endocytosed glycoproteins in isolated rat liver parenchymal cells. J Biol Chem 259:1958–1962
7. Kino T, Inamura N, Nakahara K, et al., & Imanaka H (1985) Studies of an immunomodulator, swainsonine. II. Effect of swainsonine on mouse immunodeficient system and experi-mental murine tumor. J Antibiot 38:936–940
8. Dennis JW (1986) Effects of swainsonine and polyinosinic:polycytidylic acid on murine tumor cell growth and metastasis. Cancer Res 46:5131–5136
9. Humphries MJ, Matsumoto K, White SL, et al., & Olden K (1988) Augmentation of murine natural killer cell activity by swainsonine, a new antimetastatic immunomodulator. Cancer Res 48:1410–1415
10. Powell LD, Bause E, Legler G, et al., & Hart GW (1985) Influence of asparagine-linked oligosaccharides on tumor cell recognition in the mixed lymphocyte reaction. J Immunol 135:714–724
11. Tulp A, Barnhoorn M, Bause E & Ploegh H (1986) Inhibition of N-linked oligosaccha-ride trimming mannosidases blocks human B cell development. Embo J 5:1783–1790
12. Bowlin TL & Sunkara PS (1988) Swainsonine, an inhibitor of glycoprotein processing, enhances mitogen induced interleukin 2 production and receptor expression in human lymphocytes. Biochem Biophys Res Commun 151:859–864

Tautomycin

Key Words: [PP1/PP2A inhibitor]

Structure:

Molecular Formula: $C_{41}H_{66}O_{13}$
Molecular Weight: 766
Solubility: DMSO, ++ ; H_2O, + ; MeOH, +++

Discovery/Isolation:

Tautomycin was isolated from *Streptomyces spiroverticillatus* as a new antibiotic which exhibits strong toxicity against a variety of eukaryotic cells including fungi, yeasts and animal cells [1]. The structure of tautomycin was proposed on the basis of chemical degradation and spectroscopic evidence [2–4].

Biological Studies:

Tautomycin induced bleb formation and protein phosphorylation of K562 cells, as do the protein phosphatase inhibitors, okadaic acid and dinophysistoxin-1. Tautomycin inhibited the specific binding of [³H]okadaic acid to protein phosphatases. Furthermore, tautomycin inhibited protein phosphatases isolated from mouse brain. These results indicate that tautomycin induces bleb formation through inhibition of protein phosphatases [5]. MacKintosh and Klumpp reported that tautomycin inhibited the catalytic subunits of PP1 more potently than PP2A [6]. However, Hori et al. reported that tautomycin is equally effective for PP1 and PP2A. Kurisaki et al. investigated the effects of tautomycin on cytoskeleton [7,8]. Kawamura et al. reported the structure-activity relationship in phosphatase inhibition and apoptosis induction [9].

Biological Activity:

0.1–1 µg/ml: blebbing formation of human erythroid leukemia cell K562 [1,10]
7.5 mg/kg: ID_{50} in administered orally to mice [1]

0.16 and 0.4 nM: K_1 for PP1 and PP2A, respectively [6]
22 nM, 32 nM and 6 nM: IC_{50} for PP1, PP2A, and phosphatase in chicken gizzard actomyosin, respectively [11]

References:

1. Cheng X-C, Kihara T, Kusakabe H, et al., & Isono K (1987) A new antibiotic, tautomycin. J Antibiot 40:907–909
2. Ubukata M, Cheng X & Isono K (1990) The structure of tautomycin, a regulator of eukaryotic cell growth. J Chem Soc Chem Commun:244–246
3. Cheng XC, Ubukata M & Isono K (1990) The structure of tautomycin, a dialkylmaleic anhydride antibiotic. J Antibiot 43:809–819
4. Ubukata M, Cheng X, Isobe M & Isono K (1993) Absolute configuration of tautomycin, a protein phosphatase inhibitor from a Streptomycete. J Chem Soc Perkin Trans 1:617–624
5. Magae J, Osada H, Fujiki H, et al., & Isono K (1990) Morphological changes of human myeloid leukemia K-562 cells by a protein phosphatase inhibitor, tautomycin. Proc Japan Acad 66(B):209–212
6. MacKintosh C & Klumpp S (1990) Tautomycin from the bacterium *Streptomyces verticillatus*. Another potent and specific inhibitor of protein phosphatases 1 and 2A. FEBS Lett 277:137–140
7. Kurisaki T, Magae J, Nagai K, et al., & Yamasaki M (1993) Morphological changes and reorganization of actinfilaments in human myeloid leukemia cells induced by a novel protein phosphatase inhibitor, tautomycin. Cell Struct Funct 18:33–39
8. Kurisaki T, Taylor RG & Hartshorne DJ (1995) Effects of the protein phosphatase inhibitors, tautomycin and calyculin-A, on protein phosphorylation and cytoskeleton of human platelets. Cell Struct Funct 20:331–343
9. Kawamura T, Matsuzawa S, Mizuno Y, et al., & Ichihara A (1998) Different moieties of tautomycin involved in protein phosphatase inhibition and induction of apoptosis. Biochem Pharmacol 55:995–1003
10. Magae J, Watanabe C, Osada H, et al., & Isono K (1988) Induction of morphological change of human myeloid leukemia and activation of protein kinase C by a novel antibiotic, tautomycin. J Antibiot 41:932–937
11. Hori M, Magae J, Han YG, et al., & Karaki H (1991) A novel protein phosphatase inhibitor, tautomycin. Effect on smooth muscle. FEBS Lett 285:145–148

Trapoxin A

Key Words: [Cell cycle inhibitor][Histone deacetylase inhibitor]

Structure:

Molecular Formula: $C_{34}H_{42}N_4O_6$
Molecular Weight: 602
Solubility: DMSO, +++ ; H_2O, ± ; MeOH, ++

Discovery/Isolation:

Trapoxin A was isolated from the culture broth of *Helicoma ambiens* RF-1023. This compound exhibits detransformation activities against v-*sis* oncogene-transformed NIH3T3 cells as antitumor agents. The structures were found to be cyclotetrapeptides, cyclo[(S)-phenylalanyl-(S)-phenylalanyl-(R)-pipecolinyl- (2S,9S)-2-amino-8-oxo-9,10-epoxydecanoyl-] by X-ray analysis, mass spectrometric, NMR and chemical studies [1].

Biological Studies:

Kijima et al. found that trapoxin was found to cause accumulation of highly acetylated core histones in a variety of mammalian cell lines. In vitro experiments using partially purified mouse histone deacetylase showed that a low concentration of trapoxin irreversibly inhibited deacetylation of acetylated histone molecules. Chemical reduction of an epoxide group in trapoxin completely abolished inhibitory activity, suggesting that trapoxin binds covalently to histone deacetylase via the epoxide [2].

Taunton et al. isolated two nuclear proteins that copurified with histone deacetylase activity by using a synthesized trapoxin affinity matrix. Both proteins were identified by peptide microsequencing, and a complementary DNA encoding the histone deacetylase catalytic subunit (HD1) was cloned from a human Jurkat T cell library. As the predicted protein is very similar to the yeast transcriptional regulator Rpd3p, these results support a role for histone deacetylase as a key regulator of eukaryotic transcription [3].

Biological Activity:

200 ng/ml: ID_{50} for growth of *sis*-transformed NIH3T3 cells [1]
100 nM: hyperacetylation in various cell lines [2]
2 and 50 ng/ml: MIC for FM3A and TR303, trichostatin-resistant mutant of FM3A [2]

References:

1. Itazaki H, Nagashima K, Sugita K, et al., & Nakagawa Y (1990) Isolation and structural elucidation of new cyclotetrapeptides, trapoxins A and B, having detransformation activities as antitumor agents. J Antibiot 43:1524–1532
2. Kijima M, Yoshida M, Sugita K, et al., & Beppu T (1993) Trapoxin, an antitumor cyclic tetrapeptide, is an irreversible inhibitor of mammalian histone deacetylase. J Biol Chem 268:22429–22435
3. Taunton J, Hassig CA & Schreiber SL (1996) A mammalian histone deacetylase related to the yeast transcriptional regulator Rpd3p. Science 272:408–411

Triacsins

Key Words: [Acyl-CoA synthetase inhibitor]

Structure: show triacsin C

Molecular Formula: $C_{11}H_{17}N_3O$
Molecular Weight: 207
Solubility: DMSO, ++ ; H_2O, ± ; MeOH, ++

Discovery:

WS-1228 A and B were isolated as vasodilators from the culture filtrate of a strain of *Streptomyces aureofaciens* [1] and its structures were determined [2]. Omura et al. re-isolated triacsins A, B, C, and D as acyl-CoA synthetase [3]. Triacsins C and D are identical with WS-1228 A and B.

Biological Studies:

Acid hydrolysis of triacsins results in corresponding polyenic aldehydes with no activity. This suggests that the *N*-hydroxytriazene moiety is essential for inhibitory activity against acyl-CoA synthetase. Kinetic analysis indicates that inhibition of triacsin A is noncompetitive with respect to the two substrates ATP and coenzyme A, but is competitive with respect to long-chain fatty acids [4].

Tomoda et al. revealed that the inhibition of acyl-CoA synthetase by triacsins leads to the inhibition of lipid synthesis and eventually to the inhibition of proliferation of Raji cells by using triacsins [5]. Korchak et al. demonstrated a role for acyl-CoA esters in regulating activation of O^{2-} generation [6]. Triacsin C also blocks de novo synthesis of glycerolipids and cholesterol esters but not recycling of fatty acid into phospholipid: evidence for functionally separate pools of acyl-CoA [7].

Biological Activity:

3.6 µM: IC_{50} for triacsin C for acyl-CoA synthetase from *Pseudomonas aeruginosa* [4]
8.7 µM: IC_{50} for triacsin C for acyl-CoA synthetase from rat liver [4]
8.97 µM: the apparent K_i value when oleic acid is used as substrate [4]

References:

1. Yoshida K, Okamoto M, Umehara K, et al., & Imanaka H (1982) Studies on new vasodilators, WS-1228 A and B. I. Discovery, taxonomy, isolation and characterization. J Antibiot 35:151–156
2. Tanaka H, Yoshida K, Itoh Y & Imanaka H (1982) Studies on new vasodilators, WS-1228 A and B. II. Structure and synthesis. J Antibiot 35:157–163
3. Omura S, Tomoda H, Xu QM, et al., & Iwai Y (1986) Triacsins, new inhibitors of acyl-CoA synthetase produced by *Streptomyces* sp. J Antibiot 39:1211–1218
4. Tomoda H, Igarashi K & Omura S (1987) Inhibition of acyl-CoA synthetase by triacsins. Biochim Biophys Acta 921:595–598
5. Tomoda H, Igarashi K, Cyong JC & Omura S (1991) Evidence for an essential role of long chain acyl-CoA synthetase in animal cell proliferation. Inhibition of long chain acyl-CoA synthetase by triacsins caused inhibition of Raji cell proliferation. J Biol Chem 266:4214–4219
6. Korchak HM, Kane LH, Rossi MW & Corkey BE (1994) Long chain acyl coenzyme A and signaling in neutrophils. An inhibitor of acyl coenzyme A synthetase, triacsin C, inhibits superoxide anion generation and degranulation by human neutrophils. J Biol Chem 269:30281–30287
7. Igal RA, Wang P & Coleman RA (1997) Triacsin C blocks de novo synthesis of glycerolipids and cholesterol esters but not recycling of fatty acid into phospholipid: evidence for functionally separate pools of acyl-CoA. Biochem J 324:529–534

Trichostatin A

Key Words: [Cell cycle inhibitor][Histone deacetylase inhibitor]

Structure:

Molecular Formula: $C_{17}H_{22}N_2O_3$
Molecular Weight: 302
Solubility: DMSO, ++ ; H_2O, - ; MeOH, ++

Discovery/Isolation:

Trichostatin was isolated from the metabolites of *Streptomyces hygroscopicus* as active compounds against trichophytons and some fungi. The structure was determined to be a derivative of a primary hydroxamic acid by chemical and spectroscopic evidences [1].

Biological Studies:

Trichostatin A was re-isolated as a potent inducer of differentiation in murine erythroleukemia cells [2]. Yoshida and Beppu found that trichostatin A blocked the proliferation of 3Y1 cells specifically at the G1 and G2 phases. Interestingly, when trichostatin A-arrested cells at the G2 phase were released from the inhibition, cells with 4C DNA entered a new S phase without passage through the M phase, resulting in the formation of proliferative tetraploid cells [3]. Yoshida et al. clearly indicate that trichostatin A is a potent and specific inhibitor of the histone deacetylase and that the in vivo effect of trichostatin A on cell proliferation and differentiation can be attributed to the inhibition of the enzyme [4].

Morphological reversion of *sis*- or *ras*-transformed NIH3T3, T24 and HeLa by trichostatin A was reported [5,6]. Hoshikawa et al. suggest that gelsolin is one of the putative proteins necessary for the morphological changes of human carcinoma cells induced by trichostatin A [6]. Trichostatin A showed both induction and repression of several gene expressions via chromatin remodeling [7–9]. CpG methylation was also influenced by trichostatin A treatment [10,11].

Induction of caspase-3 protease activity and apoptosis by trichostatin A were also reported [12,13]

Biological Activity:

1.5×10^{-8} M: differentiation in murine erythroleukemia cells [2]
3.4 nM: K_i for partially purified histone deacetylase from FM3A cells, noncompetitive [4]
1 ng/ml: induction of normal and flat phenotype in *sis*-transformed NIH3T3 cells [5]
73 nM: IC_{50} for luciferase expression directed by the IL-2 enhancer in Jurkat T cells [7]

References:

1. Tsuji N, Kobayashi M, Nagashima K, et al., & Koizumi K (1976) A new antifungal antibiotic, trichostatin. J Antibiot 29:1–6
2. Yoshida M, Nomura S & Beppu T (1987) Effects of trichostatins on differentiation of murine erythroleukemia cells. Cancer Res 47:3688–3691
3. Yoshida M & Beppu T (1988) Reversible arrest of proliferation of rat 3Y1 fibroblasts in both the G1 and G2 phases by trichostatin A. Exp Cell Res 177:122–131
4. Yoshida M, Kijima M, Akita M & Beppu T (1990) Potent and specific inhibition of mammalian histone deacetylase both in vivo and in vitro by trichostatin A. J Biol Chem 265:17174–17179
5. Sugita K, Koizumi K & Yoshida H (1992) Morphological reversion of sis-transformed NIH3T3 cells by trichostatin A. Cancer Res 52:168–172
6. Hoshikawa Y, Kwon HJ, Yoshida M, et al., & Beppu T (1994) Trichostatin A induces morphological changes and gelsolin expression by inhibiting histone deacetylase in human carcinoma cell lines. Exp Cell Res 214:189–197
7. Takahashi I, Miyaji H, Yoshida T, et al., & Mizukami T (1996) Selective inhibition of IL-2 gene expression by trichostatin A, a potent inhibitor of mammalian histone deacetylase. J Antibiot 49:453–457
8. Van Lint C, Emiliani S, Ott M & Verdin E (1996) Transcriptional activation and chromatin remodeling of the HIV-1 promoter in response to histone acetylation. EMBO J 15:1112–1120
9. Sowa Y, Orita T, Minamikawa S, et al., & Sakai T (1997) Histone deacetylase inhibitor activates the WAF1/Cip1 gene promoter through the Sp1 sites. Biochem Biophys Res Commun 241:142–150
10. Selker EU (1998) Trichostatin A causes selective loss of DNA methylation in *Neurospora*. Proc Natl Acad Sci USA 95:9430–9435
11. Razin A (1998) CpG methylation, chromatin structure and gene silencing-a three-way connection. EMBO J 17:4905–4908
12. Lee E, Furukubo T, Miyabe T, et al., & Kariya K (1996) Involvement of histone hyperacetylation in triggering DNA fragmentation of rat thymocytes undergoing apoptosis. FEBS Lett 395:183–187
13. McBain JA, Eastman A, Nobel CS & Mueller GC (1997) Apoptotic death in adenocarcinoma cell lines induced by butyrate and other histone deacetylase inhibitors. Biochem Pharmacol 53:1357–1368

Tryprostatin A and B

Key Words: [Cell cycle inhibitor][Microtubule binder]

Structure: shows tryprostatin A

Molecular Formula: $C_{22}H_{27}N_3O_3$
Molecular Weight: 381
Solubility: DMSO, +++ ; H_2O, - ; MeOH, +

Discovery/Isolation:

Tryprostatins A and B were isolated as cell cycle inhibitors produced by *Aspergillus fumigatus* [1,2]. The structures were determined mainly by the use of spectroscopic methods, especially by detailed analyses of their ^{1}H and ^{13}C NMR spectra with the aid of 2D NMR techniques including pulse field gradient heteronuclear multiple-bond correlation spectroscopy, and their absolute configurations were determined on the basis of the optical rotational values and CD spectra [3].

The derivatives of tryprostatins, cyclotryprostatin A-D and spirotryprostatin A, B were also isolated from the culture broth of the same strain [4–6].

Biological and Chemical Studies:

The cell cycle inhibition mechanism and primary target of tryprostatin A (TPS-A) were investigated in detail by Usui et al [7]. TPS-A induced the reversible disruption of cytoplasmic microtubules in a range of concentrations that specifically inhibited M-phase progression. Interestingly, TPS-A inhibited the assembly in vitro of microtubules, but there was little or no effect on the self-assembly of purified tubulin when polymerization was induced by glutamate, paclitaxel, or digestion of the C-terminal domain of tubulin. However, TPS-A blocked tubulin assembly induced by inducers interacting with the C-terminal domain, MAP2, tau and poly-(l-lysine). They concluded that TPS-A is a novel inhibitor of MAP-dependent microtubule assembly.

The structure-activity relationships between tryprostatin derivatives were also studied by Kondoh et al [8]. They found that the methoxy moiety on the indole ring is important for reducing cytotoxicity and that cyclotryprostatin D induces microtubule assembly.

Biological Activity:

16.4 and 4.4 µM: MIC for cell cycle progression of tsFT210 at M phase of TPS-A and -B, respectively [2]
25–50 µM: cell cycle inhibition of 3Y1 cells in M phase (TPS-A) [7]
250 µM: 40% inhibition of in vitro microtubule assembly (TPS-A) [7]

References:

1. Cui CB, Kakeya H, Okada G, et al., & Osada H (1995) Tryprostatins A and B, novel mammalian cell cycle inhibitors produced by *Aspergillus fumigatus.* J Antibiot 48:1382–1384
2. Cui CB, Kakeya H, Okada G, et al., & Osada H (1996) Novel mammalian cell cycle inhibitors, tryprostatins A, B and other diketopiperazines produced by *Aspergillus fumigatus.* I. Taxonomy, fermentation, isolation and biological properties. J Antibiot 49:527–533
3. Cui CB, Kakeya H & Osada H (1996) Novel mammalian cell cycle inhibitors, tryprostatins A, B and other diketopiperazines produced by *Aspergillus fumigatus.* II. Physico- chemical properties and structures. J Antibiot 49:534–540
4. Cui CB, Kakeya H & Osada H (1997) Novel mammalian cell cycle inhibitors, cyclotryprostatins A-D, produced by *Aspergillus fumigatus,* which inhibit mammalian cell cycle at G2/M phase. Tetrahedron 53:59–72
5. Cui CB, Kakeya H & Osada H (1996) Novel mammalian cell cycle inhibitors, spirotryprostatin A and B, produced by *Aspergillus fumigatus,* which inhibit mammalian cell cycle at G2/M phase. Tetrahedron 52:12651–12666
6. Cui CB, Kakeya H & Osada H (1996) Spirotryprostatin B, a novel mammalian cell cycle inhibitor produced by *Aspergillus fumigatus.* J Antibiot 19:832–835
7. Usui T, Kondoh M, Cui CB, et al., & Osada H (1998) Tryprostatin A, a specific and novel inhibitor of microtubule assembly. Biochem J 333:543–548
8. Kondoh M, Usui T, Mayumi T & Osada H (1998) Effects of tryprostatin derivatives on microtubule assembly in vitro and in situ. J Antibiot 51:801–804

Tunicamycin

Key Words: [N-linked glycosylation inhibitor][Secretion inhibitor]

Structure: shows tunicamycin I

Molecular Formula: $C_{36}H_{58}N_4O_{16}$
Molecular Weight: 802
Solubility: DMSO, ++ ; H_2O, ++ ; MeOH, ++

Discovery/Isolation:

Tunicamycin was isolated as an antiviral antibiotic produced by *Streptomyces lysosuperificus*, nov. sp. The antibiotic is active against animal and plant viruses, Gram-positive bacteria, yeast and fungi [1]. Analytical studies on fatty acids which were obtained by the hydrolysis of tunicamycin suggested that tunicamycin is a mixture of at least ten homologous antibiotics [2]. Ito et al. isolated tunicamycin components by HPLC and the structures of individual antibiotics were elucidated [3].

Biological Studies:

Tunicamycin showed antiviral activity in agar diffusion, cytopathic effect suppression and plaque reduction tests against the multiplication of Newcastle disease virus in cultured chick embryo fibroblasts [4]. Takatsuki and Tamura found that this antiviral activity was partially reversed with some amino sugars and their derivatives [5]. Takatsuki et al. determined the tunicamycin inhibition point as the transfer of N-acetylglucosamine from UDP-GlcNAc to a lipid intermediate [6].

   Several groups reported that tunicamycin induces apoptosis in several myeloid or malignant cell lines [7,8]. Dricu et al. suggested that N-linked glycosylation plays an important role in maintenance of viability of melanoma cells through regulating the translocation of the IGF-I receptor to the cell surface by experiments on tunicamycin treatments [8].

Biological Activity:

0.5 mg/ml: MIC for Newcastle disease virus production by the agar-diffusion plaque inhibition method [1]

0.05 µg/ml: apoptosis induction in HL-60 cells [7]

5 µg/ml: inhibition of IGF-I receptor expression in the cell membrane of SK-MEL-2 and 90VAVI cells [8]

References:

1. Takatsuki A, Arima K & Tamura G (1971) Tunicamycin, a new antibiotic. I. Isolation and characterization of tunicamycin A. J Antibiot 24:215–223
2. Takatsuki A, Kawamura K, Okina M, et al., & Tamura G (1977) The structure of tunicamycin. Agric Biol Chem 41:2307–2309
3. Ito T, Takatsuki A, Kawamura K, et al., & Tamura G (1980) Isolation and structures of components tunicamycin. Agric Biol Chem 44:695–698
4. Takatsuki A & Tamura G (1971) Tunicamycin, a new antibiotic. II. Some biological properties of the antiviral activity of tunicamycin. J Antibiot 24:224–231
5. Takatsuki A & Tamura G (1971) Tunicamycin, a new antibiotic. III. Reversal of the antiviral activity of tunicamycin by aminosugars and their derivatives. J Antibiot 24:232–238
6. Takatsuki A, Kohno K & Tamura G (1975) Inhibition of biosynthesis of polyisoprenol sugars in chick embryo microsomes by tunicamycin. Arg Biol Chem 39:2089–2091
7. Perez-Sala D & Mollinedo F (1995) Inhibition of N-Linked glycosylation induces early apoptosis in human promyelocytic HL-60 cells. J Cell Physiol 163:523–531
8. Dricu A, Carlberg M, Wang M & Larsson O (1997) Inhibition of N-linked glycosylation using tunicamycin causes cell death in malignant cells: role of down-regulation of the insulin-like growth factor 1 receptor in induction of apoptosis. Cancer Res 57:543–548

UCN-01

Key Words: [Cell cycle inhibitor][Kinase inhibitor]

Structure:

Molecular Formula: $C_{28}H_{26}N_4O_4$
Molecular Weight: 482
Solubility: DMSO, +++ ; H_2O, - ; MeOH, +

Discovery/Isolation:

UCN-01 was found in the culture broth of *Streptomyces* sp., which produces staurosporine. The structure of UCN-01 differs from staurosporine in that the C-7 carbon bears a hydroxyl group [1].

Biological and Chemical Studies:

UCN-01 is a protein kinase C specific inhibitor compared with staurosporine [1] and inhibits cPKC [2]. There are good reviews about the characteristics, biological activities and usage of UCN-01.

UCN-01 shows strong cytotoxic effects on the growth of mammalian cells and shows antitumor activity in vivo against three human tumor xenografts [3]. UCN-01 retards but does not prevent cell cycle progression through the S phase, but cells are clearly blocked from exit of G1 and entry into S phase [4]. This cell cycle arrest is accompanied with inhibition of pRB phosphorylation [5,6]. A clinical trial of UCN-01 as a cancer antitumor agent is now in progress. Recently, it was found that UCN-01 strongly binds to human α1-acid glycoprotein by in vitro protein binding experiments [7,8]. The binding leads to changes of pharmacokinetics/pharmacodynamics of UCN-01 between different species and causes a reduction in distribution and clearance, resulting in high plasma concentrations in humans [7,8].

Biological Activity:

4.1, 42, and 45 nM: IC_{50} for protein kinase C, protein kinase A and $pp60^{src}$ kinase, respectively [1]

42, 32, and 58 nM: IC_{50} for cdk2, 4, and 6, respectively [6]

$10\,\mu M$ human $\alpha 1$-acid glycoprotein: complete inhibition of the initial uptake of UCN-01 ($1\,\mu M$) into isolated rat hepatocytes [9]

References:

1. Takahashi I, Kobayashi E, Asano K, et al., & Nakano H (1987) UCN-01, a selective inhibitor of protein kinase C from *Streptomyces*. J Antibiot 40:1782–1784
2. Mizuno K, Noda K, Ueda Y, et al., & Ohno S (1995) UCN-01, an anti-tumor drug, is a selective inhibitor of the conventional PKC subfamily. FEBS Lett 359:259–261
3. Akinaga S, Gomi K, Morimoto M, et al., & Okabe M (1991) Antitumor activity of UCN-01, a selective inhibitor of protein kinase C, in murine and human tumor models. Cancer Res 51:4888–4892
4. Seynaeve CM, Stetler-Stevenson M, Sebers S, et al., & Worland PJ (1993) Cell cycle arrest and growth inhibition by the protein kinase antagonist UCN-01 in human breast carcinoma cells. Cancer Res 53:2081–2086
5. Akiyama T, Yoshida T, Tsujita T, et al., & Akinaga S (1997) G1 phase accumulation induced by UCN-01 is associated with dephosphorylation of Rb and CDK2 proteins as well as induction of CDK inhibitor p21/Cip1/WAF1/Sdi1 in p53-mutated human epidermoid carcinoma A431 cells. Cancer Res 57:1495–1501
6. Kawakami K, Futami H, Takahara J & Yamaguchi K (1996) UCN-01, 7-hydroxylstaurosporine, inhibits kinase activity of cyclin- dependent kinases and reduces the phosphorylation of the retinoblastoma susceptibility gene product in A549 human lung cancer cell line. Biochem Biophys Res Commun 219:778–783
7. Sausville EA, Lush RD, Headlee D, Set al., & Kobayashi S (1998) Clinical pharmacology of UCN-01: initial observations and comparison to preclinical models. Cancer Chemother Pharmacol 42 Suppl:S54–9
8. Fuse E, Tanii H, Kurata N, et al., & Kobayashi S (1998) Unpredicted clinical pharmacology of UCN-01 caused by specific binding to human alpha1-acid glycoprotein. Cancer Res 58:3248–3253
9. Fuse E, Tanii H, Takai K, et al., & Sugiyama Y (1999) Altered pharmacokinetics of a novel anticancer drug, UCN-01, caused by specific high affinity binding to alpha1-acid glycoprotein in humans. Cancer Res 59:1054–1060

Ustiloxin A

Key Words: [Antitumor][Microtubule inhibitor]

Structure:

Molecular Formula: $C_{28}H_{43}N_5O_{12}S$
Molecular Weight: 674
Solubility: DMSO, + ; H_2O, ++ ; MeOH, +

Discovery/Isolation:

Ustiloxin A was isolated from the water extract of false smut balls on the panicles of rice plants caused by the fungus *Ustilaginoidea virens*. The absolute structure was determined by a combination of X-ray crystallographic and amino acid analyses [1].

Biological Studies:

Bioactivities of Ustiloxin A against microtubule assembly as well as to mammal, plant and fungal cells have been studied.

Biological Activity:

1.0–6.0 μM: IC_{50} for inhibition mitosis of a variety of human tumor cell lines [2]
0.7 μM: IC_{50} for tubulin polymerization in vitro [3]
20 μM: 89% inhibition of the binding of 10 μM vinblastine [4]

References:

1. Koiso Y, Natori M, Iwasaki S, et al., & Sato Z (1992) Ustiloxin: A phytotoxin and a mycotoxin from false smut balls on rice panicles. Tetrahedron Lett 33:4157–4160
2. Koiso Y, Li Y, Iwasaki S, et al., & Sato Z (1994) Ustiloxins, antimitotic cyclic peptides from false smut balls on rice panicles caused by *Ustilaginoidea virens*. J Antibiot 47:765–773
3. Li Y, Koiso Y, Kobayashi H, et al., & Iwasaki S (1995) Ustiloxins, new antimitotic cyclic peptides: interaction with porcine brain tubulin. Biochem Pharmacol 49:1367–1372
4. Luduena RF, Roach MC, Prasad V, et al., & Iwasaki S (1994) Interaction of ustiloxin A with bovine brain tubulin. Biochem Pharmacol 47:1593–1599

Vinca Alkaloid
(Vinblastine/Vincristine)

Key Words: [Antitumor][Apoptosis][Cell cycle inhibitor][Microtubule inhibitor]

Structure:

	Vinblastine (R=CH$_3$)	Vincristine (R=CHO)
Molecular Formula:	C$_{46}$H$_{58}$N$_4$O$_9$	C$_{46}$H$_{56}$N$_4$O$_{10}$
Molecular Weight:	811	825
Solubility:	DMSO, ++ ; H$_2$O, + ; MeOH, ++	

Discovery/Isolation:

Vinblastine (vincaleukoblastine), and vincristine were isolated as new alkaloids from *Vinca rosea* Linn. Structure analyses were reported by Neuss et al. and Gorman et al [1–3]. X-ray analyses of both compounds were reported [4]. Vindesine (deacetylvinblastine amide) was synthesized from either vinblastine or deacetylvinblastine [5].

Biological Studies:

Both compounds were isolated as therapeutically useful antitumor alkaloids. The antitumor effects are due to the interference of microtubule functions by disruption of mitotic spindle and aggregation induction of microtubule protein [6,7]. The effects of two compounds on the microtubule assembly in vitro were studied by several groups [8–10]. Tubulin protein possesses two distinct binding sites for vinblastine per molecule [11]. These sites are different from the colchicine binding site and vinblastine stimulates colchicine binding [12]. The antitumor activity of *Vinca* alkaloids, containing vindesine, were investigated [5,10,13].

Biological Activity:

Vinblastine

$4 \times 10^{-5} - 2 \times 10^{-4}$ M: induction of tubulin paracrystals in vitro [8]
Affinity constant: 6.2×10^6 M^{-1} (high-affinity site), 8×10^4 M^{-1} (low-affinity site) [11]
4.3×10^{-7} mole/liter: IC_{50} for tubulin assembly (3.0 mg/ml tubulin) [9]
$0.178\ \mu M$: K_i for inhibition of net tubulin addition at the assembly ends [10]
40 nM: complete growth inhibition in L-cells [10]

Vincristine

$0.085\ \mu M$: K_i for inhibition of net tubulin addition at the assembly ends [10]
40 nM: 25% growth inhibition in L-cells [10]

References:

1. Neuss N, Gorman M, Svoboda GH, et al., & Beer CT (1959) *Vinca* alkaloids. III. characterization of leurosine and vincaleukoblastine, new alkaloids from *Vinca rosea* Linn. J Am Chem Soc 81:4754–4755
2. Neuss N, Gorman M, Hargrove W, et al., & Manning RE (1964) *Vinca* alkaloids. XXI. The structures of the oncolytic alkaloids vinblastine (VLB) and vincristine (VCR). J Am Chem Soc 86:1440–1442
3. Gorman M, Neuss N & Svoboda GH (1959) *Vinca* alkaloids. IV. Structural features of leurosine and vincaleukoblastine representatives of a new type of indole-indoline alkaloids. J Am Chem Soc 81:4745–4746
4. Moncrief JW & Lipscomb WN (1965) Structures of leurocristine (vincristine) and vincaleukoblastine. X-ray analysis of leurocristine methiodide. J Am Chem Soc 87:4963–4964
5. Barnett CJ, Cullinan GJ, Gerzon K, et al., & Nelson RL (1978) Structure-activity relationships of dimeric Catharanthus alkaloids. 1. Deacetylvinblastine amide (vindesine) sulfate. J Med Chem 21:88–96
6. Bensch K & Malawista S (1969) Microtubular crystals in mammalian cells. J Cell Biol 40:95–107
7. Malawista S, Sato H & Bensch K (1968) Vinblastine and griseofulvin reversibly disrupt the living mitotic spindle. Science 160:770–771
8. Na GC & Timasheff SN (1982) In vitro vinblastine-induced tubulin paracrystals. J Biol Chem 257:10387–10391
9. Owellen RJ, Hartke CA, Dickerson RM & Hains FO (1976) Inhibition of tubulin-microtubule polymerization by drugs of the *Vinca* alkaloid class. Cancer Res 36:1499–1502
10. Jordan MA, Himes RH & Wilson L (1985) Comparison of the effects of vinblastine, vincristine, vindesine, and vinepidine on microtubule dynamics and cell proliferation in vitro. Cancer Res 45:2741–2747
11. Bhattacharyya B & Wolff J (1976) Tubulin aggregation and disaggregation: mediation by two distinct vinblastine-binding sites. Proc Natl Acad Sci USA 73:2375–2378
12. Wilson L & Friedkin M (1967) The biochemical events of mitosis. II. The in vivo and in vitro binding of colchicine in grasshopper embryos and its possible relation to inhibition of mitosis. Biochemistry 6:3126–3135
13. Sweeney MJ, Boder GB, Cullinan GJ, et al., & Todd GC (1978) Antitumor activity of deacetyl vinblastine amide sulfate (vindesine) in rodents and mitotic accumulation studies in culture. Cancer Res 38:2886–2891

Wortmannin

Key Words: [Apoptosis] [PI-3 kinase inhibitor]

Structure:

Molecular Formula: $C_{23}H_{24}O_8$
Molecular Weight: 428
Solubility: DMSO, +++ ; H_2O, ± ; MeOH, +++

Discovery/Isolation/Structure:

Wortmannin is an antifungal antibiotic isolated from a culture of *Penicillium wortmanni* Klocker [1]. The crystal structure and absolute configuration of wortmannin was determined by Petcher et al. [2].

Biological and Chemical Studies:

Okada et al. found that wortmannin inhibited ^{32}P labeling of phosphatidylinositol trisphosphate, a product of phosphatidylinositol 3-kinase (PI 3-kinase). Wortmannin was effective on the all three of the PI-3 kinase activities. The inhibition was a non-competitive type with regard to ATP and was observed consistently when PI, PI mono-phosphate, or PI bisphosphate was used as substrate. They concluded that wortmannin abolished the formyl peptide-induced stimulation of neutrophils as a result of the inhibition of PI 3-kinase [3]. Yao and Cooper found that wortmannin inhibited the ability of NGF to prevent apoptosis in PC-12 cells and concluded that cell survival appeared to be mediated by a PI-3 kinase signaling pathway distinct from the pathway that mediates differentiation [4,5].

Biological Activity:

0.4–3.2 µg/ml: growth inhibition of *Botrytis allii, B. cincerea, B. fabae, Cladosporium herbarum and Rhizopus stolonifer*[1]

5 nM: IC_{50} for noncompetitive inhibition of all three of the PI 3-kinase activities found in the cytosol fraction of guinea pig neutrophils [3]

References:

1. Brian PW, Curtis PJ, Hemming HG & Norris GLF (1957) Wortmannin, an antibiotic produced by *Penicillium wortmanni.* Trans Brit Mycol Soc 40:365–368
2. Petcher TJ, Weber H-P & Kis Z (1972) Crystal structure and absolute configuration of wortmannin and of wortmannin *p*-bromobenzoate. J Chem Soc Comm:1061–1062
3. Okada T, Sakuma L, Fukui Y, et al., & Ui M (1994) Blockage of chemotactic peptide-induced stimulation of neutrophils by wortmannin as a result of selective inhibition of phosphatidylinositol 3-kinase. J Biol Chem 269:3563–3567
4. Yao R & Cooper GM (1995) Requirement for phosphatidylinositol-3 kinase in the prevention of apoptosis by nerve growth factor. Science 267:2003–2006
5. Yao R & Cooper GM (1996) Growth factor-dependent survival of rodent fibroblasts requires phosphatidylinositol 3-kinase but is independent of pp70[s6k] activity. Oncogene 13:343–351

Subject Index

Authors and Collaborators

Authors

HIROYUKI OSADA
   Antibiotics Laboraory, RIKEN
   Hirosawa 2-1, Wako, Saitama 351-0198, Japan
   E-mail: antibiot@postman.riken.go.jp

MINORU YOSHIDA
   Department of Biotechnology, Graduate School of Agriculture and Life Sciences
   The University of Tokyo
   Bunkyo-ku, Tokyo 113-8657, Japan
   E-mail: ayoshida@hongo.ecc.u-tokyo.ac.jp

MASAYA IMOTO
   Department of Applied Chemistry, Faculty of Science and Technology
   Keio University
   Hiyoshi, Kohoku-ku, Yokohama 223-8522, Japan
   E-mail: imoto@applc.keio.ac.jp

KAZUO NAGAI
   School of Bioscience and Biotechnology
   Tokyo Institute of Technology
   4259 Nagatsuta, Midori-ku, Yokohama 226-0026, Japan
   E-mail: knagai@bio.titech.ac.jp

TAKAO KATAOKA
   School of Bioscience and Biotechnology
   Tokyo Institute of Technology
   4259 Nagatsuta, Midori-ku, Yokohama 226-0026, Japan
   E-mail: tkataoka@bio.titech.ac.jp

TAKEO USUI
   Antibiotics Laboraory, RIKEN
   Hirosawa 2-1, Wako, Saitama 351-0198, Japan
   E-mail: usui@postman.riken.go.jp

Collaborators

TATSUHIKO SUDO, RIKEN
HIDEAKI KAKEYA, RIKEN
MASASHI UEKI, RIKEN
KEN-ICHI TOGASHI, RIKEN
SIRO SIMIZU, RIKEN
RYUJI HAMAMOTO, RIKEN
LIN NIE, RIKEN

H. OSADA

M. YOSHIDA

M. IMOTO

K. NAGAI (*left*) T. KATAOKA (*right*)

M. UEKI (*Left*) T. USUI (*right*)   H. KAKEYA

T. SUDO (*front left*)
L. NIE (*front right*)
K. TOGASHI (*rear left*)
R. HAMAMOTO (*rear center*)
S. SIMIZU (*rear right*)